中传学者文库编委会

主　任： 廖祥忠　张树庭

副主任： 蔺海波　李　众　刘守训　李新军　王　晖
　　　　　杨　懿　柴剑平

成　员（按姓氏笔画排序）：

王廷信	王栋晗	王晓红	王　雷	文春英
龙小农	付　龙	叶　龙	刘东建	刘剑波
任孟山	李怀亮	李　舒	张绍华	张　晶
张根兴	张毓强	林卫国	郑　月	金　炜
金雪涛	周建新	庞　亮	赵新利	徐红梅
贾秀清	高晓虹	隋　岩	喻　梅	熊澄宇

中传学者文库

1954-2024

主编／柴剑平 执行主编／龙小农 副主编／张毓强 周建新

探索自然 面向未来

黄志洵文集

黄志洵 著

中国传媒大学出版社

·北京·

图书在版编目（CIP）数据

探索自然　面向未来：黄志洵文集 / 黄志洵著. -- 北京：中国传媒大学出版社，2024.8.

（中传学者文库 / 柴剑平主编）.
ISBN 978-7-5657-3775-6

Ⅰ. TN91-53

中国国家版本馆 CIP 数据核字第 202431W3L8 号

探索自然　面向未来：黄志洵文集
TANSUO ZIRAN MIANXIANG WEILAI: HUANG ZHIXUN WENJI

著　　者	黄志洵
责任编辑	杨小薇
封面设计	锋尚设计
责任印制	李志鹏

出版发行	中国传媒大学出版社			
社　　址	北京市朝阳区定福庄东街 1 号	邮　编	100024
电　　话	86-10-65450528　65450532	传　真	65779405
网　　址	http://cucp.cuc.edu.cn		
经　　销	全国新华书店		
印　　刷	北京中科印刷有限公司		
开　　本	710mm×1000mm　1/16		
印　　张	20		
字　　数	233 千字		
版　　次	2024 年 8 月第 1 版		
印　　次	2024 年 8 月第 1 次印刷		
书　　号	ISBN 978-7-5657-3775-6/TN·3775	定　价	99.00 元

本社法律顾问：北京嘉润律师事务所　郭建平

总　序

媒介是人类社会交流和传播的基本工具。从口语时代到印刷时代，再经电子时代至今天的数智时代，媒介形态加速演变、融合程度深入发展，媒介已然成为现代社会运行的基础设施和操作系统。今天，人类已经迈入媒介社会，万物皆媒、人人皆媒，无媒介不社会、无传播不治理。今天，无论我们怎么用力于信息传播的研究、怎么重视信息传播人才的培养都不为过。

中国传媒大学（其前身为北京广播学院）作为新中国第一所信息传播类院校，自1954年创建伊始，即与媒介形态演变合律同拍、与国家发展同频共振，努力探索中国特色信息传播人才培养模式、构建中国信息传播类学科自主知识体系，执信息传播人才培养之牛耳、发信息传播研究之先声，被誉为"中国广播电视及传媒人才摇篮""信息传播领域知名学府"。

追溯中传肇始发轫之起源、瞩望中传砥砺跨越之未来，可谓创业维艰而其命维新。昔日中传因广播而起，因电视而兴，因网络而盛，今天和未来必乘风破浪、蓄势而上，因人工智能而强。在这期间，每一种媒介兴起，中传均吸引一批志于学、问于道、勤于术的学者汇聚于此，切磋学术、传道授业，立时代之潮头，回应社会需求，成为学界翘楚、行业中坚，遂有今日中传学术研究之森然气象，已历七秩而弦歌不断，将传百世亦风华正茂。

自新时代以来，中传坚守为党育人、为国育才初心，励精图治、勠力前行，秉承"系统治理、创新图强、交叉融合、特色发展"的办学理念，牢牢把握高等教育发展大势、传媒业态发展趋势，瞄准"智能传媒"和"国际一流"两大主攻方向，以世界为坐标、以未来为向度，完成了全面布局和系统升级，正在蹄疾步稳、高质量推动学校从传统高等教育向未来高等教育跨越、从传统传媒教育向智能传媒教育跨越、从国内一流向世界一流跨越，全力建设中国特色、世界一流传媒大学。

中国特色、世界一流，在于有大先生扎根中国大地，汇聚古今、融通中外；在于有大先生执教黉门，学高为师、身正为范；在于有大先生躬耕杏坛，敦品积学、启智润心。习近平总书记更强调，高校教师要立志成为大先生，在教书育人和科研创新上不断创造新业绩。中传广大教师素来以做大先生为毕生职志，努力成为新时代"经师"与"人师"的统一者，做真学问、立高品行，践履"立德树人"使命。

2024岁在甲辰，欣逢中传建校70华诞，学校特邀约部分学者钩玄勒要、增删批阅，遴选已公开刊发的论文汇编成集，出版"中传学者文库"，意在呈现学校在学科建设、科学研究、服务行业实践等方面的最新成果，赓续中传文脉，谱写时代新声。

文库汇聚老中青三代学者，资深学者渊渟岳峙、阐幽抉微；中年学者沉潜蓄势、厚积薄发；青年学者踌躇满志、未来可期。文库与五十周年校庆所出版的"北广学者文库"相承接，大致可勾勒中传知识生产薪火相传、三代辉映之概貌，反映中传在构建中国特色新闻传播类、传媒艺术类、传媒技术类学科体系、学术体系和话语体系方面的耕耘与收获，窥见中国特色信息传播类学科知识体系构建的发展脉络与轨迹。

这一构建过程，虽筚路蓝缕，却步履铿锵；虽垦荒拓野，亦四方辐辏。一批肇始于中传，交叉融合、具有中国特色的学科，如播音主持艺术学、广播电视艺术学、传媒艺术学、数字媒体艺术学、政治传播学等，从涓涓细流汇入滔滔江河，从中传走向全国，展现了中传学者构建中国自主知识体系的学术想象力和创新力。文库展示的虽然是历史，实则是呈现今天；看似是总结过去，实则是召唤未来。与其说这套文库的出版，是对既有学术成果的展示，毋宁说是对未来学术创新的邀约。

回首过往，七秩芳华。我们深知，唯有将马克思主义基本原理与中华优秀传统文化相结合，才能推动中华学术创造性转化和创新性发展，推动中国自主知识体系的构建。我们深知，唯有准确把握媒介形态演变的脉动、深刻认知媒介形态变革所产生的影响，才能推动中国信息传播类学科自主知识体系的构建与时俱进。

展望未来，星辰大海。我们深知，以人工智能为代表的产业和科技革命正迅疾而来，媒介生态正在加速重构，教育形态正在全面重塑，大学之使命与价值正在被重新定义；我们深知，唯有"胸怀国之大者"、面向世界科技前沿、面向经济主战场、面向国家重大需求，才能确保中传始终屹立于中国乃至世界传媒教育发展之潮头。

如何应对人工智能带来的深刻变革，对中传而言是一场要么"冲顶"、要么"灭顶"的"兴亡之战"。我们坚信，不管前方是雄关漫道，还是荆棘满途，唯有勇敢直面"教育强国，中传何为？"这一核

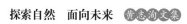

心命题，奋力书写"智能传媒教育，中传师生有为！"的精彩答卷，才能化危为机，奋力开创人工智能时代中传智能传媒教育新纪元。

功不唐捐，芳华七秩；风帆正举，赓续创新。

是为序。

第十四届全国政协委员，中国传媒大学党委书记、教授、博士生导师

In January 2021, Academician Jian Song presented his newly published English book to the author of this book. (Song was the Minister of the Ministry of Science and Technology of China and President of the Chinese Academy of Engineering.)

In the year 2004, the author of this book and Jian Song took this photo at an academic conference.

In 2013, the author of this book won the special award of the Ministry of Science and Technology System. The author took a photo with Academician Jingpei Cheng (Cheng was to be Vice Minister of Science and Technology) at the meeting.

The older generation of natural scientists had always been a model for the author to follow. This photo was taken in 1980.

(From left to right) Professor Zhigong Ren of Johns Hopkins University, Academician Tzu-Ching Huang, former President of Peking University, Academician Peyuan Zhou.

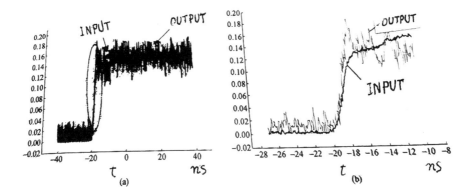

The author of this book guided doctoral students to complete a microwave experiment, the experimental frequency f=5.94GHz. This diagram shows the output waveform ahead of the input shape.

The National Physical Laboratory of the United Kingdom sent to the Chinese Academy of Metrology Science a "Newton tree" (part of the legendary wood). This is the photo of the author with it.

This is the laser experimental system built under the author's guidance. The wavelength is 632.8nm.

In 2014, the author of the book's doctoral student Rong Jiang (third from the left) passed the thesis. This is the full photo of the committee (Left 2 was Academician Zhonghua Zhang).

About the Author

Zhixun Huang, a specialist on microwave engineering and optics, was a professor of microwave engineering in Communication University of China. For many years, he had been working in the university as a doctoral tutor. And he was a visiting fellow of Electronic Institute of Chinese Academy of Sciences, member of IEEE. He had been engaged in research of electromagnetic theory, wave theory, microwave techniques, and the evanescent state in physics and electronics. And he kept doing scientific research on propagating velocity of electromagnetic waves, surface plasma waves, etc.

Prof. Huang had published many papers in various academic journals. In 1989, he got Reward of National Science and Technology Progress, due to the study of high frequency electromagnetic field strength standards. His book *An Introduction to the Theory of Waveguide Below Cut-Off* won the first National Scientific and Technology Book Award of China. In his book he first indicated that there could be the group velocity $v_g<0$ in the evanescent waves mode of the waveguide below cut-off. In 1993, he built a new characteristic equation of normal modes in coated circular waveguides including conductor loss. Moreover, in 2003, through an experiment in the coaxial photonic crystal, a group velocity of (1.5-2.4)c was observed in the stop-band of frequency. In 2013, the negative group velocity (NGV) was observed by the experiments in microwave region.

From 1999 to 2017, Prof. Huang published 6 books to discuss the superluminal propagation of light, the electromagnetic field theory, and the theory of guided waves. In 2011, he published the book *Recent Advances in Modern Physics*. In May 2013, Prof. Huang won the Special Innovation Award by the China High-tech Industry Originality Awards because of the research work on faster-than-light area.

In 2014, he published the book *Wave Sciences and Superluminal Light Physics*. Besides, the book also discussed the research of three negative physical parameters, i.e. the negative refraction index, negative wave velocity and negative Goos-Hänchen shifts. And he suggested the concept of negative characteristic motion of electromagnetic wave.

In 2017, he published the book *Study on the Superluminal Light Physics*. In this book he queried the validity of statement on light speed in vacuum(c) and the definition of meter. And he gave the concept of negative energy vacua. In 2018, he derived the Proca's wave equations (PWE). In 2022, he published the book *The Light of Physical Open Thoughts*, in which he criticized the relativity, the black hole physics, and the concept of gravitational waves.

Preface

I'm a natural scientist. Basically, I am a person with wings of thoughts inside me. In my decades-long research, I have published more than 100 papers and 15 books, monographs, demonstrating a strong creativity. However, the nine English papers included in this book, published between 2021 and 2023, are a sign that my scientific thinking is maturing, and these articles prove that I have a deep understanding of the natural world. Nature is rich, beautiful and full of mysteries. Nine articles recorded my repeated thinking and reflected my inner feelings by which you can clearly see the picture of my mental state. Of course, the descriptions used have not only words, but also a lot of mathematical physics equations. It is by careful mathematical analysis, as well as numerous experimental results, that I have reached the conclusions in this book.

I have loved science all my life. At the same time, I also love literature and music. Of all the western classical music, I admire Beethoven the most. Now I imagine that the total score of the nine symphonies he created can be compiled into a collection, so the book I contribute now may be comparable to that. This is not to say that I am a master, but to express the understanding and yearning of scientists for musicians. ... The beautiful nature let me spend my life in its arms. What a happy thing!

China's scientific career has advanced by leaps and bounds in recent years, in aerospace, high-speed railways, supercomputers and quantum communications, as well as in the theory in scientific papers. The theoretical research of Chinese scientists is gaining more and more international attention, as evidenced by the popularity of my paper in many countries. I hope to be a scientist who can bring honor to the motherland, explore nature, face the future, and be worthy of the times.

Finally, I would like to thank the Communication University of China for its firm support of my research work over the years.

Zhixun Huang

Introduction

At present, the international scientific community is confronted with many academic problems and scientific doubts, which shows that human's understanding of nature is still superficial. This book is a collection of long English essays published by the author in recent years abroad (the USA, the UK, India), with substantial logical thinking and profound mathematical analysis. This book has a broad vision and captures the essence of things. It is a comprehensive work of natural science. In fact, this book is a summary of the author's long thinking and research on the basic theories of physics.

For example, the use of quantum theory is to understand and analyze the nature of vacuum, and to pursue the mystery of quantum entanglement. For light, it is pointed out that the uncertainty of the speed of light leads to the irrationality of the existing definition of meters in metrology. The analysis of the nature of photons is particularly in-depth, which in turn involves the problem of wave-particle duality of light. There are many papers in this book, and the study of aster-than-light is the author's most comfortable subject. It not only contains insightful insights, but also discusses the outstanding contributions of Chinese scientists.

For example, this book analyzes the completeness of the basic equations of electromagnetism (Maxwell equations). The covariance of Maxwell equations is also studied on the basis of non-relativistic space-time view. This makes the traditional theory full of new vitality.

In addition, the book deeply analyzes the problem of negative group velocity in physics, and believes that the general success of negative group velocity experiments is the proof and symbol of the existence of advance waves, thus linking the research in different subject areas.

It is no accident that all the papers in this book are listed as "original innovative papers" when published abroad. For example, a new concept of negative energy vacuum is proposed. The current definition of meter in metrology is questioned. The western concept of faster-than-light space travel by wormhole and warp drive propulsion is criticized. The Maxwell equations in electromagnetism are completely discussed. The necessity and research significance of the existence of advance wave are deeply analyzed. The nature of quantum entanglement is analyzed in detail. The nature of photon and the duality of wave and particle are different from the traditional concepts.

A number of new theoretical equations in cut-off waveguide theory are given, which solve the problem of accurate calculation when the national attenuation standard is established. This book also represents a unique new contribution by the author after decades of faster-than-light research.

This book is a highly theoretical work, reflecting the author's creativity. It has introduced many achievements of the Chinese scientific community to the world, and its publication will further arouse great interest of the international scientific community.

Contents

Two Kinds of Vacuum in Casimir Effect ·· 001
A Study and Discussion of the 1983 Meter Definition ························ 026
To Achieve Faster than Light Astronautic Travel Whether Human Beings can Use
    Wormholes or Warp Drive Propulsion ······································ 046
The Completeness of Classical Electromagnetic Theory ······················ 080
On the Non-Relativistic Space-Time View and the Covariation of Maxwell's
    Equations ·· 115
Negative Velocity Characteristics in Electromagnetism ······················ 153
Research and Discussion of Quantum Theory ·································· 194
Study on the Essence of Photons and the Wave-Particle Duality of Light ······ 233
Circular Section Metal Wall Cut-Off Waveguide Accurate Calculation when Used as a
    Standard Attenuator ·· 261

Two Kinds of Vacuum in Casimir Effect[*]

INTRODUCTION

In 1948, Dutch physicist Hendrik Casimir discovered a physical phenomenon known as Casimir effect; It has been widely valued and studied for a long time. This is because it provides a direct demonstration and proof that "quantum vacuum is a physical reality". The basic way to demonstrate the Casimir effect is to take two neutral (uncharged) metal plates, place them parallel to each other, and the two will attract each other. The two plates form an electromagnetic boundary environment, but in classical physics, there is no electrodynamic reason for mutual attraction to actually occur. Therefore, Casimir effect is a quantum effect. In quantum field theory (QFT), there is the so-called zero point energy (ZPE), which represents the infinite energy of a vacuum state; Casimir first proposed a method to produce a finite result for this infinite vacuum, namely Casimir force; And it can be traced back to the famous van der Waals force.

In fact, Casimir began his theoretical work by studying van der Waals forces, and it was N. Bohr who advised him to consider the effects of ZPE. Interestingly, it was not until 1997 that the Casimir force was actually measured experimentally, and in the following 20 years, the Casimir effect was studied and applied. From the perspective of scientific innovation, it has brought important enlightenment and instructive benefit to people.

THE VACUUM VIEW OF MODERN PHYSICS

Matter always exists in various forms, independent of our consciousness; even distinct forms of matter must be rational. Vacuum is an objective reality, a form of matter. In a broad sense, vacuum is the largest and universal existence in the universe. In narrow engineering terms, a vacuum is the result of a human quest to live at the bottom of the atmosphere, where pressure is highest. The essence of this quest is the acquisition of local, airless environments in a small container by vacuum pump equipment and the measurement of such environments by sensing means. This is the practice of engineering vacuum.

For physical vacuum, the British physicist Dirac put forward the following viewpoint

---

[*] The paper was originally published in *Current Journal of Applied Science and Technology*, 2021, 40 (35), and has since been revised with new information.

in 1928: vacuum is a state in which the negative energy level is occupied by all electrons; Or in other words, the vacuum is filled with electrons in the negative energy level. In the past, vacuum was regarded only as an empty region of space. Dirac adopted a new concept that a vacuum is a region of space with the lowest possible energy; there are only two ways to get the non-vacuum state: one is to fill an electron into the positive energy state; the other is to create a hole in the distribution of negative energy states. The hole actually means a particle with the same mass as the electron, but with a positive charge. His theory received little attention at first, but positrons were discovered in 1932. It was Dirac that started the idea of a physical vacuum.

In 1974, the idea emerged that if there was an extremely strong force field, the energy stored in the force field could produce particles in a vacuum. For example, when a strong magnetic field above 10^6 Gs is applied to a vacuum, a kind of particle with photonic properties but with rest mass—heavy photon is produced. It has a short lifetime and will decay into an electron-positron pair.

There are signs that the nature of vacuum is not a simple problem. Nobel Laureate Dr Lee Chung-do said in 1979: "Vacuum is a real thing, a medium with Lorentz invariance. Its physical properties can be expressed through the interaction of elementary particles."

Therefore, from the perspective of modern physics, vacuum should no longer be called "space without any matter". Logically, such a space does not exist. The exact definition of vacuum (physical vacuum) is: the lowest energy state (ground state) of a system of quantum field, which is the basic form of matter's existence; Matter particles are merely the product of the excited state of the vacuum; the ground state contains no real particles. However, the generation, disappearance and transformation of virtual particles in the vacuum reflect that all modes in the quantum field are still oscillating in the ground state (zero-point vacuum oscillation). The complexity of physical vacuum can be seen from the following research topics: vacuum condensation, vacuum polarization, vacuum symmetry breaking, vacuum phase transition, etc. Some people suggest that we should discuss it in two general directions: first, the nature of vacuum itself; the other is the nature of the interaction between vacuum and particles.

Back to engineering vacuum. Here, the engineering definition of a vacuum is: "A rarefied gas state with a pressure less than 1 atm." The vacuum degree is mainly expressed in pressure units: the lower the pressure, the higher the vacuum degree, a unit of pressure expressed by the height of the mercury, i.e, the pressure of a gas in terms of length, based on Torricelli's 1644 mercury barometer experiment. This unit has long been used in industry. From a practical point of view, it is very convenient to have a scale to measure the pressure. But millimeters of mercury are not a very deterministic unit. Mercury variety, purity, temperature and other factors affect its density value, thus affecting the definition. The Torr, a unit adopted in honor of the Torricelli, is similar in size to millimeters of mercury (the Torr is 7×10^{-6} smaller than mmHg). In the International System of Units (SI), the unit of pressure is Pascal (symbol Pa), whose magnitude is $1Pa=1N/m^2= 1kg/(m\cdot s^2)$; However, 1 atm=101325Pa, so 1 Torr=1 atm/760=133.332Pa. Despite this, the Torr unit is still used in many scientific laboratories. From 1953 to 1970, humans ability to obtain a vacuum had reached 10^{-8} to 10^{-13} Torr, which is called ultra-high vacuum (UHV). By the end of the 20th century, this ability had increased to $p \leqslant 10^{-14}$ Torr, which is called extremely high vacuum (EHV).

When physicists talk about ZPE, the vacuum is, of course, a physical vacuum, not an engineering vacuum. The Casimir effect is the expression engineering vacuum condition (the pressure in the experimental environment) has on effect on the experiment, that is, the measurement of Casimir force should be carried out in the engineering vacuum environment, and even the UHV condition should be guaranteed. But past experimenters in the literature have not done so, which is puzzling. In any case, there is still plenty of room for improvement in existing experiments.

THEORETICAL BASIS OF ZERO POINT ENERGY

Does a vacuum have energy? It's not an easy question to answer. We know that the engineering definition of "vacuum" is "vacuum is a rarefied gas with a pressure of less than 1 atm (101325Pa)". However, the long-standing philosophical and scientific understanding of a vacuum is that it is an empty space (or state), which equates "vacuum" with "no matter". Scientific developments in the 20th century led to the realization that there is no such thing as a "complete absence of matter". Since matter is related to energy, the vacuum energy is zero prime no longer makes sense. This is a popular statement of the idea that a vacuum has energy.

However, in the vocabulary used by people, energy comes from energy and has the characteristics of output and application. It is in this understanding that the terms "electricity energy" "wind energy" "solar energy" "nuclear energy" are used. If the term "vacuum energy" is put forward alongside them, it looks very suspicious on the face of it. Therefore, the scientific community is not unanimous on the concept of "vacuum energy". One view is that ZPE cannot be used at all because it represents the lowest quantum state and is not an "acquired energy". Another view puts forward the slogan of "ask for energy from vacuum", and thinks that the zero-point energy (ZPE) measured by Los Alamos laboratory in 1997 is at least measurable.

Let us review the basic theory of quantum mechanics (QM) combined with statistical mechanics. The energy of a single quantum of energy (photon) amount for

$$E = \hbar\omega = hf \tag{1}$$

where $\hbar = h/2$, h is a Planck constant. In long time the experiments have shown that a photoelectric receiver based on the photoelectric effect always receives only the quantity of energy mentioned above formula or a multiple there of; none of the instruments that attempted to "split the photon" succeeded, i.e. the "partial photon" did not exist. The conclusion is clear—it is impossible for monochromatic wave photons to split into two photons of the same frequency, each carrying only part of the original photon, i.e., no photon of energy $hf/2$ exists (it is impossible to split a photon into two photons with half the original energy).

But QM tells us something else. As a result of the uncertainty principle, it is impossible to leave a microscopic particle at rest at its lowest potential energy. A quantum field can be regarded by Fourier analysis as a superposition of harmonic oscillators of different frequencies, so that it has energy even in the ground state (i.e. a physical vacuum). That is, in the lowest energy state (ground state), the energy of a harmonic oscillator at a frequency f is $hf/2$. Although it cannot be output, it is indeed energy. Quantum electrodynamics (QED) is used to quantize electromagnetic field. Firstly, vector potential is introduced as the regular coordinate, then the regular momentum is derived, then the regular coordinate and momentum are converted into operators, and finally the Hamilton operator of single-mode electromagnetic field is given. Thus, the electromagnetic field is transformed into a photon field, and the state of the electromagnetic field is represented by the photon number state (the eigenstate of the Hamilton operator).

Using the Coulomb guide, vector potential **A** satisfies the wave equation:

$$\nabla^2 A = \frac{1}{c^2} \frac{\partial A^2}{\partial t^2}$$

The Hamiltonian quantity is obtained from the canonical momentum, i.e.

$$H = \int \frac{1}{2}\left(\varepsilon_0 E^2 - \mu_0 H^2\right) dv$$

H in the integral sign is the magnetic field strength. And now we are going to expand it with orthogonal modular functions, we obtain:

$$A = \sum_{i=1}^{2} \sum_{k} \sqrt{\frac{\hbar}{2\varepsilon_0 \omega_k}} \left[a_{ki} U_{ki}(r) e^{-j\omega_k t} + a_{ki}^* U_{ki}^*(r) e^{j\omega_k t} \right] \tag{2}$$

where, i is two polarization directions; Is wave vector, wave number $k_0 = \omega\sqrt{\varepsilon_0 \mu_0} = \omega/c$, so we have:

$$\omega = ck \tag{3}$$

However, the momentum (vector) of the electromagnetic wave quantum of wave vector **k** is

$$\boldsymbol{p} = \hbar \boldsymbol{k}$$

the amplitude is

$$p = \hbar k_0 = \frac{hf}{c}$$

The energy (scalar) of this quantum is $E = \hbar \omega = hf$.

Now the generalized momentum can be obtained from $\Pi = \varepsilon_0 A$ and become an operator, so

$$A = \sum_{ki} \sqrt{\frac{\hbar}{2\varepsilon_0 \omega_k}} \left[\hat{a}_{ki} U_{ki}(r) e^{-j\omega_k t} + \hat{a}_{ki}^+ U_{ki}^*(r) e^{j\omega_k t} \right] \tag{4}$$

$$\Pi = \sum_{ki} (-j) \sqrt{\frac{\hbar \varepsilon_0 \omega_k}{2}} \left[\hat{a}_{ki} U_{ki}(r) e^{-j\omega_k t} + \hat{a}_{ki}^+ U_{ki}^*(r) e^{j\omega_k t} \right] \tag{5}$$

After quantization, the Hamilton operator is

$$\hat{H} = \sum_{ki} \hbar \omega_k \left[\hat{a}_{ki}^+ \hat{a}_{ki} + \frac{1}{2} \right]$$

where, $\hat{a}_{ki}^+$ is the photon's generation operator; $\hat{a}_{ki}$ is the annihilation operator of photons. It's easier to write:

$$\hat{H} = \sum_k \hbar \omega_k \left[\hat{a}_{ki}^+ \hat{a}_{ki} + \frac{1}{2} \right] \tag{6}$$

The commutation relation is

$$\left[\hat{a}_k \cdot \hat{a}_{k'}^+ \right] = \delta_{kk'}$$

The photon number state representing the photon field is

$$\hat{H} | n_k \rangle = \hbar \omega_k \left[n_k + \frac{1}{2} \right]_{n_k} \tag{7}$$

The photon number operator (k mode) is

$$\hat{n} = \hat{a}_k^+ \cdot \hat{a}_k$$

In the above types, $|n_k\rangle$ represents the state of n_k photons, and the average value of the light field is 0.

The same conclusion as equation (7) can be obtained by using the method of quantization of harmonic oscillator. In general, the quantized electromagnetic field is described by the eigenstate $|n_k\rangle$ of the photon number operator, which represents the states containing n_k photons (k mode). The single-mode electromagnetic field can be expressed by the simpler formula below:

$$\hat{H} = \hbar \omega \left[n + \frac{1}{2} \right] \tag{8}$$

where, n is the number of photons. Thus, the energy of k-mode electromagnetic field is not

$n\hbar\omega$, but one more term. When there are no photons in space (n=0), the energy of k-mode is not zero, but $\hbar\omega/2$, which is ZPE; Its discovery is the achievement of the quantization theory of electromagnetic field itself. Vacuum is now regarded in quantum theory as the ground state, denoted as $|0\rangle$. The energy of the ground state is

$$\langle 0|H|0\rangle = \frac{1}{2}\sum_k \hbar\omega_k \qquad (9)$$

It's actually saying that the ZPE quantity is:

$$E_0 = \frac{1}{2}hf \qquad (10)$$

In Equation (8), when n=0 (vacuum without photon), there is still a minimum energy E_0. Therefore, a quantum subsystem without quantum still has a minimum energy, whose value is exactly 1/2 of the energy carried by a quantum, namely ZPE. It must be noted that ZPE is independent of temperature; ZPE is independent of n (the state of the quantum system); ZPF is consistent with the spontaneous emission of the system (ZPE causes spontaneous emission). $hf/2$ only means that each existing mode has a radiation density equivalent to half a quantum energy, but does not mean that there can be "half a quantum" or "half a photon".

Now let's look at the harmonic oscillator analysis when statistical mechanics is combined with QM. If the temperature drops to T=0K (thermodynamic zero), it is impossible for microscopic particles not to have any motion. Otherwise, its momentum and position can be determined precisely at the same time, thus violating the Heisenberg uncertainty relation. In fact, when T=0K, the microscopic particles still vibrate. This phenomenon can be obtained by calculating and expounding the average energy of several harmonic oscillators by means of statistical mechanics, so the result is:

$$\bar{E} = \frac{hf}{e^{hf/kT}-1} + \frac{1}{2}hf \qquad (11)$$

Equation (11) is the Planck black-body radiation formula. It is the result of combining statistical mechanics and quantum theory, and also the basis of quantum noise theory and stimulated radiation theory. The unit of E is J or W s, or W/Hz. In electronics terms, it's the spectral power density, which is power per unit of bandwidth. The first term at the right end of equation (11) is the average energy of an oscillating mode at frequency f, and the second term is ZPE. That's because when we take T equals 0K, the first term is 0, and we're left with the second term.

In short, the average energy, or the power spectral density of thermal fluctuation and thermal radiation, caused first term on the right side of Equation (11). The second term is independent of temperature T, it's still there even by vibration (oscillation) of any substance at equilibrium temperature T above absolute zero is expressed by the if T is equal to 0K, which means there's still one term of energy at absolute zero. Make

$$p(f) = \frac{hf/kT}{e^{hf/kT} - 1} \quad (12)$$

There are

$$E = kT\left[p(f) + \frac{1}{2}\frac{hf}{kT}\right] \quad (13)$$

So E depends on the ratio of hf to kT. In fact, this ratio reflects the ratio of quantum effect to classical effect, and the larger the ratio is, the larger the quantum effect is (it cannot be ignored). The author calculated the relationship between the value of p(f) and hf/kT (Table 1). In fact hf/kT>0, so always p(f)<1; The question is how much less than one. Obviously, the higher the frequency and the lower the temperature, the lower the p(f).

Table 1 Relationship data of p(f) and hf/kT

hf/kT	0	0.1	0.2	0.5	1	2	3	4
p(f)	1	0.9506	0.9033	0.7708	0.5820	0.3130	0.1572	0.075

The above is the basic understanding of zero point energy (ZPE), but more macroscopic discussion is needed.

ABOUT VACUUM ENERGY

For single mode electromagnetic field, we already know that the Hamillon operator is $\hat{H} = \hbar\omega\left[n + \frac{1}{2}\right]$; therefore, when there is no photon (n=0) in space, the mode energy is not zero, but $E_0 = hf/2$, which was originally proven by P. Dirac and has since become fundamental knowledge in quantum mechanics.

The key now is to understand the nature of a vacuum; vacuum has energy fluctuations. This fluctuation of energy can be understood as virtual matter. In short, there must be energy fluctuation caused by interaction in vacuum without concrete matter according to uncertainty principle. In this sense, vacuum is a physical medium or complex system, which is one of the important achievements of quantum theory.

It must be noted that the vacuum fluctuations cannot be stopped by taking away their energy, because they have no energy. Sometimes there may be positive energy borrowed from somewhere else, and as a result there may be negative energy, which in turn rapidly absorbs energy from the positive energy region, thus reducing it to zero or maintaining some positive energy. The vacuum fluctuation is stimulated and driven by this continuous energy borrowing process. In the field of electromagnetic and electromagnetic waves, as well as in the field of laser technology, vacuum fluctuations have an experimental basis, and a term "spontaneous emission" has been used to call them.

If the vacuum energy density is assumed to be ρ_0, ρ_0 will be infinite if integrated over all frequencies according to the ZPE formula. Even if the integration of all frequencies is not reasonable, ρ_0 must be very large (J. Wheeler estimates ρ_0 up to 10^{35} J/m^3). Assuming that the vacuum in the universe has energy, it becomes a source of gravity, creating a gravitational field. Thus, cosmological studies on the large scale are linked to quantum field theory studies on the small scale. Years ago, astronomers reported finding that the expansion of the universe was accelerating. One readily available explanation is Einstein's cosmological constant. Sure, a vacuum can produce a repulsive force, but how much? Quantum field theory holds that random fluctuations in energy produce short-lived virtual particles at a constant rate. However, the vacuum energy is so great that the repulsive force applied is 10^{120} times greater than known. Some particles have negative energy, so excess is spared, leaving only a tiny residual energy that could explain the acceleration seen.

But the scientific community is divided on "vacuum energy". One view is that zero point energy cannot be exploited at all because it represents the lowest quantum state and is not an "acquired" energy. Another view puts forward the slogan of "energy from vacuum", and thinks that the (10-15) J energy obtained by the measurement of Casimir force by Los Alamos laboratory in 1998 is zero point energy or vacuum energy. Reports were mixed, however, and some described the measurement as proof of "negative energy."

As is well known, Newton's gravitational field equation takes the form of Poisson's equation, the 2nd order linear partial differential equation of the gravitational potential:

$$\nabla^2 \Phi = 4\pi G \rho \tag{14}$$

where, ρ is the density of matter and represents the source of the gravitational field. In 1915, Einstein proposed the equation of gravitational field [1]:

$$R_{\mu\nu} - \frac{1}{2} R^g_{\mu\nu} - \lambda g_{\mu\nu} = -8\pi G T_{\mu\nu} \tag{15}$$

Here, T is the momentum-energy tensor of the gravitational source, $g_{\mu\nu}$ is the space-time metric, and $R_{\mu\nu}$ and R form the tensor composed of the metric and its derivative. λ is the cosmological term and is known as the cosmological constant. In recent years, λ has been referred to as the repulsive factor. Scientific developments in the 20th century have continued into the early 21st century, and many things are related to λ.

From purely mathematical point of view, it's easy to solve this equation by taking λ minus 0. Therefore, The Einstein field equation is also written as:

$$R_{\mu\nu} - \frac{1}{2} g_{\mu\nu} R = -\kappa T_{\mu\nu} \tag{16}$$

where, g is the metric tensor; R is the curvature tensor; K is a coefficient proportional to Newton's gravitational constant:

$$K = \frac{8\pi G}{c^2} \tag{17}$$

In 1921, Einstein gave a lecture at Princeton University in the United States using formula (16).

In 1917, astronomers discovered that most spiral nebulae were receding from the Milky Way at great speeds. In fact, the action that a vacuum can produce is a repulsive force, which is consistent with the term "repulsive factor". This can be proved mathematically, and only through the equations of general relativity (GR) can we see it more clearly. GR theory gives the dynamic equation of uniform isotropic universe as

$$\ddot{R} = -\frac{4\pi}{3} G \left(\rho - 2\rho_{\mathit{eff}} R \right) \tag{18}$$

where, ρ is the average density of the universe; ρ_{eff} is the effective cosmic density considering cosmological constant and vacuum energy density; ρ_{eff} is defined as

$$\rho_{\mathit{eff}} = \rho_0 + \frac{1}{8\pi} \frac{\lambda}{G} \tag{19}$$

These mathematical formulas provide the entrance to the basic theory of cosmology. The negative sign before the second term in parentheses on the right of equation (18) indicates that the action of ρ_{eff} is opposite to that of ρ that is repulsion; the source of repulsion can be obtained from equation (19): the first is the positive cosmological constant($\lambda > 0$), and the second is the vacuum energy density ($\rho_0 > 0$).

That's just a summary. The real cosmological study (which will determine) is very complex. In the discussion of these problems, scientists have many puzzles and problems to be solved. Moreover, if GR is incorrect, the above discussion is invalid. To be clear, the fact that we have introduced Einstein's theories does not mean that I agree with Einstein.

The vacuum energy problem has not been settled in the scientific community. However, scientists have put forward many suggestions and made various efforts. In 1984, R. Forword suggested using the phenomenon of cohesion of charged thin film conductors to extract electrical energy from a vacuum. In the early 20th century, Cartan and Myshkin independently proposed that there is a long range interaction in nature — torsion field. Later this idea was widely studied. In combination with ZPE, vacuum ZPE is considered to be the energy of the torsion field. The torsion field is considered to be caused by the spin of the object and the disturbance caused by spin transverse polarization of vacuum. In 1997, A. Akimov and G. Shipov proposed in their paper on the torsion field that it is possible to extract energy from the vacuum by eddy perturbation of the physical vacuum. Interestingly, according to Dubrovsky's research, the propagation speed of the torsion field is considered to be superluminal (v>10⁹c). In 2000, it was reported that someone found evidence of the existence of the torsion field in electrolysis experiments. In early 2001, an international conference on "field propulsion technology" was held in the UK. One of the topics on the agenda was "the possibility of using ZPE to propel spacecraft". Such spacecraft, if realized, could fly freely in space for long periods of time without carrying fuel. This idea is based on the understanding of vacuum (a physical vacuum is an ocean of infinitely large fluctuations of energy) and the idea that energy can be extracted anywhere in space as long as the dynamic Casimir effect is coherent with the torsion field. Researchers believe that the 21st century may be the century of successful implementation of ZPE.

The research of Chinese scientists shows that the introduction of torsion field theory and the in-depth study of physical vacuum will help to understand the abnormal exothermic and nuclear phenomena in electrochemical processes. In the electrolysis process, microbubbles constantly appear at the electrode tip or micro-bulge. The formation, growth and collapse of bubbles are dynamic processes at the cavity boundary. Under resonance conditions, dynamic Casimir effect will occur and ZPE will be absorbed. The supernormal exotherm observed during electrolysis is not mainly due to the exotherm of nuclear reaction, but by extracting ZPE. The phenomenon of superheat is caused by two mechanisms: the torsion field generated by vortex plasma and the vacuum ZPE coherence and the dynamic Casimir effect.

CASIMIR FORCE OF A PAIR OF PARALLEL METAL PLATES

Casimir was a Dutch physicist. What led to his famous discovery? It all started with van der Waals forces. By the 1940s, H. Casimir (1906-2000), a scientist at Philips Laboratories in the Netherlands, and his colleagues carried out further research, the results of which were collected in three papers published in 1948 and 1949. For example, two interacting atoms can be regarded as two oscillators with a distance of d. In this case, the degenerated natural frequency, 0 decomposed into $\omega = \omega_0\sqrt{1+k}$, where k is the coupling strength of the oscillator, which is proportional to d^{-3}. If a portion of ZPE ($\hbar\omega/2$) is allocated to each frequency, the interaction energy is obtained:

$$E \propto \frac{\hbar\omega_0}{d^6} \tag{20}$$

However, T. Overbeek of the Philips Institute found some problems in his experiments, arguing that London's calculations were improper in assuming that electromagnetic interactions were instantaneous. If the finiteness of the speed of light (the speed of electromagnetic action) is taken into account, the van der Waals potential should be corrected over large distances. Overbeek approached Casimir and D. Polder, who devoted themselves to the problem. First analyze a simpler system, in which a single atom is in an electromagnetic cavity made of an ideal conducting wall. Calculate the interaction between the atom and the nearest cavity wall and find out how it relates to distance. After quantum electrodynamics (QED) processing of the field inside the cavity, they concluded that atoms are attracted to the wall of the cavity. This force came to be known as the Casimir-Polder force, with an associated energy of:

$$E_c = -\frac{3}{8\pi} \frac{\hbar c \alpha}{d^4} \quad (21)$$

where: α is electrostatic polarization; Casimir and Polder went on to calculate the attraction energy between two atoms, and found that if two atoms were identical, it would be

$$E_c = \frac{23}{4\pi} \frac{\hbar c \alpha^2}{d^7} \quad (22)$$

Casimir later said: "I mentioned my work to Bohr on a walk. He said it was good; it was creative work. I told him that I was puzzled by the simplicity of the formula for action over large distances; he pointed out that this may be related to ZPE. That was all it was, but it was a new road for me."

Casimir recalculated, assigning a ZPE ($\hbar\omega/2$) to each mode of the cavity, again using the well-known perturbation formula for cavity mode frequency:

$$\frac{\delta\omega}{\omega} = -2\pi\alpha \frac{|E_0(x_0, y_0, z_0)|^2}{\int |E_0(x, y, z)|^2 \, dv} \quad (23)$$

Here (x_0, y_0, z_0) is the position of the particle in the cavity, and the integration is performed over the total volume of the cavity. The energy is now determined by the sum of all the modes in the cavity, and the energy of attraction can be obtained by taking the energy difference between the two particles and the wall. Casimir's method can be said to reduce the problem of quantum electrodynamics to that of classical electromagnetism. Casimir finally gets the force between the walls of the cavity, which is caused by the zero-point field of the cavity.

This is the theoretical background that led to Casimir's discovery, but it does not yet map out the exact path to success. The key element of Casimir's scientific thinking is that only certain virtual photons are counted when calculating the energy between two uncharged conductor plates, each mode contributes a pressure to the plate, and the external infinite mode has a higher pressure than the internal infinite mode, which creates a force that brings the plates together. Experimental evidence of this force has been evidence of a vacuum electromagnetic field, improving our understanding of the vacuum. Now we say that is a quantum vacuum effect.

Imagine that two metal plates are placed in parallel at a distance of d (Fig.1). According to the analysis of electromagnetic field theory, there are two transverse polarization modes,

TE and TM, for each wave number (k) of the bowl. From the point of view of guided wave theory, it is a parallel plate waveguide structure, which may have TEM modes and a series of TE_{mn} and TM_{mn} higher-order modes. Subscripts m and n are called indexes of modes, which can generally be understood as discrete, but do not exclude the continuous change of mode indexes in some cases, unless the index is zero.

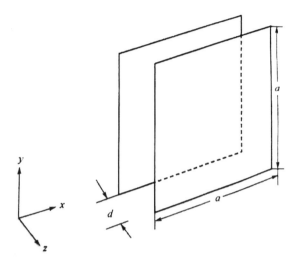

Fig. 1 Schematic diagram of Casimir force

In Casimir's case, the situation was different: he was talking about the macroscopic quantum phenomenon of ZPE changes caused by the presence or absence of a boundary, so he was comparing the absence of parallel plates (free space) with the presence of plates (boundary). The wave vector (k ⊥) parallel to the plate does not differ with or without the plate; But the wave vector (k ⊥) perpendicular to the plate is discrete (n is a discrete positive integer) with the plate and continuous (n changes continuously) without the plate. One dimensional analysis is easier to do, where the coordinates are taken as follows: the x and y axes are close to the surface of one of the plates, and the z axes are perpendicular to the plate (another metal plate is placed at z=d). In this case, the wave vector is perpendicular to the material surface and its magnitude is

$$k = \frac{n\pi}{d} \tag{24}$$

Let the same space have two cases respectively: free space without plate (no cavity), ZPE energy is E_{fr} (subscript fr is the first two letters of free); A space with a plate (cavity) and an energy of E_{cav} (subscript cav is the first three letters of cavity). The energy difference between the two cases is

$$\Delta E = E_c = E_{cav} - E_{fr} \tag{25}$$

The subscript c stands for Casimir, that is, the difference represents a particular kind of Casimir energy, i.e. E_c. If $E_c \neq 0$, it means that ZPE changes after the insertion of the cavity (such as two metal plates). If $E_c < 0$, it can be known that $E_{cav} < E_{fr}$, that is, ZPE becomes smaller after the insertion of the cavity. Such as E_c (absolute value) is related to cavity wall spacing (d), then the occurrence of force can be judged, which is the cause of Casimir force. In classical physical theories, such as Maxwell's electromagnetic field theory, there is neither the concept of ZPE nor the generation of Casimir forces. Therefore, this energy and this force are both products of Quantum Field Theory. If it can be proved by experiments, the correctness of relevant theories can be expressed in reverse.

In 2000, R.London provided a simple and straightforward derivation. The following will be elaborated according to his ideas. When the electromagnetic field to be quantanization, the radiative Hamiitonian is

$$\hat{H}_r = \sum_k \sum_\lambda \hbar \omega_k \left[\hat{a}^+_{k\lambda} \hat{a}_{k\lambda} + \frac{1}{2} \right] \tag{26}$$

where, the subscript r represents radiation. Introducing the above vacuum definition, the energy eigenvalue equation is

$$\frac{1}{2} \sum_k \sum_\lambda \hbar \omega_k \|0\rangle = E_0 \|0\rangle$$

So we get the vacuum energy

$$E_0 = \frac{1}{2} \sum_k \sum_\lambda \hbar \omega_k = \sum_k \hbar \omega_k \tag{27}$$

The two polarities (TE and TM) have been summed here.

Consider that space has a one-dimensional system (parallel double conductor plates) whose existence only gives rise to discrete modes. The boundary conditions require the mode group density k=nπ/d, so the cavity energy (ZPE) can be written as

$$E_{cav} = \sum_k \hbar c k = \frac{\pi \hbar c}{d} \sum_{n=1}^{\infty} n \tag{28}$$

where: c is the speed of light. When there is no board in the same space, discrete n is replaced by a continuous variable:

$$E_{fr} = \frac{\pi \hbar c}{d} \int_0^\infty n \, dn \tag{29}$$

Therefore, a

$$E_c = \frac{\pi \hbar c}{d} \left[\sum_{n=1}^{\infty} n - \int_0^\infty n \, dn \right] \tag{30}$$

Euler-Maclaurin summation formula is used to obtain

$$E_c = -\frac{\pi \hbar c}{12d} \tag{31}$$

Find the attraction between the two plates is

$$F_c = \frac{\partial E_c}{\partial d} = \frac{\pi \hbar c}{12d^2} \tag{32}$$

The above derivation may seem banal, but it was the first to show (in the simple-dimensional case) how a finite (small) change can occur in infinite vacuum energy (electromagnetic vacuum energy), and was previously unknown to physics. So some of the scientific literature says this is a fascinating result.

However, the real world is three-dimensional, and the analysis must include those modes whose wave vectors are not perpendicular to the plate. Suppose there is a rectangular cubic cavity the plate length is a, the plate distance is d, and d ≪ a, then the sum in the x and y directions is replaced by integral, and the formula is as follows:

$$E_c = -\frac{\hbar c a^2}{\pi^2} \left\{ \sum_n \int_0^\infty dk_x \int_0^\infty \sqrt{k_x^2 + k_y^2 + \frac{n^2 \pi^2}{d^2}} dk_y - \frac{d}{\pi} \int_0^\infty dk_x \int_0^\infty dk_y \int_0^\infty \sqrt{k_x^2 + k_y^2 + \frac{n^2 \pi^2}{d^2}} dk_z \right\} \tag{33}$$

In this case, the third derivative is used in the Euler-Maclaurin summation formula, we obtained:

$$E_c = -\frac{\pi^2 \hbar c a^2}{720 d^3} \tag{34}$$

and then

$$F_c = \frac{\partial E_c}{\partial d} = -\frac{\pi^2 \hbar c a^2}{240 d^4} \tag{35}$$

This is the Casimir force formula; It can be seen that $E_c \propto a^2$ (the greater the force of the plate area), and $F \propto d^{-4}$ (the smaller the spacing, the greater the force). The reason for the former can be understood as the larger the plate area is, the more significant the difference between internal and external photon numbers is; the reason for the latter can be understood as the smaller d is, the smaller the allowable module between plates is (the fewer photons are), resulting in the larger force. Note the difference between Casimir force and Newton gravitation, where the law is $F \propto d^{-2}$ and the magnitude of the force is proportional to the product of the masses of the two plates ($F \propto m^2$). But for Casimir force, F has nothing to do with the mass of the metal plate.

Since this is now a square conductor plate with area a^2, the pressure (force per unit area) is

$$P = \frac{F_c}{a^2} = -\frac{\pi^2 \hbar c}{240 d^4} \tag{36}$$

When the dimension of F is dyne, the dimension of a is cm, and the dimension of d is om, quation (36) is

$$P = -\frac{0.013}{240 d^4} \text{ dyne/cm}^2 \tag{37}$$

If d=1μm and the plate area is a^2=1cm², F_c=-1.3×10⁻² dyne=-1.3×10⁻⁷ N, it is a very small force, approximately equivalent to the gravity of a water drop with a diameter of 0.5mm. The force is small, but may be considerable if d is greatly reduced (it must be noted that when d→0, F_c→∞.) In fact, if d=10nm, p=1atm; So in the nanoworld this force is not negligible, perhaps far more important than Newtonian gravity. In addition, when d=1μm, the Coulomb force between the two plates will be greater than the Casimir force (when the voltage between the two plates is 17mV). This number can give people an idea of the size of the Casimir force. The derivation of the formula assumes that the material is an ideal reflection at all frequencies, that is, the conductivity of the plate is infinite, so the actual material formula needs to be modified.

Casimir's formula is now part of the history of physics. But in the late 1940s, when Casimir told the famous W. Pauli that there was "attraction between two conducting plates," he dismissed it as "nonsense." But after Casimir insisted, Pauli finally accepted the result.

In physics, the Casimir effect (or Casimi force) may at first seem a bit strange—a pair of metal conducting plates (parallel to each other) in a vacuum can act as if they were attracted to each other, but not gravitation. This force is small, but it is much stronger than gravity; Its existence can be measured, but the measurable effect is required at very small distances (microns or even nanometers) between the plates, which is technically very difficult.

The quantized electromagnetic field is a quantum system with infinite number of harmonic oscillators. The ground state has zero vibration and corresponding ZPE, and the zero vibration of all modes is the vacuum fluctuation of the quantum electromagnetic field — although the mean value is zero, the mean square value is not zero. Therefore, quantum theory believes that vacuum has energy, and the total size is $\frac{1}{2}\sum_i \hbar \omega_i$. Since the degree of freedom i (i.e., vibrational modulus i) is infinite and the upper limit of ω is infinite, this vacuum energy is divergent and unobtainable. However, it is possible to calculate and measure the change in the vacuum energy by placing two parallel metal plates into an open cavity, which changes the boundary conditions of the field and thus the frequency of the harmonic oscillator, causing the energy of the vacuum state to change. Although ZPE is still divergent and cannot be observed after implantation, its energy difference before and after implantation can be calculated and observed. This is the Casimir energy, I'll call it E_c; The corresponding force acting on the metal plate, the Casimir force, is denoted as F_c. Now, E_c is equal to the difference between the ZPE in the vacuum between the plates and the ZPE in the absence of the plates:

$$E_c = \left[\sum \frac{1}{2}\hbar\omega_i\right]_{wp} - \left[\sum \frac{1}{2}\hbar\omega_i\right]_{np} \tag{38}$$

This way we can get a clearer physical idea. In equation (38), wp is "with plates", np is "no plates".

But what is the meaning of negative sign? First, the plates is attracted to each other. Next, "Negative energy" can be understood as "the emptiness between the plates is more empty than the vacuum", which must produce inward force to bring the plates closer, Because of this, Lamoreaaux's measurements in 1997 were considered to "measure negative energy"; But the question is debated.

THEORETICAL RESEARCH PROGRESS

Casimir effect is a macroscopic quantum phenomenon deduced from quantum field theory, which cannot be understood from the perspective of classical physics. But this does not mean that classical Maxwell field theory (and its associated mathematical methods) are meaningless. Of course, the main analysis method is to rely on QFT and QED. Now some situation and progress of theoretical research are given.

Casimir's articles, the main two are hard to find. Easily searchable is an article he co-authored with Polder entitled "The effect of delay on London-van der Waals forces". In this paper, it first analyzes the interaction between a neutral atom and an ideal conducting plate, and then the interaction between two neutral atoms, using quantum electrodynamics (QED) to deal with the effect of delay on the interaction energy. This article does not cite the concept of ZPE.

In 1956, E. Lifshitz proposed his own theory: using the fluctuation dissipation theorem, he derived general formulas for free energy and dispersion interaction. His Fluctuating electromagnetic field is a classical simulation of Casimir's theory. Lifshitz's method has since been used by several authors. In addition, he supported an experiment to study the attraction between dielectric materials, which promoted scientific attention.

Lifschitz derived Casimir formula using van der Waals force only; Later R, Jaffe argued that Casimir forces need not be explained by vacuum fluctuations. Van der Waals force is the force between two neutral spherical atoms at rest due to the instantaneous electric dipole moment (due to the instantaneous center difference between positive and negative charges). It should be zero at 0K, but it's not zero because of the zero vibration. In the case of Casimir effect, the physical phenomenon may be caused by ZPE or van der Waals forces. In 1993, C.Sukenik et al. conducted experiments with a cavity in which the distance between plates could be adjusted at 0.5~8μm and a sodium atom beam passed through the cavity placed in a vacuum. Experiments show that it is consistent with quantum electrodynamics (QED) calculations and not with van der Waals forces.

In 2006, M Bordag used the path integral in quantum field theory (QFT) to give the Green function expression of E_c, which is caused by the Casimir between plates being treated as a massless scalar field, actually combining classical field theory with QFT. In 2010, Qiu Weigang pointed out that there are four types of boundary conditions for double plates:(D, D), (D, N), (N, D) and (N, N), where D represents Dirichlet boundary conditions of the first type and N represents Neumann boundary conditions of the second type. The negative energy ($E_c<0$) generated by the combined action of (D, D) and (N, N) is gravitational action; the positive energy ($E_c>0$) generated by the combined action of (D, N) and (N, D) is repulsion action. In 2007, Zeng Ran et al. derived and calculated the Casimir force between plates of negatively refracted materials, and discussed the influence of dispersion relation of negatively refracted materials on Casimir effect.

In 2009, M. Bordag et al. published *Advanced in the Casimir Effect* (745 pages) in Oxford University Press, which is a tour de force Study of on Casimir effect. The book is divided into three parts: I. Physical and mathematical basis of ideal boundary Casimir effect; II. Casimir force between entities; III. Casimir force measurement and its applications in fundamental physics and nanotechnology. There are about 700 references at the end of the book, among which 5 papers written by Casimir are listed in the references for the convenience of readers. This book co-authored by four authors is the most advanced and fruitful book in theoretical research so far, which has a landmark significance. In 2007, S. Lamoreaux published a paper "Casimir forces: still surprising after 60 years". It pointed that between quantum fluctuations and forces have now pervades all areas of physics, and that experimenters and theorists have also found the Casimir force problem challenging. In this paper, so my applications related to fine structure constant, electronic structure, black hole theory are discussed. 2008 marks the 60th anniversary of the discovery of Casimir forces, and several other articles have highlighted its contributions to theoretical physics.

CASIMIR EFFECT CAUSES UNIQUE NEGATIVE ENERGY VACUUM AND SUPERLUMINAL PHENOMENA

Previous theoretical work has profoundly revealed the nature of Casimir effect causing quantum vacuum and leading to faster-than-light (superluminal) phenomenon. The uniqueness of these works requires a special discussion.

In 1990, K. Scharnhorst from Germany and G. Barton from Britain published an article in the same journal, claiming to discover the phenomenon of faster-than-light speed in Casimir effect . This was in the first half of 1990. In July of the same year, S.Ben-Menahem from the United States published a paper entitled "Causality between dual conductor plates" in the same journal, commenting on the work of the aforementioned two men. These papers are high level, for example using quantum electrodynamics (QED) concepts and Feynman diagrams for analysis. As a result of Scharmhorst's work, the author concludes that the structure of vacuum is changed by placing two plates in the vacuum, so there are two kinds of vacuum; normal vacuum or free vacuum outside the plate, and negative energy vacuum between the plates. For electromagnetic wave propagation perpendicular to the plate, the speed of light in vacuum is not the same, and the variation ($\Delta c/c$) is about 1.6×10^{-60} d^{-4}, so when $d=10^{-9}$m, $\Delta c=10^{-24}$c. Therefore, due to the double loop effect of quantum electrodynamics, Scharnhorst concluded that this would make the phase and group velocity of electromagnetic waves greater than the speed of light in vacuum.

Now that this Casimir effect has been experimentally demonstrated, we have to accept that the "two vacuums" statement is true and that it is logical that the speed of light inside and outside the plate may be different. Thus, it is the change in boundary conditions that affects the vacuum and thus the propagation speed of electromagnetic waves. In other words, the propagation of light depends on the structure of the vacuum, which is the basic idea of quantum physics. It is due to the Casimir effect that we can distinguish between the following: ① normal vacua (also called free vacua); ② vacuum between the plates; The latter is characterized by vacuum energy density and can also be called negative energy vacua, which is the physical basis for faster-than-light phenomenon.

Scharnhorst first calculated the refractive index in the direction perpendicular to the board surface:

$$n_p = \sqrt{\varepsilon_{11}\mu_{11}} \tag{39}$$

where, ε_{11} and μ_{11} represent the permittivity tensor component and the permeability tensor component; The subscript p of n stands perpendicular. Eventually the export

$$n_p = 1 - \frac{11}{2^6 45^2} \frac{e^4}{(md)^4} \tag{40}$$

where, d is the distance between two ideal conductive plates; m is mass, c is the speed of light in normal vacuum (free vacuum), then

$$c = \left\{1 + \frac{11}{2^6 45^2} \frac{e^4}{(md)^4}\right\} c_0 \tag{41}$$

where, c is the speed of light in the vertical direction of the plate surface under the condition of interplate vacuum. The difference between c and c_0 is due to the change in the vacuum structure, which is caused by placing two plates. The result is $c > c_0$, where $c_0 = 299792458$ m/s, and the c is greater than c_0, i.e., faster than light. Further calculation gives:

$$\frac{\Delta c}{c} = \frac{c - c_0}{c} \tag{42}$$

Take $d = 1\mu m$, $\Delta c / c = 1.6 \times 10^{-36}$, which is very small; But even this is at odds with special relativity (SR)—SR theory does not allow anything to exceed the speed of light, however small. Try reducing d again—for a 1nm gap (d=1nm), increment $\Delta c = 10^{-24} c$. This number is also very small, but can be significant in some cases. In short, Scharhorst did not calculate "the speed of a photon traveling between two metal plates," but rather the speed of a wave traveling vertically between two plates, and found that the phase velocity was slightly higher than the speed of light ($v_p > c$). When the frequency is not high, the dispersion can be ignored and the group velocity is equal to the phase velocity, so the group velocity is slightly higher than the speed of light ($v_g > c$).

In 1993, G. Barton and K. Scharnhorst reinterpreted the problem by calling two metal plates "parallel mirrors". Paper title, or by "OED between parallel mirrors: light signals faster than c, or amplified by the vacuum"); The abstract of the paper states: "Due to the scattering of the quantized field, the light traveling vertically between two parallel double mirrors with a frequency of ω experiences a vacuum as a dispersive medium of refractive index n(0). Our earlier low-frequency results, representing n(0)<1, combine the Kramers-Kroning dispersion relationship with the classic Sommerfeld-Brillouin argument to declare either case: ① $n(\infty) < 1$ and $c/n(\infty) > c$; ② The imaginary part of n is negative, and the vacuum between the mirrors is insufficient to respond to light detection as a normal passive medium. It is clear, therefore, that the authors are concerned with the properties of vacuum; They think that under the physical conditions and conditions of Casimir effect, the refractive index of vacuum is no longer equal to 1, but may be smaller than 1. This, of course, is still the QFT's view, as opposed to classical physics. In addition, it should be noted that Scharnhorst's "group velocity hyperluminal" has two conditions: one is the wave perpendicular to the plate; Second, the frequency is not too high ($\omega \ll m_e$).

Why does the electromagnetic wave velocity ratio c between two metal plates depend on Casimir effect occur? Conceptually, considering the Maxwell electromagnetic field at zero absolute temperature between two parallel plate mirrors (distance d), the board is assumed to be an ideal conductor at any frequency. The outer boundary condition of the plate is $E_{||} = 0$ and $B_\perp = 0$. if the field is quantized, the vacuum structure of the plate is different from that of the unbounded space. In particular, the square of the field component and the energy density differ—the latter being lower, as demonstrated by the Casimir effect.

It is well known that even without mirrors, the zero-point vibration of Dirac's electron/proton field profoundly changes the vacuum properties, which is what distinguishes QED from classical physics. For example, they introduce nonlinearity into Maxwell's equations, and light scattering occurs. These nonlinear combination mirrors alter the zero Maxwell field, resulting in light velocities between mirrors perpendicular to the mirrors that may exceed c; Between the two mirrors, with respect to unbounded space, the plane wave detection propagation is changed (inductive coupling from the Fermion of the detection field to the zero oscillation of the quantized Maxwell field). When $\omega \ll m_e$, the nonlinear correction of Maxwell equation can be concluded as Euler-Heisenberg effective Laplace quantity density:

$$\Delta L = \frac{1}{2^3 3^2 5 \pi^2} \frac{e^4}{m^4} \left\{ \left(\boldsymbol{E}^2 - \boldsymbol{B}^2 \right) + 7 \left(\boldsymbol{E} \cdot \boldsymbol{B} \right)^2 \right\} \tag{43}$$

The results show that for vertical propagation between mirrors, the effective refractive index becomes:

$$n = 1 + \Delta n \tag{44}$$

And

$$\Delta n = \frac{11 \pi^2}{2^3 3^4 5^2} \frac{e^4}{(md)^4} \tag{45}$$

The refractive index of propagation parallel to the mirror is still 1, the same as that of unbounded space.

There are now

$$\frac{1}{n} = \frac{1}{1 + \Delta n} \approx 1 - \Delta n \tag{46}$$

$$v_p = \frac{\omega}{k} = c \frac{1}{n} = c(1 - \Delta n) = c(1 + |\Delta n|) \tag{47}$$

Group velocity is (in case of no dispersion):

$$v_g = \frac{d\omega}{dk} = v_p = c(1 + |\Delta n|) \tag{48}$$

Δn is too small for real measurements; It's really just

$$|\Delta n| = \frac{\Delta c}{c} \tag{49}$$

Considering that $n = \sqrt{\varepsilon \mu}$, so

$$\Delta n = \frac{1}{2} (\Delta \varepsilon + \Delta \mu) \tag{50}$$

$\Delta\varepsilon$ and $\Delta\mu$ may be understood as the changes of permittivity and permeability with position.

In 1998, Scharnhorst published his last paper on this topic. It is worth noting that he proposed the phrase "modified QED vacua", which is consistent with our discussion of the two vacuums in this section.

EXPERIMENTAL RESEARCH PROGRESS

Speaking of experiment, it is the only means of making people accept a new theory. Because Casimir forces are weak, experiments are difficult. The first person to do an experimental verification of theoretical expectations for the attraction of parallel plates was M. Sparnaay , and although with a great deal of error (up to 100% uncertainty), it is widely cited. The experiment is still interesting today. First, it measures the Casimir force between the conductor plates. It uses springs and cleverly constructed capacitors—a change in force is a change in capacitance, which can be measured. This proves once again that experimental design in scientific research is a very innovative work. After 1973 various experiments were carried out. Since 1994, S. Lamoreaux began an experimental study of Casimir forces and worked to improve the accuracy, with experiments ranging from d=0.6 to 6μm, In fact, this was the beginning of the verification of Casimir forces with sufficient accuracy, which was published in 1997, half a century after Casimir's paper was published. Lamoreaux measures the Casimir force between a gold-coated spherical lens and a conductor plate connected to a torsion balance. The Casimir force makes it torsion, and the measurement accuracy is 5%~10% at the distance of d=1μm, and the result is equivalent to E=(10-15)J. The earlier theory was finally well proved (though not by two plates but by one between a plate and a gold-plated surface), and in 1998, U. Mohideen and A. Roy measured the Casimir force more accurately at d=0.1-0.9μm, using aluminized plates and small spheres, the latter with a diameter of 200±4μm. Laser technology was used in the measurement, and the experimental results confirmed the Casimir theory with an accuracy of 1%. Knowing that Casimir himself did not die until 2000 (at the age of 91), and that he had reached an advanced age in 1997 and 1998, it must have been a mixed feeling to hear that the Casimir force had been accurately measured.

Mohideen's experiment is of great significance. His measurement result is: F_c=-(160~2) pN, that is, the absolute value of F_c, from 1.6×10^{-10} N to 2×10^{-12}N, are very small; This is between two aluminum surfaces. The force is so small that the difficulty of measuring it is conceivable; Moreover, the influence of skin depth and surface roughness cannot be ignored and must be reflected in the data correction. Fig. 2 is an example of this measurement (quoted from Klimchitskaya et al.), where d represents the distance between two aluminum surfaces, and the vertical coordinate F_c is in units of 10^{-12}N; the dotted line is the ideal metal surface, the actual line is the non-ideal metal surface.

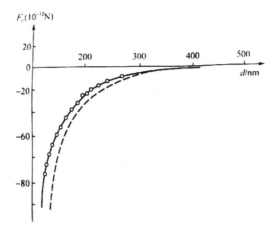

Fig. 2 Schematic diagram of Casimir force measurement results

Literature is another example of measurement; There are many more articles of this kind, and it is impossible to describe in detail the relevant research work. G. Klimchitskaya of St. Petersburg published a long article "Casimir Forces between practical materials: experiments and theory" in 2009, with a total of 340 references at the end. It can be seen that in recent years there has been more and more in-depth and meticulous research. In addition, Chinese scientists have published theoretical work in this area, but no one seems to have carried out experiments, possibly because of the difficulty of doing experiments.

CONCLUSION

Casimir effect can also be regarded as a pillar of QED. It can be seen from the foregoing that the concept of ZPE is used in Casimir force, so the experimental proof of Casimir force (F_c) is also the experimental proof of ZPE theory. This work later led to a great discussion on "whether vacuum really has energy", and in concept radiated to the recognition of virtual photons and negative energy. It can be said that it is "connected with the whole body"! The improvement and deepening of understanding of basic scientific problems will certainly lead to advances in applied science, for example, the Casimir effect has been proposed in nanotechnology to be considered.

The focus of this paper is to improve the understanding of the nature of quantum vacuum. In the past, to say that "vacuum is not empty" was already a criticism and subversion of classical physics. Now it seems doubly strange to say that there is a negative energy vacuum that is "empty" than the normal physical vacuum. But these theories are rigorously justified; Casimir effect can create an environment with refractive index less than 1(n<1) and lead to the appearance of superluminality, which is one of the representations of "quantum superluminality". These advances in basic science will certainly open up new fields of application. In short, it is not the Casimir structure that creates the quantum vacuum, but the structure that makes the quantum vacuum "emerge" in a clever way as a perceptible physical reality. This is truly a scientific achievement.

But much remains unclear. People ask, for example, is ZPE a physical reality or an auxiliary analytical tool? Is it an energy that can be extracted and applied in practice? Can we really get energy from a vacuum? There is a great deal of disagreement and debate about this. For example, although Lamoreaux's results are in line with ZPE's expectations, Jaffe argues that the Casimir effect does not provide a measure of ZPE, and therefore cannot measure energy in a vacuum. S. Catroll said that vacuum fluctuations are real and it is the Casimir effect that makes them manifest. In short, the scientific community is still divided on the Casimir effect.

The Casimir effect can be derived (explained) by classical physics? I do not agree with this statement, because there is no "(weak) attraction between parallel plates" in classical electrodynamics anyway; Otherwise, the great scientist Wolfgang Paul (1900-1958) would not have said at first (but later changed his mind) that Casimir's paper was nonsense. Some theoretical derivations do not lead us to believe that "classical physics is valid and therefore quantum physics is not involved", and we firmly believe that quantum theory is the most convincing way to understand Casimir effect.

COMPETING INTERESTS

Author has declared that no competing interests exist.

REFERENCES

[1] Casimir H, Polder D. The influence of retardation on the London—van der Waals forees[J]. Phys. Rev., 1948,73(4): 360–372.

[2] Casimir H. On the attraction between two perfectly conducting plates[J]. Proc. K. Ned. Akad. Wet., 1948, B51: 793–795.

[3] Casimir H. Introductory remarks on quantum electrodynamics[J]. Physica, 1953, 19: 846–849.

[4] Casimir H, Ubbink J. The skin effect at high frequencies[J]. Philips Tech. Rev., 1967, 28: 300–315.

[5] Casimir H. Some remarks on the history of the so called Casimir effect. Bordag(ed). The

Casimir Effect 50 Years Later[M]. Singapore: World Sci. , 1999.

[6] Ford L. Casimir force between a directric sphere and a wall[J]. Phys Rev A, 1998, 58(6): 4279–4286.

[7] Dodonov V. Dynamical Casimir effeet in a nondegenerate cavity with losses and detuning [J]. Phys Rev, 1998, A58: 4147–4150.

[8] Zeng R, et al. Effect of negative refractive index materials on Casimir effect[J]. Acta Physica Sinica, 2007, 56(11): 446–6450.

[9] Huang Z X. On Zero point vibration energy and Casimir force[J]. Engineering Science, 2008, 10(5): 63–69.

[10] Klimchitskaya G. The casimir force between real materials: experiment and theory[J]. Rev Mod Phys, 2009, 81: 1827–1885.

[11] Huang Z X. Some problems of quantum noise[J]. Journal of Electronic Measurement and Instrumentation, 1987, 1(3): 1–10.

[12] Puthoff H. Source of vacuum electromagnetic zero point energy[J]. Phys Rev A, 1989, 40: 4857–4862.

[13] Germelin A. ZPE and Casmir–Birdler effect [C]. Advances in Quantum Mechanics, Vol. 1, 2000: 232–248.

[14] Cole D, Puthoff H. Extracting energy and heat from the vacuum[J]. Phys Rev E, 1993, 48: 1562–1567.

[15] Pinto F. Engine cycle of an optically controlled vacuum energy transducer[J]. Phys Rev B, 1993, 60: 14740–14752.

[16] Yu Y. Introduction to General Relativity[M]. 2nd Ed. Beijing: Peking University Press, 1997.

[17] Hehl F. General relativity with spin and torsion: foundations and prospects[J]. Rev Mod Phys, 1976, 3: 393.

[18] Akimov A, Shipov G. Torsion field and experimental manifestations[J]. Journal of New Eenergy, 1997, 2(2): 67–81.

[19] Reed D. Excitation and extraction of vacuum energy via EM torsion field coupling theoretical model[J]. Journal of New Energy, 1998, 3(2/3): 130–140.

[20] Jiang X, Lei J, Han L. Dynamic Casimir effect in a electrochemical system[J]. Journal of New Energy, 1999, 3(4): 47–49.

[21] Jiang X, Lei J, Han L. Torsion field and tapping the zero point energy in an electrochemical system[J]. Journal of New Energy, 1999, 4(2): 93–95.

[22] Lei J, Jiang X. Casimir effect and extraction of ZPE[J]. Science and Technology Review, 1999, 4: 10–12.

[23] Lei J, Jiang X. Electrochemical anomalies and flexible field theory[J]. Science & Technology Review, 2000, 6: 3–5.

[24] Ni G, Chen S. Advanced Quantum Mechanics[M]. Shanghai: Fudan University Press, 2000.

[25] Larrimore L. Effects of particle size on the structure of the particle[J]. Journal of Physics, 2002, 115: 1–4.

[26] London R. The quantum theory of light[M]. New York: Oxford University Press, 2000.

[27] Bordag M. The Casimir offect for a sphere and a cylinder in front of plane and corrections

to the proximity force theorem[J]. Phys Rev D, 2006, 73: 125018.
[28] 邱为钢. 卡西米尔效应的格林函数计算方法 [J]. 大学物理, 2010, 29(03): 33-34+49.
[29] Bordag M, et al. Advanced in the Casimir effect[M]. New York: Oxford Univ. Press, 2009.
[30] Lamoreaux S. Casimir forces: still surprising after 60 years[J]. Physics Today, 2007, (Feb.): 40–45.
[31] Scharnborst K. On propagation of light in the vacuum between plates[J]. Phys Lett, B, 1990, 236(3): 354–359.
[32] Ben Menahem S. Causality between conducting plates[J]. Phys Lett, B, 1990, 250: 133, 1–13.
[33] Barton G, Scharnhorst K. QED between parallel mirrors: light signals faster than light, or amplified by the vacuum[J]. J Phys A: Math Gen, 1993, 26: 2037–2046.
[34] Scharnhorst K. The velocities of light in modified QED vacua[J]. Ann. d. Phys., 1998, 7: 700 – 709.
[35] Lamoreaux S. Demonstration of the Casimir in the 0.6 to 6μm range[J]. Phys Rev Lett, 1997, 78: 5–8.
[36] Mohideen U, Roy A. Precision measurement of the Casimir force from 0.1 to 0.9μm[J]. Phys Rev Lett, 1998, 81: 4549–52.
[37] Harris B W, Chen F, Mohideen U. Precision measurement of the Casimir force using gold surfaces[J]. Phys. Rev. A, 2000, 62(5): 052109.
[38] Huang Z X. On the quantum faster-than-light property[J]. Journal of Communication University of China (natural science edition), 2012, 19(3): 1–17.

A Study and Discussion of the 1983 Meter Definition*

INTRODUCTION

In 1960, the 10th International Conference on Metrology (CGPM 10) decided to name the Metric Convention established in 1875 the International System of Units (SI). It has seven basic units, they are: length unit "meter" (m), time unit "second" (s), current unit "ampere" (A), temperature unit "Kelvin" (K), mass unit "kilogram" (kg), material quantity unit "mole" (mol), luminous intensity unit "candela" (cd). These basic units and many derived units make up the entire system of units of measurement. The earliest definition of the meter was approved by the French Academy of Sciences in 1799: $1/4 \times 10^7$ of the earth's meridian is called a meter, which was defined in 1875. Later, it was found that it could not meet the needs of industrial development for measurement accuracy, so in 1889, the International Congress of Metrology adopted the distance between the two lines of the platinum-iridium alloy meter ruler as the definition value of 1m. A platinum-iridium meter No.6 is called the "International Meter Original". Each country participating in the Metric Convention has an identical platinum-dependent alloy meter, which is regularly compared with the international meter original instrument.

The relative accuracy of the international original meter prototype is 10^{-7}. After World War II, the German Federal Bureau of Physical Technology (PTB) successfully developed the Krypton-86 low-pressure gas discharge lamp. The vacuum wavelength of the orange line radiated from the Krypton-86 isotope is a fixed value. So in 1960, the International Metrological Conference adopted a new definition of the meter: "the meter is 1650763.73 times the length of the vacuum wavelength of the $2p_2$-$5d_5$ transition radiation of the Krypton-86 atom"... The above historical situation shows that the definition of the basic unit is not static and will change constantly with the progress of science and technology and the needs of industrial development.

In 1983, the international metrology community took a new step by adopting fundamental physical constants as the basis for establishing a new definition of the meter. The reason for this situation, is due to the invention of laser in 1960, the rapid development of laser technology, including the measurement of optical frequency technology to achieve

---

* The paper was originally published in *International Journal of Science Academic Research*, 2021, 2 (12), and has since been revised with new information.

a very high precision. In 1972, the National Bureau of Standards (NBS) scientist K. Evenson published the research work of his team—to achieve the frequency measurement of methane (CH_4) laser with highly complex technology, and obtain accurate frequency value f_{CH_4}, which has never been done before. Since the wavelength of the laser had been measured with considerable accuracy, it was possible to multiply this by the wavelength of the methane (λ_{CH_4}) to get the speed of light in vacuum. On this occasion, the international metrology circle tried to formulate a new definition of the meter "based on the basic physical constants" (in fact, based on the speed of light in vacuum), which we called "the 1983 definition of the meter" or "the current definition of the meter". Of course, there was a transition period from 1972 to 1983, and the current definition of the meter was not immediately decided.

There are two outstanding problems with the definition of the meter using Kr-86 spectral line (wavelength $\lambda = 605.7$nm) as the basic unit. First of all, there is a contour asymmetry in the spectral line, resulting in a wavelength difference of 1×10^{-8} between the center and the maximum light intensity. Secondly, the new laser frequency stabilization technology makes the frequency stability and reproducibility better than 1×10^{-9}, which is more than 100 times higher than that of Kr-86 orange line. Thus, the sheer technical appeal of metrology prompted the international metrology community to abandon the 1960 metre definition and switch to the 1983 metre definition. We stress that this is not a rational decision based on basic scientific principles. Problems with the current meter definition have been exposed since 1983, which is why we are writing this article.

THE ESTABLISHMENT OF THE DEFINITION OF METER IN 1983 AND ITS SPIRITUAL ESSENCE

Physics has long known that light has wave-particle duality, which has the characteristics of particle (the photon), but light wave is also a kind of electromagnetic wave; in fact, there is a broad electromagnetic spectrum. Therefore, any idea of that light is simple is wrong. If the experiment is carried out in an engineering vacuum without air, the following formula holds:

$$c = f\lambda \tag{1}$$

where, f and λ are respectively the frequency and wavelength of light wave, and c is the speed of light wave (the speed of light in vacuum). This thinking is entirely based on the understanding that light is a wave, and has nothing to do with the particle nature of light. In fact, no one has ever directly measured the speed of photons.

Now consider the work of Evenson's team in 1972 and how things have evolved in the years since 1972 to 1975. Evenson built a complex optical frequency measurement system using a laser frequency chain starting from the cesium atom frequency standard, including six different lasers and five microwave klystrons, the results were obtained

$$f_{CH_4} = 88.376181627 \times 10^{12} \text{ Hz} \tag{2}$$

The measurement accuracy is 6×10^{-10}; The known wavelength value of methane is about 3.39μm, which can be calculated using the best value at that time, then we obtained:

$$c = \lambda_{CH_4} f_{CH_4} = (299792456.2 \pm 1.1) \text{ m/s} \tag{3}$$

That is, the accuracy is 3.6×10^{-9}. For that alone, the accuracy of measuring the speed of light in vacuum has improved by a factor of 100. This created a great attraction for the International Bureau of Metrology. So what is the measurement of λ_{CH_4}? From 1972 to 1973, the following precise measurements were obtained by the international famous metrological institutions :

American Bureau of Standards (NBS): 3.392 231376(12) μm

International Bureau of Measurement (IBS): 3.392 231 376(8) μm

National Research Council of Canada (CNRC): 3.392 231 40(2) μm

The first two are defined in terms of barycentric points, and the last is defined in terms of intermediate points. The International Advisory Committee on Definition of Meters (CCDM) decided in June 1973 to use the following data as standard values (recommended values) for methane spectral line wavelengths

$$\lambda_{CH_4} = 3.392\ 231\ 40 \text{ μm} \tag{4}$$

The uncertainty is 4×10^{-9}. Therefore, the standard value was determined by CCDM in 1973:

$$c = (299792458 \pm 1.2) \text{ m/s} \tag{5}$$

The uncertainty is 4×10^{-9}. Later (1972~1974), several new measurements appeared, but they were all within the uncertainty range of the above standard values. This value was thus endorsed by the International Astronomical Union (August 1973) and the International Metrology Conference (1975). In 1983, the CGPM-17 made the following statement on the unit of length: "The travel length of light in vacuum in the period of $(299792458)^{-1}$'s is called 1 meter." Obviously, this is defined by taking the result of formula (5) as the most accurate value of the speed of light in vacuum.

However, the meter definition adopted and promulgated by CGPM-17 in 1983 must be understood as a universal physical constant without error, i.e.

$$c = 299792458 \text{ m/s} \tag{5a}$$

In this statement, ±1.2m/s is removed, which means that the uncertainty of the value c is zero. Such coercion is questionable; Moreover, since 1983, for nearly 20 years, the situation in the international metrology community is that it is very difficult to indirectly realize the definition of meter according to the formula $c = \lambda f$. In order to achieve the definition of meter with certainty, the wavelength value of the specified frequency stabilized laser is required as the standard spectral line. At that time, director of International Bureau of Metrology, Dr. T. Quinn, personally issued a "Notice on the realization of the definition of meters" (Metrologia, Vol.36, No.2, 211) in 1999, indicating that there are finally 12 kinds of lasers available. This situation shows that the implementation of the 1983 meter definition is not smooth. Dr. Quinn later elaborated on the indirect realization of the metre definition several times.

Let's consider the essence of the 1983 definition. Write formula (1) as follows:

$$\lambda = \frac{c}{f} \tag{1a}$$

If c immobilized, units of length can be derived from frequency (that is, time). Then, the measurement technology can depend not on (not pursue) reducing the uncertainty of the wavelength, but on the high level of light frequency measurement. Therefore, the International Bureau of Metrology is actively promoting the 1983 meter definition, not out of scientific considerations, but for the convenience and need of measurement technology. After the definition of meter was published in 1983, many metrologists in the world said, "The measurement of the speed of light that has lasted for 300 years can come to the end." They also said that "this was a perfect full stop". The author thinks that such view and practice are wrong. Science has no limits and endless development. No one can "ban research" or "ban testing" on a certain academic topic or direction. This is decided by the essence of natural science. When it comes to the speed of light in vacuum, measurements that have been going on for more than 300 years should not stop. This is not only because of the never-ending nature of scientific development, but also because of the existing problems in the definition of meter...We would even go so far as to say that the 1983 "ban" has done science a disservice. The author's view is clear: the measurement and research of the speed of light should not stop after more than 300 years.

This paper emphasizes that the seven basic units of metrology should be independent of each other and should not cross influence each other. This is a fundamental principle of modern metrology. The current definition of the meter violates this principle by using the unit of "seconds". This means that the meter definition depends on the second definition. Some people think it's good, but we can't laugh at that. Each of the base units should exist independently of other units. Many metrologists have sadly overlooked this. Another problem with the current meter definition is the confusion of the relationship between the basic unit and derived unit. According to formula $\lambda = c/f$, since frequency (corresponding time) is the basic unit, wavelength can only be derived unit. Thus, this definition effectively makes length lose its status as the fundamental unit and become the derived unit; This is very inappropriate. A closer look at the 1983 meter definition reveals more problems. If it is a light wave (light is, firstly an electromagnetic wave), then the definition should specify that it is a plane wave. But the ideal plane wave is not technically available, so what to do?

In other words, the speed of light should be the ideal velocity of a plane wave; If not, there will be effects such as curvature effect; and so on. In addition, there are some theoretical and experimental problems in the current definition of meter, which will be discussed one by one.

THE SPEED OF LIGHT CANNOT BE CONSTANT IN A REAL PHYSICAL VACUUM

"A vacuum is empty space without matter", this is an old saying in classical physics. In fact, we can never be sure if a space is really empty, even if the air is pumped out of it first to achieve the so-called "ultra-high vacuum". That's because there are plenty of photons that are constantly being created and then annihilated, albeit briefly, but virtual photons can do just as much physical action as ordinary photons. Evidence has long been available, such as Spanish scientists who found in 2011 that rotating bodies (graphite particles with a diameter of 100nm) slow down in an engineered vacuum, indicating that the vacuum also has friction. In fact, there are plenty of photons in space that are constantly being created and annihilated before we can measure them directly. Although they appear only briefly, these "virtual photons" can exert electromagnetic effects on objects just like ordinary photons. Scientists at the Institute of Optics of Spain's National Research Council say this electromagnetic action can slow down the rotation of objects. Just as two cars collide head-on with more force than rear-end, a "virtual photon" colliding with a rotating object in the opposite direction produces more force than it does in the same direction. The degree of deceleration also depends on temperature, because the higher temperature, the more "virtual photons" are created and annihilated, creating more friction. At room temperature, it takes about 10 years for a 100nm diameter graphite particle, which is abundant in interstellar dust, to spin down to about a third of its initial speed; At 700°C (the average temperature in the hot region of the universe), the process takes just 90 days.

The findings reported in *New Scientist* suggest that a vacuum does not guarantee constant values for precise measurements. Now, there are three situations when we are faced with a physical vacuum:

1. The effect of quantum vacuum oscillations is that the speed of light in a vacuum may not be a constant, but rather fluctuate, albeit slightly, around an average value.
2. Quantum vacuum polarization also has a similar effect and is periodic.
3. Casimir effect not only shows the correctness of quantum vacuum view, but also brings the diversity of vacuum and the possibility of faster-than-light speed (superluminality).

First look at the effects of quantum vacuum oscillations, which are related to the physical effects of virtual particles. Quantum field theory (QFT) considers that all quantum fields in the vacuum state are still moving, that is, all modes are still oscillating in the ground state, which is called vacuum zero-point oscillation. Virtual particles appear, disappear and transform into each other constantly in vacuum because of the interaction between quantum fields. The Website of Science Daily reported that French scientists and German scientists respectively put forward their research results, the content is that the speed of light is a real characteristic constant, and the quantum theory holds that the vacuum is not empty, but a flickering particle. This causes the speed of light not to be fixed, but to have fluctuating values. So today physicists are starting to get it right thinking.

However, when the interaction between particles and vacuum is considered, the physical phenomenon of vacuum polarization appears. For example, positively charged particles attract virtual electrons in vacuum and repel virtual positrons in vacuum. That changes the way the virtual cloud's charge is distributed. This situation is similar to the phenomenon of dielectric polarization in classical physics. There are four physical interactions in nature; electromagnetic interaction and weak interaction belong to the same mechanism and are described by the same equation, so it is called weak-electric unified theory. But in the vacuum polarization effect of electromagnetic action (also known as the electron field Dirac vacuum polarization effect), photons polarize the vacuum, creating pairs of electrons (electron e^-, positron e^+) that create charges and currents, which then return to photons. In the weak action vacuum polarization effect (also known as the neutrino field Dirac vacuum polarization effect), Z^0 bosons polarize the vacuum, producing neutrino pairs, resulting in weak charges and weak flows, and then returning to Z^0 bosons. Feynman diagram can be drawn in both cases. The difference is that the former has no static mass and the latter has static mass. This comparative study can deepen the understanding of vacuum polarization. American physicist J. Franson published a paper in June 2014, which attracted wide attention in the physics circle. The paper claimed that it had been proved that the speed of light was slower than the value thought in the past. His argument is based on observations of supernova SN1987A in 1987, when photons and neutrinos were detected on Earth from the explosion. Photons arrived 4.7 hours later than neutrinos, a phenomenon that had previously been only vaguely explained. Franson thinks this may be caused by the vacuum polarization of the Photon—it splits into a positron and an electron and recombines into a photon in a very short time. Under the gravitational potential, the particle energy changes slightly during the recombination, making the speed slow. As the particles travel 168,000 light-years (SN1987A to Earth), this constant merging and splitting will cause the photons to arrive late.

Another factor is the Casimir effect on the speed of light. If two parallel metal plates are put in a vacuum, the inner and outer states of the plates are not the same. The vacuum degree between the two plates is higher and deeper, so it has the force to make the two plates close to each other. This Casimir effect has been experimentally demonstrated, so the above statement of "two vacuums" is correct. This makes it logical that the speed of light inside and outside the plate may be different. Thus, it is the change in boundary conditions that affects the vacuum and thus the propagation speed of electromagnetic waves. In other words, the propagation of light depends on the structure of the vacuum, which is the basic idea of quantum physics. Due to the Casimir effect, we can distinguish between the following two: (1) normal vacuum (also known as free vacuum); (2) The vacuum between the plates with plates is characterized by a reduced vacuum energy density, so the author believes that it can also be called negative energy vacuum.

Now, considering vacuum as a unique medium, its refractive index and wave velocity can be calculated:

Phase velocity

$$v_p = \frac{c}{n} \tag{6}$$

Group velocity

$$v_g = \frac{c}{n_g} \tag{7}$$

where, n is the phase refractive index, referred to as the refractive index; n_g is the group refractive index. The relation between phase refractive index and group refractive index is

$$n_g = n + f\frac{dn}{df} \tag{8}$$

For non-dispersive media, $dn/df = 0$, so $v_g = v_p$, group velocity is consistent with phase velocity.

In 1990, K. Scharnhorst published the paper "Light propagation in vacuum between bimetallic plates". The Casimir effect structure is analysed. Two metal plates close together; This imposes certain boundary conditions on the photon vacuum fluctuation. Scharnhorst calculated by quantum electrodynamics (QED) method, and obtained that the refractive index perpendicular to the direction of the plate surface is:

$$n_p = 1 - \frac{11}{2^6 \times 45^2}\frac{e^4}{(md)^4} \tag{9}$$

Note that the interplate is in vacuum state, and the above formula represents $n_p < 1$; In formula (9), d is the distance between two ideal conductive plates, and m is the mass; m is defined as the speed of light in normal vacuum or free vacuum, then the c is

$$c = \left\{1 + \frac{11}{2^6 \times 45^2} \frac{e^4}{(md)^4}\right\} c_0 \tag{10}$$

where, c is the speed of light in the vertical direction of the plate surface under the condition of interplate vacuum, and the difference is of c and c_0 due to the change in the vacuum structure, which is caused by the placement of double plates. The result is $c > c_0$, here $c_0 = 299792458$m/s, c is faster than the speed of light. Further calculation gives:

$$\frac{\Delta c}{c} = \frac{c - c_0}{c} = 1.6 \times 10^{-60} d^{-4} \tag{11}$$

If $d = 1\mu m$, $\Delta c / c = 1.6 \times 10^{-36}$, it is very small; but even this it is not consistent with special relativity (SR). d can be reduced again, for the 1nm gap (d=1nm), the increment $\Delta c = 10^{-24} c$; This data is also very small, but theoretically important. In short, Scharnhorst did not calculate "the speed of a photon traveling between two metal plates," but the speed of a wave traveling vertically between two plates, and found that the phase velocity was slightly higher than the speed of light ($v_p > c$). When the frequency is not high, the dispersion can be ignored and the group velocity is equal to the phase velocity, so the group velocity is slightly higher than the speed of light ($v_g > c$). To sum up, "vacuum" changes the speed of light through a variety of physical processes. Therefore, how to understand and define the "vacuum" of "the speed of light in vacuum" becomes a problem.

ON THE THEORETICAL BASIS OF THE CURRENT DEFINITION OF METRE

The International Bureau of Metrology did not say that the 1983 definition of the meter was based on the theory of special relativity (SR), but we can conclude that this is the case because SR has a principle of invariance of light speed. This paper points out two important points: first, the principle of constant speed of light has its own shortcomings, that is, it is not satisfactory in logic self-consistency; Secondly, as a postulate of SR, "the principle of invariance of light speed" lacks real experimental proof. In recent years, however, some experimental results may falsify the invariance of light speed. This undermines the theoretical basis of the 1983 definition of metre. SR is based on two postulates and a transformation. The first postulate states that "the laws of physics are the same in all inertial systems", that is, in all inertial systems, not only the laws of mechanics are equally true, but also the laws of electromagnetic and optics. The second postulate states that "light in vacuum always has a certain speed, independent of the motion of the observer or the light source, and independent of the colour of the light". This is what Einstein called the L principle. In order to eliminate the apparent contradiction of the above two postulates (relativity of motion and absoluteness of optical propagation), SR holds that "principle L is true for all inertial systems". In other words, the coordinate transformation between different inertial frames must be Lorentz transformation (LT). On the second postulate, Einstein said in 1905 that "light in empty space always travels at a certain speed, independent of the motion of the emitter". The 1921 statement reads: "At least for a certain inertial system K, the hypothesis that light travels at speed in vacuum is also confirmed. According to the principle of special relativity, we must also assume that this principle is true for any other inertial system". In 1949, it was stated that "light always travels at a constant speed in vacuum, independent of the colour of light and the motion of the light source".

Another core concept associated with the second postulate is the relativity of simultaneity. If clock at point A can define the time t_A of an event at A, and clock at point B can define the time t_B of an event at B. But how does the compare of t_A and t_B? A definition of simultaneity is needed. For this reason, Einstein proposed the assumption that the speed of light is constant. If an optical pulse is being sent at t_A, the time indicated by the clock at B is

$$t_B = t_A + \frac{L}{c_{AB}} \tag{12}$$

where L is the distance between two points, and c_{AB} is the one-way speed of light from A to B. But c_{AB} is unobservable, because it depends on the prior synchronization of clocks A and B (one-way speed of light is related to the definition of simultaneity). Einstein now defines simultaneity in terms of $c_{AB} = c_{BA} = c$, as opposed to the principle of constant speed of light in the loop (experiments so far have only shown constant speed of light in the loop, not in one direction). If the principle of invariance of the speed of light is correct, time and simultaneity are not absolute, and length measurements lose their absoluteness (they give different results in different inertial systems). It must be pointed out that the invariable absoluteness of the speed of light is incompatible with the principle of relativity in a narrow sense, which emphasizes the relativity of motion. There is an irreconcilable contradiction between the two basic assumptions of SR, which was demonstrated by E. Silvertooth in the 1970s. Einstein himself had doubts about this and tried to prove that there was only an apparent contradiction, but it did not solve the compatibility. Einstein actually put the cart before the horse and looped logic when he proved compatibility by using two inferences derived from postulates: relativity and length contraction. Einstein asserts that there is no absolute motion to adhere to the principle of relativity, and introduces light, which has no rest system and therefore is absolute motion, to construct a second postulate. The two postulates are extremely incompatible.

More people think that the current statement of the principle of the invariable speed of light is a hypothesis, so far the lack of real experimental proof. Even relativistic scholars acknowledge this, for example, as Prof. Y. Zhang pointed out, saying that "the invariable speed of light has been experimentally proved" is not true. Einstein's principle of invariance of light speed refers to the one-way speed of light, that is, the speed at which light travels in any direction. But many experiments measure not the isotropy of one-way light but the invariance of loop light speed. In addition, the 1994 reprint of [12] emphasizes the unpredictability of one-way speed of light because "we have no prior definition of simultaneity, and the definition of speed of light depends on the definition of simultaneity." Zhang believed that Newton's absolute simultaneity could not be realized in reality. Einstein proposed the assumption that the speed of light is constant, that is, the optical signal against the clock; ... It is a hypothesis because it is not an empirical result, because the isotropy of the unidirectional speed of light has not (and cannot) be proved experimentally. To measure the speed of light in one direction, one has to check two clocks in different places, and to do this one has to know the exact value of the speed of light in one direction. This is a logical cycle, so attempts to test the speed of light in one direction are futile. (Many experiments listed in reference [12] are to prove the principle of constant speed of light in the loop.)

In terms of experiments, literature [12] lists 12 experiments on "invariance of light speed" (from 1881 to 1972) and 16 experiments on "independence of light speed and motion of light source" (from 1813 to 1966). But the former only shows the loop speed of light invariable principle, the latter only applies to $v \ll c$ case. Some people believe that SR theory has been firmly established over the centuries and can't be wrong. This is not true. In fact, there have been people in the scientific community for many years who have put forward ideas that are different from the invariable principle of the speed of light. In 1936, A. Proca proposed the modification of Maxwell equations considering the rest mass of photon ($m_0 \neq 0$). However, in the theoretical system of Proca equations, the invariable principle of light speed is no longer correct, and the speed of light will be related to the frequency of electromagnetic waves.

CHINESE SCIENTISTS REFUTE THE PRINCIPLE OF INVARIABLE SPEED OF LIGHT WITH EXPERIMENTS

It must be emphasized that in recent years, Chinese scientists have made unique contribution of using well-designed experiments to obtain reliable data after long-term study, thus falsifying the principle of the invariable speed of light. We are only going to talk about two things here; First, Prof. R. Wang used modern technology to reproduce the Sagnac type experiment, using moving fiber, hollow fiber, zigzag moving fiber and segmentation fiber, at different speeds. It is proved that the speed has an effect on the propagation of light in the fiber moving back and forth, and that the propagation time of light is different. "Our results falsify the principle of the light-speed constancy", Wang said in 2005. Now we will focus on the large distance deterministic experiment of Chinese scientists on the assumption that the speed of light is constant. Now we review the theory first; since the principle of invariance of the speed of light comes from the static ether, and the Michelson-Morley experiment denied the ether, should the principle of invariance of light speed still exist? Einstein's approach not only preserved the hypothesis that the speed of light does not change, but enhanced it. He said: "The first step is to reject the ether hypothesis: then the second step is to make the principle of relativity accommodate the fundamental lemma of Lorentz's theory, because to reject this lemma is to reject the basis of the theory." The following is the lemma: "The speed of light in vacuum is constant, and light is independent of the motion of the luminous body. " We raise this lemma to principle. For simplicity we'll call it the principle of invariance of the speed of light. In Lorentz's theory, this principle is only true for a system in a special state of motion: that is, the system must be static relative to the ether. If we want to preserve the principle of relativity, we must allow the invariable principle of the speed of light to hold for any system of non-acceleration motion.

Einstein added, "As a rule of thumb, we also put the following values

$$\frac{2\overline{AB}}{t'_A - t_A} = c \tag{13}$$

as a universal constant, the speed of light in empty space. It is essential to define time by means of a stationary clock in the stationary system. We call the time now suitable for the stationary system definition 'stationary time'."

Obviously, there are some hypotheses that need to be tested experimentally. Einstein's 1905 paper had no such experimental proof, so Einstein called his approach a hypothesis "aided by some physical experience." For a hundred years people have mostly accepted it immediately, without considering whether there is a problem with it. To summarize, Einstein stated in 1905: place an identical clock in two places (A, B) in space, and the event at A corresponds the time t_A, event at B corresponds the time t_B. But there is still no definition of public time. It is now stipulated that the time required for the optical signal A→B is ($t_B - t_A$), which is equal to the time required for the optical signal to reflect back to point A, i.e.($t'_A - t_B$); then

$$t_B - t_A = (t'_A - t_B) \tag{14}$$

so these two clocks are in synchronization. ... Einstein's above "regulation" is in fact a description of his second postulate (the invariable principle of the speed of light), since Eq.

(14) is actually equal to

$$\frac{L}{c_{AB}} = \frac{L}{c_{BA}} \tag{15}$$

namely, $c_{AB} = c_{BA}$; But this hypothesis required experimental proof, and Einstein could not come up with one. In 2004, Prof. J. Lin pointed out that modern technology is capable of measuring t_A, t_B and t'_A, using space technology. The problem can be looked at another way, Einstein's equation is actually:

$$t_B = \frac{1}{2}(t'_A + t_A) \tag{16}$$

This is the time definition of the arithmetic mean, based on the assumption that the speed of light does not change. If formula ($t_B - t_A = t'_A - t_B$) is correct, that light takes the same oneway time of "forward" and "back" the same way, then, "the speed of light is independent of the direction in which it travels". Thus, the "principle of invariance of the speed of light" (or "principle of constancy of the speed of light") becomes an essential theoretical assumption. But that, of course, is a thing of necessary experimental proof. In short, the principle of invariance of the speed of light, one of the two cornerstones of SR, was simply Einstein's way of preserving the mathematical form of the original physical equations based on the static ether; That is, the time of light signal passing through "forward" and "back" is defined to be equal, and the concepts of "stationary system" and "stationary clock" are introduced. So how did a team of Chinese scientists design a large-distance experiment using satellite technology to falsify the principle that the speed of light does not change? First, the research team, led by J. Lin, he is a distinguished scientist at the Chinese Academy of Launch Vehicle Technology, and he is Member of the International Academy of Astronautics. Prof. Lin is a famous expert in satellite navigation technology. His original and novel ideas and methods of redefining space and time based on rocket measurements have received attention and praise in the scientific community. The team also includes experts from the National Time Research Center of the Chinese Academy of Sciences. The team did not specialize in physics and did not deliberately "find faults with relativity". They decided to do the research after discovering problems with SR theory during their long spaceflight practice. The relevant work received financial support from the state, and finally the achievement appraisal meeting was held under the auspices of China Aerospace Science and Technology Corporation, which was approved. A paper on the results was published in January 2009. The title of the paper is "The Crucial Experiment for Checking Einstein's Postulate of the Constancy of light Speed".

It is a unique work for Lin's team to carry out experiments on large distances with the help of satellites by using high-tech aerospace technology. As we know, the world entered the space age in 1957. Time technology (atomic clocks and time signals travel over long distances) and satellite communications (navigation messages) make one-way optical (electromagnetic) signals a reality. Experimental conditions were then available to test whether Einstein's 1905 paper's hypothetical defining equation $t_B - t_A = t'_A - t_B$ was true or not. In 2008, Prof. Lin completed a decisive experiment on Einstein's 1905 definition of simultaneity on the TWSTT (Two-way Satellite Time Transmission) facility of national Timing Research Center, Chinese Academy of Sciences. (Former Shanxi Observatory). Experimental observations show that Einstein's postulated equations are not valid in the presence of relative motion! The principle of experimental verification is based on the principle of special relativity and the definition of one-way optical (electromagnetic) signal simultaneity. By comparing the measurement mechanism of the simultaneity definition of one-way optical signal with that of Einstein's two-path optical signal, it is proved that, under the condition of relative motion between A and B, the signal transmission time of the two-path optical signal is necessarily not equal when it is decomposed into two one-way optical signals "forward" and "back". In the experiment conducted by Prof. Lin, the cesium atomic clocks of Xi'an Lintong Ground Observation Station and Urumqi Ground Observation Station carried out bidirectional time transmission through Sinosat and Zhongwei I satellite respectively. The observation data proved that although the relative speed between the satellite and the ground station was only 1m/s, the distance of signal transmission through the synchronous satellite reached 72,000 km, resulting in the time difference between the "forward" and "back" one-way signals between Xi'an Lintong station and Urumqi station, with a difference of 1.5ns. The observational results confirm the conclusion of Lin's theoretical analysis, and the uncertainty in the experiment is ±0.01ns. The results of this decisive experiment, carried out by the Space System and impossible to carry out in a ground-based laboratory, have shaken one of the cornerstones of SR. Therefore, Lin believes that the traditional time and space theory should be reconsidered from the perspective of satellite system and inertial navigation measurement principle. From the perspective of one-way light (electromagnetic wave) signal characteristic of satellite navigation, the Galilei transform should be restored to its position. On the face of it, all it takes a ground station (for point A) and a satellite (for point B). We can to do the experiment. But they don't work that way; The development of modern atomic clock technology and aerospace technology makes it possible to synchronize time using one-way optical signals. The TWSTT concept synchronizes the time of two atomic clocks at a distance by sending electromagnetic signals (pulses of seconds from different clocks at the same time) to each other at the same time. Now, Lin and his colleagues are using two atomic clocks that A_I, A_k, in principle, should send light signals to each other at the same time. In fact, the observation stations which are far away from the earth and rotate with the earth in the geocentric inertial system cannot realize the direct line of sight observation

A Study and Discussion of the 1983 Meter Definition

and communication. Therefore, the time synchronization observation model of the two direction one-way optical signal of the clock and clock is realized by the transmission of the geosynchronous fixed-point communication satellite S_n. FIG. 1 is a schematic diagram of the decision experiment arrangement.

In practical experiments, there are many factors to consider, such as the influence of the motion of ground stations and satellites in the geocentric inertial system on the observation equation, and other complex problems. There's even a Sagnac effect to consider. Lin's team finally obtained the observation equation of two-way satellite time transfer. In principle, one-way signal observation consists of clock difference, Sagnac effect and signal transfer time. The clock difference and Sagnac effect are corrected for the actual one-way observations before the two basic elements of Einstein's one-way optical signal simultaneity definition can be obtained: the optical signal's arrival reading on the time clock and the time required for the signal to travel this distance.

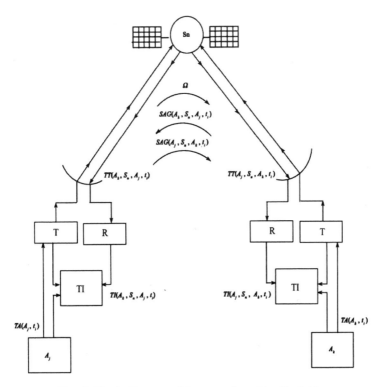

Fig. 1 Block diagram of the experiment by Prof. Lin
(S_n—satellite, R—receiver, T—transmitter,
TI—time interval computer, A—atomic clock)

However, in the two-way satellite time transmission, both parties have mastered the observation values of the two direction one-way signals sent by both parties through communication means. The two direction one-way signal observation sum clock difference and Sagnac effect cancel automatically due to the asymmetry in principle, so the relationship between the one-way signal transmission time and the reading on the clock of both parties is finally obtained. Finally, the conclusion of Lin is: "The crucial experiment for checking Einstein's postulate of the constancy of the speed of light was performed at the high precision TWSTT (Two Way Satellite Time Transfer) facility of the National Time Service Center, Chinese Academy of Sciences. The principle of the crucial experiment was based on the principle of special relativity and the definition of simultaneity by means of one way light signal. By comparison the measurement mechanisms of one way light signal simultaneity and 'to-and-fro' two way light signal simultaneity, the principle of the crucial experiment has proved: if there exists relative motion, the 'uplink' and 'downlink' light signal passage times of the 'to-and-fro' two way light signal are not equal. The cesium atomic clocks at Xi'an station and Urumuqi station transferred and exchanged pps time signals via Sino satellite and China Sat-1 satellite. The observation uncertainty is the order of 0.01 ns. The observed data have proved the equality $t_B - t_A = t'_A - t_B$, which was introduced by definition in Einsteins 1905 paper, is not valid in case if there exists relative motion between A and B."

But, is the time difference measured by Lin due to the different "forward" and "back" distances of optical signals caused by satellite drift? Prof. Q.Ma is an expert on SR theory, and his two monographs (Chinese book published in Shanghai and English book published in New York). The author suggested that he study this problem, and after reading the relevant materials, he wrote an article "On the Significance of Lin Jin's Experiment" (to be published), in which he said:

In the dual-path signal transmission experiment conducted by Lin's team with satellites published in 2009, it was found that the one-way signals "forward" and "back" did not take the same time to pass through. We analyze the significance of Lin's experiment. Many past experimental measurements have shown that the speed of light varies in rotating systems, and that the speed of light varies from east to west on the Earth's surface, known as the Sagnac effect. After excluding Sagnac effect, Lin's experiment measured the time difference between "forward" and "back" one-way optical signals caused by the drift speed of 1m/s satellite. Some people think that the time difference is caused by the difference in the transmission distance of the forward and back light signals caused by the satellite drift. However, the change of relative velocity is the change of actual distance caused by the movement of relative objects, so the time difference measured by Lin experiment is caused by the difference of relative velocity of forward and back light signals. Based on the estimation of the theoretical value of the Lin's experiment based on the different relative velocities of the forward and back light signals caused by the satellite drift, we accurately obtain the measured value of this experiment (1.6ns). Therefore, it is concluded that the Lin's experiment clearly does not support Einstein's assumption of constant speed of light. Therefore, after detailed analysis, Prof. Ma concluded that the time difference measured by Lin's team was not caused by the satellite's drifting motion. He said that the experiment proved that electromagnetic signals transfer satellites drift under the condition of one-way variable speed of light. The time difference calculated by Ma is 1.6ns, which is very consistent with the 1.5ns measured by Lin. He also suggested better experiments in space.

Our conclusion is that Chinese scientists, standing at the starting point of a new era, have deepened and interpreted the problem. The title of the paper of Lin et al. indicated that he wanted to do a deterministic experimental test of Einstein's hypothesis that the speed of light is constant. They did successful experiments on a very large distance (72 000km) to test whether the one-way speed of light is isotropic, and came to a negative conclusion, answering the long-standing question. Therefore, the author believes that the Lin's experiment may have shaken the foundation of SR. The author also believes that this is a bold experiment, and very important; The space powers, such as the United States and Russia, have not done. At the same time, it also proves the thesis of this paper that the current definition of meter needs to be improved. Einstein rejected the invariable principle of the speed of light as early as 1911 on the grounds that gravitational potential slows the speed of light down. Einstein is often incongruous and confusing; the speed of light is just one example.

ON "INCONSTANT PHYSICAL CONSTANTS"

Physical constants arise because new theories and laws are constantly emerging in physics. Some constants are very famous and are associated with the names of great physicists. For example, the universal gravitational constant (G) reminds us of I. Newton, who discovered the universal gravitational force. The mention of electron charge (e) reminds us of J. Thomson, who discovered electrons; Mention of the Planck constant (h) reminds us of M. Planck, the inventor of quantum theory; and so on. The metrology circle has always attached great importance to the basic physical constants, and in recent years it has been advocated to establish datum based on the basic physical constants. On November 16, 2018, the member states of the International Bureau of Metrology (more than 60 countries) voted to adopt a resolution to revise the International System of units (SI); According to the resolution, four basic units of SI will be defined instead by fundamental physical constants:

Kilogram (kg) — Planck constant h

Amperes (A) — electron charge e

Kelvin (K) — Boltzman constant k

Mol — Avogadro number N_A

As for the other three basic units, the meter (m) was first defined in 1983 in terms of the speed of light in vacuum (c). There are also two units, the second (s) and the candela (cd), which are not yet defined in terms of fundamental physical constants.

This is the most significant change in metrology since 1960. The International Bureau of Metrology hopes to bring a new atmosphere to metrology, and also hopes that the SI system will be stable in the future... However, in recent years, there have been some new discussions about the fundamental physical constants in the international physics circle, among which two things are very interesting. First of all, can all measurable dimensions that characterize the physical world be extrapolated as one parameter in principle? In fact, no one yet knows why these constants are these value. Although scientists have been able to determine these constants with great precision in the laboratory using highly sophisticated techniques, their origins remain unknown. One theory is that these values determine the conditions under which galaxies, stars and other cosmic formations can exist, and create the conditions for life to emerge and develop. But this is a bit like the anthropic principle, which sees everything as being in service of human birth and existence. But neither earth nor men is not in the center of universe, and their existence or absence is of little consequence to the cosmic universe. We can imagine god creating the world. Right before the big bang, God (the nature) is sitting at the console wondering: "What should I set for the speed of light?" "How much charge should I put on the electrons?" "How do I value the parameter h that determines the quantum size?" It is not clear whether "God" created everything in the universe after careful consideration or by grabbing at random numbers.

In addition, research suggests that the term "constant" may not be the right word, as they may vary over time and space. Since 1930, researchers have speculated that some constants are not constant. Two terms are now popular internationally: Inconstant constants, and Not so-constancy constancy. These are based on related studies. In 2001, the international research team led by J.Webb used the world's largest astronomical telescope set in Hawaii to observe and study some of the most distant quasars in the deep space of the universe, and found that the microscopic structure constant hundreds of millions of years ago was smaller than the current value, thus judging that the speed of light in the early stage of the evolution of the universe was larger than the current value. The fine structure constant (FSC) is defined as

$$\alpha = 2\pi \frac{e^2}{hc} \tag{17}$$

The standard value given by international metrology circle is $\alpha^{-1}=137.03599761$. Physicist J. Barrow was a member of the Webb's team and participated in the study for two years, he says: "we look at the spacing of absorption lines for different chemical elements, which depends on any small change in the red shift as absorption occurs. Since the light leaves these stars (5~11) Ga earlier, it is possible to determine whether there α has been a change in the past 11Ga by comparing the observed line interval with the current laboratory line interval. In two years, 147 quasars were observed and the results were quite unexpected—the early values were about $|\Delta\alpha/\alpha|=7\times10^{-6}$ smaller than they are now." His words indicate that if averaged over a year, it is equivalent to $|\Delta\alpha/\alpha|=5\times10^{-16}a^{-1}$ (the observation interval is 3Ga to 11Ga ago). That may not seem like much, but being a physical constant doesn't allow for such annual variations. So some in the international scientific community say this is one of the amazing discoveries of experimental physics in the last 50 years. New scientist reported on 3 July 2004, a reanalysis of data from the Oklo reactor in West Africa a month ago showed that it had grown 4.5×10^{-8} in the past 2Ga, so it was slightly smaller in the past than it is now, this conclusion is consistent with that of Webb's group. There are three elements that constitute the fine structure constant, namely h, e and c. So who is responsible for the change? For simplicity, consider only one possible case, where one of e, h, or c fails to hold constant. Even so, there are different views. In August 2002, physicist P. Davies described the results of his team's study of Webb's paper in *Nature*. To determine which physical constants are likely to change, Davies' team applied analytical techniques such as the second law of thermodynamics. It turns out that the speed of light c was not constant and slowed down over billions of years. If so, he thought relativity and the $E=mc^2$ formula might have to be abandoned! Webb's observations show that the atomic structure that emits quasar light is slightly different from that seen by humans, but the difference is significant... Recently, there are many other literatures on variable speed of light, such as [22] and [23]. It should be noted that we are not saying that these fundamental physical constants are not to be trusted; Nor is the leadership of the International Bureau of Metrology, which is at the helm of global metrology, all wrong. In this paper, the speed of light in vacuum is taken as an example to show that there are many factors affecting the constancy and stability of the constant, and they are not negligible. Fundamental physical constants cannot be oversimplified or idealized, because there is no absolute constant in practice. Whether the isotropy of the value is guaranteed is also a big problem, reflected in the time difference has been ns class.

CONCLUSION

Length measurement is the basic measurement which is closely related to human life and its importance is beyond doubt. In ancient China, due to the need of measuring land, building houses, and building bridges, there were not only length measuring tools (rulers), but also length measuring standard. The latter takes its law tube resonant frequency (equivalent to wavelength) as the ruler reference. The ancient Egyptians built great structures—pyramids and temples—that could not be carried out without geometrical measurements of length and angles. In modern times, early weights and measures have MKS system, CGS system, the first letter of the former stands for meters (m), the first letter of the latter stands for centimeters (cm); This is all proof that the length measure is the head... Modern industrial production and science and technology advance rapidly, Planck length in micro aspect, light year (ly) in macro aspect, both reflect the importance of length measurement.

In 1960, the meter was defined in terms of atomic radiative transitions; and in 1983, the fundamental physical constant (the speed of light in vacuum) was used to define the meter. These are all valuable efforts and have yielded considerable achievements. But science is constantly evolving, and definitions are not permanent. What new, advanced and rational definitions are likely to emerge? This article does not provide a credible solution. We only discuss the problems existing in the current definition of meter, hoping to attract the attention of the physical and metrological circles and find new solutions in extensive and in-depth exploration. Future methods must be logically consistent, experimentally feasible and reproducible. To this end, the primary task is to develop basic science.

If we dig into the basic theory of physics, we find a strange phenomenon. For example, the invariable principle of the speed of light, which Einstein solemnly proposed in 1905 when he proposed SR, was refuted by Einstein himself from 1907 to 1911. For example, in Einstein's 1911 paper, the core idea was that the gravitational potential slows down the speed of light. This view was carried on even 103 years later (in 1914) by the physicist J.Franson. Franson was a proponent of general relativity (GR), so where did he leave SR? Such a situation, so that we have to suspect that the principle of the invariance of the speed of light is fundamentally wrong, with it as the theoretical basis of the current definition of the meter has collapsed!

REFERENCES

[1] Shi C Y. An introduction to modern metrology [M].Beijing: China Metrology Press, 2003.

[2] Giacomo P. Documents Concerning the New Definition of the Metre[J]. Metrologia, 1984, 19: 163.

[3] Shen N. Optical Frequency Standard[M]. Beijing: Peking University Press, 2012.

[4] Einstein A. The Meaning of Relativity[M]. Princeton, N J, USA: Princeton University Press, 1922.

[5] Zhang Y. Experimental Basis of Special Relativity[M]. Beijing: Science Press, 1994.

[6] Ma Q. Question the Logic Consistent of Relativity[M]. Shanghai: Science and Technology Literature Press, 2004.

[7] Ma Q. The Theory of Relativity[M]. New York: Nova Pub.Co., 2014.

[8] Webb J K, Murphy M T, Flambaum V V, et al. Further evidence for cosmological evolution of the fine structure constant[J]. Physical Review Letters, 2001, 87(9): 091301.

[9] Albrecht A , Magueijo J .Time varying speed of light as a solution to cosmological puzzles – art. no. 043516[J].Physical review. D Particles and fields, 1999(4):59.

[10] Franson J D. Apparent correction to the speed of light in a gravitational potential[J]. New Journal of Physics, 2014, 16(6): 065008.

[11] Qiu G. A Study on Metrology in Chinese History[M]. Beijing: China Metrology Press, 1992.

To Achieve Faster than Light Astronautic Travel Whether Human Beings can Use Wormholes or Warp Drive Propulsion*

INTRODUCTION

Flying outside the Earth system is called space flight, and beyond the solar system is called astronautics. The universe is just so big—Proxima Centauri is 4.3ly from Earth and Sirius is 8.8ly (ly is light-year, 1ly=9.5×10^{12}km). The speed of light in vacuum c =299792458m/s $\cong 3\times10^5$km/s, it would take years to get there at the speed of light! Einstein's 1905 paper "On the electrodynamics of moving bodies" developed the theory of special relativity (SR), which asserted that faster-than-light motion was impossible. However, Einstein's 1915 paper proposed the theory of general relativity(GR), and some physicists think it is possible to have faster-than-light motion without violating SR, based on GR's theory of spacetime bending. In 1988, M. Morris and K. Thorne published a paper in PRL entitled "Wormholes, time machines, and weak energy conditions"; Worm holes, the paper says, could solve the difficulties of interstellar travel by building a very large curvature for spacetime. Try to build and maintain the wormholes needed for interstellar travel; And wormholes can even create time machines and shock causality. In 2000, L. Ford et al. said in their paper that if wormholes could exist, they would become spherical portals to distant places. It doesn't violate the laws of physics, but it does require a lot of negative energy. Since the force of negative energy is repulsive, it prevents the wormhole from collapsing. Moreover, unlike space bending, which acts as a converging lens, negative energy acts as a diverging lens on light, which is necessary for entry into and exit from the wormhole.

Relativistic scholars proposed a faster-than-light spaceflight scheme called Warp drive based on GR without violating SR. In 1994, M. Alcubierre envisioned a method of faster-than-light spaceflight based on GR theory: a "spacetime bubble" (or "warp bubble") of curved spacetime, with a spaceship inside it. If the spacetime in front of the bubble contracts and the space time behind it expands, the distance to the destination will be shortened and the distance from the departure point will be increased. This would mean that the spacecraft was moving, but it was actually stationary relative to the neighboring world, and of course there was no need to worry about the "upper limit of light speed" of SR theory, which meant that the spacecraft could travel faster than the speed of light without violating the

---

* The paper was originally published in *Current Journal of Applied Science and Technology*, 2022, 41 (16), and has since been revised with new information.

prevailing theory of physics. But calculations show that this requires negative energy to surround the spacecraft. Since the law of conservation of energy cannot be violated, great negative energy is accompanied by an equal amount of great positive energy. No one knows how to get at this enormous power, and the solution is impossible even without error. So the concept has been ignored by the scientific community, except in science fiction movies.

Wormhole and warp drive propulsion are the products of GR theory, based on the concept of spacetime integration and curved spacetime. Although it is well known that SR asserts that "faster-than-light motion is impossible", GR denies this statement in practice, indicating that relativity has internal inconsistencies. But these studies all assert that "exotic matter with negative energy is needed".

The author did not accept the integration of space and time and curved spacetime; Now, personal views aside for the moment. We note that while many physicists espouse the theory of relativity, they firmly believe that faster-than-light space travel must be pursued in the future so that humans can explore beyond the solar system (or even the Milky Way), which is consistent with the author's view. Research trends in the United States (e.g., K. Thorne's insistence on wormholes and H.White's insistence on warp drive propulsion) have brought the issue back to the public. Therefore, I decided to participate in the discussion and contribute some of my own views.

For example, in the year of 2012, NASA scientist H.White said at a conference that the shape of the warp bubble could be improved to reduce the unrealistic energy requirements. His comments, and the meetings NASA held about them, were encouraging. In early 2022, the journal *European Phys. Jour. C* reported that White and his team had made a nanoscale warp bubble; Interest has also increased dramatically.

WORMHOLES AND SPACETIME BENDING

In 2014, the American science fiction film *Interstellar* is hot in all countries. The story imagines that in a future where humans are on the verge of extinction, NASA is trying to get people to leave Earth and fly to Saturn, and then travel through a wormhole to another galaxy. The journey from Saturn's orbit to the center of Andromeda was made at faster-than-light speed because it took so little time and the distance was so great (5×10^6 ly). The film was made with the help of physicist Kip Thorne of the California Institute of Technology (CIT), one of the authors of the wormhole theory. Thanks to special spacetime tunnel like wormholes, with one end in the orbit of Saturn and the other in the center of Andromeda, a hero can make the trip in seconds that would take light more than five million years to make. This is a fascinating part of the film.

In 1916, at the beginning of GR, L.Flamm pointed out that Schwarzschild solutions to Einstein's gravitational field equation described empty spherical wormholes if the topology was properly chosen. This was the earliest discovery, only a few months after the equations of the gravitational field were published. Many years ago, scientists proposed that wormholes were actually warped, deformed spaces, that could connect two different points in cosmic spacetime. The result is a tunnel-like structure that can be straight or bent. Thus, a wormhole is a tunnel through time and space that allows almost instant travel between distant locations. In the 1950s, J.Wheeler et al., and in the 1980s, K.Thorne et al, the meaning of their papers all like this situation.

Another imagines a "superspace" in which the curved spaces of our universe and those of other universes can be drawn as 2D images embeded in a higher-dimensional hyperspace. Hyperspace is just an imaginary tool, but it's useful for explaining wormholes. Thorne imagines a wormhole that goes through hyperspace. It could have two holes, like one on Earth and other on Vega. The two openings are connected by a hyperspace tunnel, perhaps only 1km long. In this way, one can enter from an opening near the Earth and exit from an opening near Vega, 26ly away. There's a famous diagram, that imagines our universe as a 2D curved surface, strongly curved, with a distance of 26ly along the surface, but a distance of 1km between the two holes. Wheeler's young assistant professor, M. Kruskal discovered the evolution of globular wormholes from solving the equation of the gravitational field—starting with no wormhole, with a singularity near Earth and a singularity near Vega; The two may then grow in hyperspace, meet, and annihilate, creating wormholes when they do. Then it shrinks and disappears. The time between creation and disappearance of a wormhole is very short.

K.Thorne calculations based on the equation of the gravitational field lead to the following finding: (1)Some exotic materials is needed to provide the gravitational force that pushes open the walls of wormhole; (2)The exotic matter penetrating the wormhole should have a negative energy density; But this is true from view of the light beam passing through the wormhole, and the energy density is still positive for the wormhole reference frame. As mentioned earlier, Wheeler suggested that wormholes could be self-destructing based on his quantum bubble hypothesis, in which virtual particles mysteriously appear and disappear all the time. Unfortunately, Wheeler's theory suggests that these flickering wormholes are tiny, on the Planck order of magnitude, or about 10^{-33}cm in length. In other words, wormholes are almost impossible to measure. To make them big, they have to have exotic matter; Because the negative properties of exotic matter might push the perimeter of the wormhole outward, making it large and stable enough for a person or spacecraft to pass through.

If we hope a wormhole can use to travel, it must at least allow the signal to pass through as light. The light at the entrance is convergent and the light at the exit is divergent. In order for the converging light to become astigmatism somewhere in the middle of the wormhole, this conversion must be done with negative energy. And since the force of gravity of negative energy is actually a repulsive force, it prevents the wormhole from collapsing. ... So everything depends on the possibility of generating negative energy. In

1978, physicists proposed a theory called "Quantum inequality", saying that "the amount of allowable negative energy is inversely proportional to its time and space scale", that is, the larger the negative energy is, the shorter the duration of its existence. Conversely, if the negative energy is weak, it can last for a long time. Moreover, for the greater negative energy, the corresponding positive energy comes closer to it. In the Casimir effect, the plates have to be very close together to get a lot of negative energy. In 1996, Ford et al. proved that the wormhole radius was less than 10^{-30}cm. To get a macroscopic wormhole, the negative energy needs to be concentrated in a very thin (say, 10^{-19}cm) area. ... In short, although quantum theory allows negative energy, it imposes strict limits on its value and duration, making wormholes a moot idea at this stage.

Ford and Roman explain the laws of nature associated with negative energy. Some mechanisms, such as making black holes thermodynamically compatible, suggest that negative energy can exist. But it's impossible to produce negative energy without limit, because that would contradict the second law of thermodynamics. Imagine, for example, a exotic matter generator that steadily supplies a negative energy flow outwards; However, due to the law of conservation of energy, it must have a positive energy flow as a by-product. If the two are directed in different directions, they become an inexhaustible source of energy for the positive energy region, and thus can be made into perpetual motion machines. But that's impossible according to the second law of thermodynamics.

Ford and Roman's paper, "Constraints on Negative Energy Density in Flat Spacetime," states that, unlike classical physics, in quantum field theory energy density can become negative at a point in spacetime without limitation. This violates known classical energy conditions, such as the weak energy condition. Specific examples, such as the Casimir effect and the squeezed states of light, are supported by practical observations. The theoretical expectation of black hole evaporation also includes negative energy density. On the other hand, if the laws of quantum field theory are not limited to negative energy, they may produce significant macroscopic effects that violate the second law of thermodynamics, such as wormholes, warp drive propulsion, and time machines.

Ford and Roman derived the negative energy bound as seen by an inertial frame observer for the free massless scalar field of 4-dimensional Minkowski spacetime (flat spacetime). Limits on the amplitude and duration of negative energy; Quantum inequality is written in a similar form to the uncertainty principle:

$$\hat{\rho} \geq -\frac{k}{t_0^4} \tag{1}$$

where $\hat{\rho}$ is the energy density integral and t_0 is the time characteristic width. Therefore, the longer the time required to maintain, the less negative energy can be obtained. In summary, analytical calculations show that wormhole sizes are strictly limited.

Go back to the 1988 paper by M.Morris and K.Thorne, a wormhole is a short "handle" of a spatial topology that connects widely separated regions of the universe. The Schwarzschild metric describes a wormhole if the topology is chosen correctly. However, the wormhole's boundaries prevented two-way operation, and its throat snapped so fast that it could not pass even in one direction. To prevent pinching (singularities) and boundaries, crossing the neck must be at non-zero pressure and energy. Then two questions arise: (i) does quantum field theory allow the type of pressure-energy tensor required to maintain a bidirectionally passable wormhole? (ii) Do the laws of physics allow the establishment of wormholes in the universe where the space parts are initially simply connected? If the laws of physics allow a traversable wormhole to exist, might they also allow such a wormhole to be transformed into a "time machine" that defies causality?

Wormhole creation must be accompanied by the choice of closed time-like curves and/or a discontinuous future light cone, which also defies weak energy conditions. The spacetimes created by such wormholes are known, but what is not known is whether the pressure-energy tensor specified by Einstein's equation in those spacetimes is allowed by quantum field theory.

Wormhole creation with great spacetime curvature will be controlled by the laws of quantum gravity. A plausible case requires quantum foam. For a topological type with a length scale of the order of Planck-Wheeler length, there is a finite probability amplitude:

$$\sqrt{\frac{G\hbar}{c^3}} = 1.3 \times 10^{-33} \text{ cm} \tag{2}$$

Imagine an advanced civilization pulling the wormhole out of the quantum bubble and magnifying it to classical size. This may be analyzed by developing calculations of spontaneous wormholes produced by quantum tunneling.

In addition, for any passable wormhole, a double ball encircled one entrance (but its outer spacetime was almost flat), as seen from the other entrance through the wormhole, is an outer bound surface. This means that without an event horizon, the pressure-energy tensor $T_{\mu\nu}$ of the wormhole must violate the average weak energy condition (AWEC); That is, to go through the wormhole we will to have a zero geodesic, with a tangent vector equals theta, along which we will have theta $k^\mu = dk^\mu / d\zeta$, and then

$$\int_0^\infty T_{\mu\nu} k^\mu k^\nu d\zeta < 0 \tag{3}$$

Thus, if it can be shown that quantum field theory prohibits violations of AWEC, it may rule out the possibility of advanced civilizations maintaining permeable wormholes.

ALCUBIERRE SPACETIME METRIC ANALYSIS

Miguel Alcubierre, a Mexican physicist who worked in the UK for a long time, wrote his famous paper entitled "The Warp Drive Hyper-Fast Travel Within General Relativity". Firstly, he used the language of the 3+1 formalism of GR, because it will permit a clear interpret ation of the results. In this formalism, spacetime is described by a foliation of spacelike hypersufaces of constant cordinate time t. The geometry of spacetime is then given in terms of the following quantities: the 3 metric γ_{ij} of the hype-surfaces, the lapse function α that gives the interval of proper time between nearby hypersufaces as measured by the Eulerian observers, and the shift vector β^i that relates the spatial coordinate systems on different hypersurfaces. Using these quantities the metric of spacetime can be written as:

$$ds^2 = -d\tau^2 = g_{\alpha\beta}dx^\alpha dx^\beta$$
$$= -(\alpha^2 - \beta_i\beta^i)dt^2 + 2\beta_i dx^i dt + \gamma_{ij}dx^i dx^j \tag{4}$$

As long as the metric γ_{ij} is positive definite for t all values, spacetime is guaranteed to be completely hyperbolic modal.

Suppose the ship is moving in x coordinates; Want to find a metric that "pushes" the spacecraft along an orbit described as a random function $x_s(t)$ of time; Therefore it is written under normalized conditions:

$$\alpha = 1$$
$$\beta_x = -v_s(t)f(r_s), \quad \beta_y = \beta_z = 0$$
$$\gamma_{ij} = \delta_{ij}$$

where $v_s(t) = dx_s(t)/dt$, $r_s(t) = \left[(x - x_s(t))^2 + y^2 + z^2\right]^{\frac{1}{2}}$, and the $f(r_s)$ is:

$$f(r_s) = \frac{\tanh[\sigma(r_s + R)] - \tanh[\sigma(r_s - R)]}{2\tanh(\sigma R)} \tag{5}$$

So the metric can be written

$$ds^2 = -dt^2 + \left[dx - v_s f(r_s)dt\right]^2 + dy^2 + dz^2 \tag{6}$$

Therefore, the spacetime structure can be described as follows: The 3D geometry of hypersurfaces is always flat. A time-like curve perpendicular to a hypersurface is geodesic, and spacetime is always flat except for a few regions.

Consider the external curvature tensor K_{ij}, which describes how 3D hypersurfaces are embedded into 4D spacetime:

$$K_{ij} = \frac{1}{2\alpha}\left(D_i\beta_j + D_j\beta_i - \frac{\partial g_{ij}}{\partial t}\right) \cong \frac{1}{2}\left(\partial_i\beta_j + \partial_j\beta_i\right) \tag{7}$$

Introduce the concept of volume expansion angle:

$$\theta = -\alpha TrK$$

It can also be illustrated that the volume expands behind the spacecraft and compresses in front. The trajectory of the spacecraft is actually a time-like curve, with eigentime equal to coordinate time. The spacecraft flies on geodesic and has no time dilation.

It can be shown that $d\tau = dt$, so the ship is moving on a time-like curve. Therefore, it can be concluded that when the spacecraft is flying, it does not experience time dilation. This also directly proves that the spacecraft flies on geodesic. This means that even if the coordinated acceleration is an arbitrary function of time, the intrinsic acceleration along the flight path of the spacecraft will be zero.

To see how one can use this metric to make a round trip to a distant star in an arbitrary small time, let us consider the following situation: Two stars A and B are separated by a distance D in flat spacetime. At time t_0, a spaceship starts to move away from A at a speed $v<1$ using its rocket engines. The spaceship then stops at a distance d away from A. We will assume that d is such that $R \ll d \ll D$. It is at this point that a disturbance of spacetime of the type described, centered at the spaceship's position, first appears. This disturbance is such that the spaceship is pushed away from A with a coordinate acceleration that changes rapidly from 0 to a constant value a. Since the spaceship is initially at rest ($v_s = 0$), the disturbance will develop smoothly from flat spacetime.

When the spaceship is halfway between A and B, the disturbance is modified in such a way that the coordinate acceleration changes rapidly from a to $(-a)$. If the coordinate acceleration in the second part of the trip is arranged in such a way as to be the opposite to the one we had in the first part, then the spaceship will eventually find itself at rest at a distance d away from B, at which time the disturbance of spacetime will disappear (since again $v_s = 0$). The journey is now completed by moving again through flat spacetime at a speed v.

Moreover, we see that when a spacecraft travels in flat spacetime, time dilates only from the beginning and end of the flight. Since the round-trip is only twice the distance, we can get back to planet A in any small intrinsic time, both from the ship's point of view and from the planet's point of view. The spacecraft will be able to travel faster than light. However, it will always maintain a time-like trajectory; That is, in its local light cone, the light itself is also pushed along by the warping of spacetime.

The above analysis is based on: the three-dimensional geometry of hypersurfaces is always flat; Information about spacetime curvature will determine the time lapse indicating that the time-like curve perpendicular to the hypersurface is geodesic; The latter is contained in the external curvature tensor K_{ij}, which describes how 3D hypersurfaces are embedded into 4D spacetime. ... The problem is that the metric described above has an important shortcoming: it violates three major energy conditions (weak energy condition, principal energy condition, and strong energy condition). The weak energy condition and the main energy condition require the energy density to be positive for all observers. If one could calculate the Einstein tensor from the metric and use the 4-dimensional velocities of Euler observers, it would be shown that those observers would see the following energy density:

$$T^{\alpha\beta}n_{\alpha}n_{\beta} = \alpha^2 T^{00} = \frac{1}{8\pi}G^{00} = -\frac{1}{8\pi}\frac{v_s^2\rho^2}{4r_s^2}\left(\frac{df}{dr_s}\right)^2 \tag{8}$$

The fact that this expression is always negative indicates that it violates both the weak energy condition and the principal energy condition, and it can also be proved that it violates the strong energy condition using similar methods.

We see that when it happens in a wormhole, it takes exotic matter to do faster-than-light (FTL). But even if we believe that exotic matter is forbidden, it is well known that quantum field theory allows the existence of negative regions in some special environments (such as the Casimir effect). The need for exotic matter does not eliminate the possibility of using spacetime warps like those described above for super-fast interstellar travel.

Anyway, the calculations show that there has to be negative energy, and it has to surround the starship. It's very difficult to do that; The first is to require vast quantities of exotic matter. For example, it was later proved that a warp bubble moving at $v=10c$ has a wall thickness of 10^{-30}cm. Assuming a starship size of 200m, the negative energy required is equivalent to ten billion times the mass of the observable universe! Other improvements have been proposed, but not much in terms of implementability. ... According to astrophysicists, Warp drive could be a reality within the next 100 years, making Star Trek-style space travel possible. That's according to J. Lewis, a professor at Sydney University in Australia, who told ABC news that the futuristic concept is part of the theory of relativity, which describes how we can warp time and space. So you can travel through the universe as fast as you want. It's theoretically possible, but can we build a warp drive? We have clues that there are materials out there, but whether we can assemble them all and build a warp drive remains to be seen.

The physicist's view—that future space flight will inevitably require FTL travel —has long been advanced by Chinese scientists, including Song Jian, Lin Jin and me. And we've already noticed inconsistencies in relativity—SR says that FTL motion is impossible, GR says it is possible (otherwise theory of wormholes and Warp drive wouldn't have happened). This inherent contradiction has led to the bizarre claim that future FTL spaceflight is still based on relativity.

DISCUSSION ON WARP DRIVE PROPULSION SCHEME

It must be admitted that if GR is an airtight and perfectly correct theory, Alcubierre uses it to the extreme. He was followed by a number of similar papers, which seemed more to satisfy mathematical interests. And we just focus on one thing, is there any way we can do FTL space travel in the future? Is it just talk?

Warp drive is a theoretical prediction or vision for FTL space travel. The basic principle is that a spacecraft in a warp bubble can travel to the far reaches of the universe at an arbitrarily large speed while remaining stationary relative to the nearby space (frame of reference). This is because the spacetime in front of the warp bubble has been managed to contract, shortening the distance from its destination; At the same time, the spacetime behind the warp bubble expands, increasing the distance between the warp bubble and the starting point. Observers outside the disturbance region may observe the spacecraft traveling at FTL speed.

SR's belief that nothing can move at faster-than-light speeds "remains true" in GR, Alcubierre said. But the description would be more accurate—In GR, nothing can travel locally faster than the speed of light.

Since human daily life is based on Euclidean space, it is natural to assume that if no matter can travel locally at FTL speed, then the two places are separated by an appropriate distance D, as measured by the observer who always stays in the starting position, the round-trip time between the two places cannot be less than 2D/c. Of course, we know from our study of SR that if a person is flying back and forth at close to the speed of light, his measured time is smaller. But within the framework of GR, and without introducing exotic topologies (wormholes), it may come as a surprise to many that one can actually make this round trip in less time than measured by a stationary observer. But the idea is easy to understand if we think about the early expansion of the universe and consider the relative separation velocities of two observers moving together. If the relative velocity is defined as the rate of change of the proper space distance in intrinsic time, a value much larger than the speed of light is obtained. This does not mean that the observer will travel at FTL speed, they will always move within a local cone of light; The great speed of separation comes from the expansion of spacetime itself. The above example shows how to use the expansion of spacetime to move away from a place at an arbitrarily large speed. You can also use the contraction of spacetime to get to a place at any speed. In this way, spacecraft would be propelled away from Earth and toward distant stars through "spacetime" itself. We can reverse this process and return to Earth.

After the paper of Alcubierre was published, the response seemed to be more enthusiastic than wormhole (see [14-20]). However, most of them are still mathematical analysis, and there are few articles discussing how to design and realize them physically. But some papers are novel, such as Santa-Pereira's, "Using fluid dynamics to deal with curved phases to advance spacetime geometry", which is similar to the Chinese scientists use of aeromechanics to deal with problems related to relativity. In addition, reference [20] suggests that the use of negative energy density may be avoided. The Alcubierre Warp drive metric is a kind of spacetime geometry with spacetime distortion called warp bubble.

One of the giant particles gets global FTL. The solution of the field equation of Alcubierre metric with fluid material as gravity source is given. The energy-momentum tensor under consideration has two types of fluid: perfect fluid and parameterized perfect fluid (PPF), which is an exploratory and more flexible model aimed at exploring the possibility of Warp drive solutions with positive matter density components. It has been shown that the Alcubierre metric relates this geometry to Burgers' equation, which describes the motion of shock waves in an inviscous fluid but brings the solution back to a vacuum. The same thing happens in two quarters of a perfect fluid. Other solutions for ideal fluids show that flexural drive of positive matter density is possible, but at the cost of complex drive regulation function solutions. With regard to PPF, a solution is also obtained, showing that positive matter density can produce velocity. The weak energy, principal energy, strong energy and zero energy conditions for all subclasses of study were calculated to satisfy the requirements of perfect fluid and to generate constraints in the PPF, thus making it possible for the density of positive matter to also produce warp bubbles. Combining all the results, the energy-momentum tensor describing a more complex form of matter or field distribution yields solutions to field equations with Warp drive metric, where negative matter density may not be a strict prerequisite for obtaining "Warp velocity."

However, the author believes that negative energy density can not be invoked, and this argument still needs to be tested over time, because the above situation is only the result of fluid dynamics analysis.

THE CONCEPTUAL CONFUSION OF SPACE AND TIME INTEGRATION

We have discussed in detail the basic theory of wormhole and warp drive propulsion. This is contrary to the author's intention, because we do not fundamentally agree with relativity, especially with its two foundations: the integration of space and time and the curvature of spacetime. As is known to all, both SR and GR take space and time integration as the starting point. All the relativity literature talks about spacetime, but what does spacetime really mean? People don't actually know. The textbook description gives the impression that Minkowski's approach, while having some mathematical expressiveness benefits, but violates physical reality. It is practically impossible and meaningless to "add" a space vector to a time vector. Fundamentally, time and space should not be confused. We believe that space is continuous, infinite, three-dimensional, isotropic; Time is the symbol of the continuity and sequence of matter's movement. It is continuous, unidirectional, and passes evenly without beginning or end. Space and time exist independently of human consciousness. Moreover, space is space and time is time; they are fundamental quantities that describe the physical world. The so-called spacetime does not exist in Metrology and the International System of Units (SI), nor does it have measurable properties. It is unreasonable to confuse time and space, two completely different physical concepts, by artificially constructing a new parameter (so-called 4D spacetime) with physical quantities of different dimensions.

Therefore, in writing this paper, the author, contrary to his original intention, tries to take the position of relativists (including Thorne and Alcubierre), and consider the possibility of "FTL travel". Instead, what we of a basic concepts; This paper takes literature [8] as an example for illustration. Of course, we are not here to blame [8] authors and interviewees, but to say that relativity does have such conceptual holes and confusion. In [8], the Japanese physicist Michio Kakai is quoted as saying that there are two ways to move faster than light: extending space (warp drive, for example) and curving space (wormhole, for example); here he was say "space", not "spacetime". Reference [8] also says, "spacetime near the warp bubble is very warped"; But later it says: "The FTL drive of a warp drive can be attributed to space expansion." Now, it is spacetime warping, or is space warping?! Relativity itself (and those men who believe in it) does often confuse spacetime with space. This is inadmissible, because since the whole theory of relativity is based on the integration of time and space, how can use space replace spacetime?! However, in terms of the physical meaning of warp drive, it is possible to explain what is actually happening by saying that space in front of the ship is shrinking while space behind the ship is expanding. In that case, space is enough without spacetime; But is it still relativity?!

It is a shame that this article about Warp dive had to stop to discuss the integration of time and space. The mark of space is length, the unit is meter; Time is measured in seconds. But spacetime cannot exist and stand as an independent physical quantity, nor can it be assigned its own unit. In astronomy and cosmology, the GR term "matter bends spacetime" often means "matter bends space". The idea of time bending has always been elusive—how can time be bent? Bending is a kind of geometrical description. For invisible time, it is meaningless to say whether time is curved or not. This concept has no physical reality... These old problems now affect our discussion of Warp drive, making "turning to GR for FTL travel" being a hollow and impractical idea.

According to reference [5], Alcubierre said: "We clear see how the volume elements are expanding behind the spaceship, and contracting in front of it." At here he uses the word "volume" (i.e. space), and he didn't uses the word "spacetime"!

HOW TO OBTAIN THE NEGATIVE ENERGY

The concept of negative energy was first put forward by scientist in the 1920s, and in accordance with his ideas, positron was discovered, which was also the earliest antimatter. But Dirac is just using this concept to analyze the problem. He doesn't say that "there may be negative energy in nature" or that "negative energy can be artificially created," and the positrons carry positive energy. It is necessary to discuss the problem of negative energy because of the study of superluminal velocity.

The core of GR theory is spacetime bending. Normally matter gives spacetime a positive curvature, like a sphere. Wormhole theory led to the time machine, which states that in order to travel through time, spacetime needs to have a negative curvature, like a saddle. But the imagery may be hard to understand; Warp drive also seems to require negative curvature spacetime... The key point is that quantum laws (which are based on the uncertainty principle) allow the energy density to be negative in some places, as long as it can be compensated for by the positive energy density elsewhere, keeping the total energy density positive.

To his credit, G. Feinberg addressed negative energy as early as 1968 in his paper "The Possibility of Tachyon particles." He says: "Another problem with tachyon particles arises from the fact that, for momentum vectors in phase space, the sign of the energy can be changed by Lorentz transformations." There is a more direct relation between the positive and negative energy solutions of the wave equation than the phase time momentum. This connection, to include tachyon particles, must include the presence of negative energy states. However, we shall see that the negative energy solutions for tachyon particles are very similar to those for ordinary particles, i.e. in quantum field theory these solutions are associated with creation operators instead of annihilation operators. In fact, Feinberg was the first to show that the FTL problem was related to the concept of negative energy. But we should also see that his ideas are contradictory—if the tachyon carries negative energy, it leads to negative mass. However, the core of Feinberg tachyon theory is virtual mass (take $m_0 = j\mu, \mu > 0$), and the two are not compatible.

Later, many scientists pointed out that any FTL space program would use negative energy. For example, Russian astronomer S. Krasnikov (who proposed a one-way FTL spacetime channel, but not the same as a wormhole), as well as K. Olum in the United States, B. Bassett in the United Kingdom. Their thoughts are the same.

Now, we'll discuss a few possible ways to obtain negative energy, but they are not guaranteed success and are mentioned just to keep the mind active.

Use the Compressed Vacuum State

The concept of quantum vacuum is different from that of classical vacuum. Because there are fluctuations, there is a process of virtual particles that spontaneously emerge and disappear. Vacuums with fluctuations correspond to an average energy density of zero, but the scientists have come up with a way to make it less than zero. One example is a "compressed vacuum state" in which positive energy is present in one part of space and negative energy in another, but the two are balanced throughout space. This effect is said to be achieved by passing a laser beam through a non-linear optical material. Alternating positive and negative energies are created as photons enhance or suppress vacuum fluctuations.

Use Strong Gravity to Produce Negative Energy Particles

S. Hawking, a British physicist, suggested that pairs of virtual particles exist in a vacuum, including cases where one is a particle and the other is an antiparticle; They were first created together, then separated, then came together and annihilated. Since energy cannot be created out of zero, one of the particle—antiparticle pairs has positive energy and the other has negative energy, but the latter has a short life and is a virtual particle. It has to find a partner and annihilate with it. ... Hawking thus proposed the concept of negative energy particles; In his view virtual particles (particles whose existence cannot be proved by measurement) have both positive and negative energies. If there is a black hole, where gravity is too strong, the energy of the real particle may also be negative. It is also possible that a negative energy virtual particle falls into a black hole and becomes a real particle or antiparticle. However, Hawking radiation is a positive energy particle.

Hawking's theory is based on quantum field theory, in line with the "quantum view of vacuum," while the idea that black holes evaporate through radiation was proposed in 1974. According to general relativity (GR), the curvature of spacetime near black holes is so large that it disturbs the vacuum fluctuations. It is this extreme curvature of spacetime that requires negative energy to come out and enter the black hole, which means positive energy will come out of the black hole. So the conservation of energy cannot always be violated, and black hole physics must be consistent with thermodynamics. This shows that classical physics is not bad, the law of conservation of energy, the law of thermodynamics are to be observed.... However, the existence of negative energy particles is still a hypothesis in the author's opinion and lacks experimental proof.

Search for Negative Energy from Quasars

In the section of "Energy and momentum distribution of gravitational field" published by Hu Ning in 2004, the complex derivation even showed the result that "the total energy of gravitational field is negative". The reason of the negative energy density is not accepted is that it leads to "negative mass". And we can't explain the attraction between stars without the concept of negative energy density. The gravitational energy and the energy stored in the gravitational field must be negative, which Hu Ning said is an insurmountable contradiction existing in linear gravitational field equation. The implication seems to be that more rigorous theoretical equations will overcome this paradox.

Use Casimir Effect

It is generally believed that Casimir effect can be used to obtain negative energy. In 1948, Dutch physicist H. Casimir proposed in the paper, take two flat smooth metal plates, put them parallel to each other in a vacuum environment, if the distance d is very small, it will be found that there is a mutual attraction between them (later called Casimir force), it is not Newton's universal gravitation, the real existence is very strange. The physicist had dismissed this as nonsense, but gradually accepted it at Casimir's insistence.

On the "negative energy vacuum" in Casimir effect, I published an English article in 2021. Now summarize the main point. The quantized electromagnetic field is a quantum system with infinitely many harmonic oscillators. The ground state has zero-point vibration and corresponding zero-point energy (ZPE), and the zero-point vibration of all modes is the vacuum fluctuation of the quantum electromagnetic field—although the mean value is zero, the mean square value is not zero. Therefore, quantum theory believes that vacuum has energy and the overall value is $\frac{1}{2}\sum_i \hbar\omega_i$. Since the degree of freedom i (that is, the modulus of vibration i) is infinite and the upper limit of ω is infinite, this vacuum energy is divergent and unobtainable. However, it is possible to calculate and measure the change in the vacuum energy by placing two parallel metal plates, constructed an open cavity. The boundary conditions of the field change, and the harmonic oscillator frequency changes, causing the energy of the vacuum state to change. Although ZPE is still divergent after cavity placement, it cannot be observed. But the difference in energy before and after it can be calculated and observed. So this is the Casimir energy, which is called E_c. The corresponding force on the metal plate which is called the Casimir force F_c. Now, E_c equal to the difference between the ZPE of the vacuum between the plates and the ZPE of the absence of the plates:

$$E_c = \left[\sum \frac{1}{2}\hbar\omega_i\right]_Y - \left[\sum \frac{1}{2}\hbar\omega_i\right]_N \tag{9}$$

We can get a clearer idea of physics by putting it this way; In the above formula Y, for "Yes", is after the plate is inserted; N for "No", is no board case.

It is worth noting that the expressions of E_c and F_c given by the above derivation, and finally have a minus sign; What does it mean physically? There is a view that the Casimir energy is negative energy and the Casimir force between the two conducting plates is attracted to each other. "Negative energy" can be understood as "the emptiness between the plates is more empty than the vacuum", which must produce inward force to bring the plates closer. Because of this, Lamoreaaux's measurements in 1997 were considered to "measure negative energy", but the question is debated.

Why do many people always think that Casimir energy E_c is negative energy? When two parallel metal plates are placed in engineering vacuum environment, the energy density between the plates is lower than that outside the plates due to quantum effect. It is generally believed that the energy density outside the plate is zero, so the energy density inside the plate is negative. This assertion led to the belief that Lamoreaaux measured negative energy. However, the experimenter himself did not say that he measured the negative energy. Instead, he designed a sophisticated device to measure the Casimir force F_c, thus calculating that the Casimir energy $E_c=(10\sim15)$J and the spacing value d=1μm. This experiment was published in 1997, half a century after H. Casimir's paper. Casimir himself did not die until 2000 (91 years old), and was very old in 1997 and 1998. It must have been a mixed feeling to hear that Casimir force was accurately measured.

Use Negative Wave Velocity

The WKD experiment, which was published in *Nature* in 2000 under the title "Gain Assisted Superluminal Light Propagation", is a negative wave velocity (actually a negative group velocity) experiment. After the results were announced, heated discussion ensued. Some people do not agree that it is a FTL experiment, in order to prove their point of view, the analysis and calculation, but there is a negative energy density result. Because of the interaction between the light pulses and the cesium (Cs) atomic gas in the experiment, the energy density of the electromagnetic field they calculated from electromagnetic theory was only the electromagnetic component of the internal energy of the cesium atomic gas. It is calculated that the negative energy density moves from the exit to the entrance at approximately negative group velocity $v_g=-c/310$. They found it difficult to interpret the negative energy density and speculated that it might mean "extracting energy from caesium atomic gas."

The final results are given in literature [28]. When the parameters in WKD experiment are substituted, the result is $w<0$ (w is the energy density). If considered as the electromagnetic energy density of the pulse in the medium, then it is negative! Furthermore, the energy density increases in the opposite direction of $A(z,t)$. Since it can be proved that the energy transmission velocity v_e is approximately equal to the group velocity v_g, it is known that $v_g \cong v_g <0$, the energy density propagates in the opposite direction to the incident wave.

Use Metamaterials

Metamaterials are known as super-materials. The broad concept includes photonic crystals (PC), left-handed materials (LHM), absorbing materials, stealth materials, etc., and the range is very wide. The narrow understanding is negative refractive index (NRI) materials, that is, left-handed materials character.

In 1964, V. Veselago published the first paper on this subject, entitled "Electrodynamics of ε, μ simultaneously negative". This paper proposes a new concept—although no matter with $\varepsilon<0$ and $\mu<0$ occurring at the same time has been found in the past, its existence does not conflict with the existing laws of physics. Clearly, Veselago believes that negative refraction propagation occurs only if the permittivity and permeability are both negative.

Here we are concerned with energy relations; In isotropic media, the electric field energy density and magnetic field energy density are respectively:

$$w_e = \frac{1}{2} \varepsilon E^2 \tag{10}$$

$$w_m = \frac{1}{2} \mu H^2 \tag{11}$$

where E is electric field intensity, H is magnetic field; So ε or μ negative will lead to negative energy density. If space has both electric and magnetic fields, the total energy density is

$$w = w_e + w_m = \frac{1}{2} \varepsilon E^2 + \frac{1}{2} \mu H^2 \tag{12}$$

If both $\varepsilon<0$ and $\mu<0$ are true, then $w<0$; But if one of them is positive, then the total energy density is not necessarily negative.

The possibility of negative energy upsets Vesselago, because it has no place in classical physics. He used the concept of frequency dispersion to explain it; Under the condition of no dispersion and no absorption, if the constraint that the total energy cannot be zero is kept (which is consistent with the traditional understanding of energy), it is impossible for, ε, μ to be negative, that is, there is no possibility to satisfy both $\varepsilon<0$ and $\mu<0$ at the same time. However, the following expression is more general:

$$w = \frac{1}{2}\left[\frac{\partial(\varepsilon\omega)}{\partial\omega} E^2 + \frac{\partial(\mu\omega)}{\partial\omega} H^2\right] \tag{13}$$

To keep the value w greater than 0, the value must be met:

$$\frac{\partial(\varepsilon\omega)}{\partial\omega} > 0 \tag{14}$$

$$\frac{\partial(\mu\omega)}{\partial\omega} > 0 \tag{15}$$

Both inequalities do not mean that ε and μ cannot be negative at the same time; However, to satisfy the requirement of the above inequality, it ultimately depends on the relationship between ε, μ and frequency.

Although Vesselago ruled out $w<0$ based on classical electrodynamics, experimental techniques developed decades after the publication of his paper brought the question to a conclusion. In March 2000, at a meeting of the American Physical Society, scientist D. Smith of the University of California, San Diego (UC-SD) announced the completion of negative refractive index experiments in microwaves, making the LHM predicted many years ago.

FUNDAMENTAL FTL RESEARCH THAT DOES NOT RELY ON "SPACETIME BENDING"

Relativity is based on the integration of space and time and the curvature of spacetime, but both can be problematic. As it turned out, SR prohibited FTL speed, and some physicists bypassed SR and turn to used GR for develop FTL theory, propagating wormholes or warp drive propulsion. Recent trends seem to have made warp bubble a buzzword, thanks to the persistence of the American physicist H.White in warp drive and the continued publication of papers in the *European Physics Journal C*.

In February 2022, China's *Science and Technology Daily* published an article by reporter Tang Fang entitled "Warp Bubbles Found in the Real World? Don't Worry, Faster than Light Travel is Too Early". The article said:

"Science fiction heroes often use a tool called a warp drive for faster than light interstellar travel. In the virtual universe of *Star Trek*, warp drive is a faster than light propulsion device. Inspired by this, and based on general relativity, physicist M. Alcubierre proposed the scientific concept of warp drive in 1994.

A team led by H. White, a physicist at the Defense Advanced Research Projects Agency (DARPA), has discovered a warp bubble in the real world, the *European Journal of Physics, Series C* reported recently. White's nanoscale warp bubble is thought to open the door to creating warp engines and ushering in the faster than light age.

In response, Li Li, an associate researcher at the Institute of Theoretical Physics, Chinese Academy of Sciences, told *Science and Technology Daily* in February that White's team did some numerical simulations and predicted that a certain microstructure could give a negative energy density distribution, which is similar to the negative energy needed to maintain the Alcubierre spacetime structure (warp bubble). But whether it can be linked to warp bubbles or warp engines remains to be seen."

When the author saw this report, the feeling is joy and sorrow. The good news is that FTL research is seeing a resurgence in the new century, even though we may not like the projects mentioned; The worrying is that the public is not aware of the fact that scientists in many countries (including China) have been working on FTL problems for decades, but they still seem to be "relying on GR", which we do not approve of or like. If GR does have a problem on correctness, then we what to do? !

If we start with G. Feinberg's serious FTL research paper, the history of related research is 55 years. Scientists from the United States, the United Kingdom, China, Russia, Italy and other countries participated in the research. Theoretical work and experimental work are abundant. In this case, it is well established that " any thing can't travel faster than the speed of light" "No credible evidence of faster than light experiments has been found";

It all doesn't seem realistic. The 1993 SKC experiment, for example, sped up photons by 70% to $v=1.7c$ in a human lab — does this count as the faster than light motion of "anything"? Is it too much to erase the efforts and explorations of so many countries and workers in a few words?... Although I have been doing FTL research for more than 20 years and have published many articles and books, I am still willing to start from scratch. Here are a few highlights we'd like to address... Here is a list of some of the ideas and work that Chinese scientists have done in FTL research—it's probably better to do basic research than talks only about "human travel to the universe by FTL".

We do not Agree with the Research Approach of "Bypassing SR and Resorting to GR"

Many physicists, yearning for FTL spaceflight but fearful of the SR ban, found GR to be useful, and went in to tinker with mathematical formulae to turn physical problems into mathematical problems. Could this approach bring us closer to FTL space travel? That's impossible! For example, the negative energy problem, although the author also suggested some possible way means to obtain negative energy, but it is very difficult to have practical effect.

Interesting researchers are advised to first see if SR's ban on FTL speed is correct or not. According to the mass-velocity formula, the mass of the particle in motion is:

$$m = \frac{m_0}{\sqrt{1-\beta^2}} \qquad (16)$$

where m_0 is the rest mass of the particle, and $\beta = v/c$, v is the particle velocity, and c is the speed of light. So the energy of the particle is:

$$E = \frac{m_0 c^2}{\sqrt{1-\beta^2}} \qquad (17)$$

Therefore, when the value of v gradually increases from a low value, E gradually increases; When v is equal to c, E is infinite. If $v > c$, E becomes imaginary. Both infinite and imaginary energies are meaningless in practice, so SR theory determines that "faster than light cannot exist".

These views, it must be said, are superficial. First of all, Lorentz's mass-velocity formula is derived for the electromagnetic mass of electrons. Even if it applies to charged particles such as electrons and protons, it cannot be generalized to all motion bodies like SR. In fact, there is a lack of experimental proof that the mass-velocity formula can applies to neutral particles and objects, then the so-called "light barrier" may not really exist.

In addition, the electron is not an ordinary motion of particle, but a special charge of the particle motion. So the energy is not infinite even if v is equal to c. In addition, it can be proved that when the velocity v increases, both the charge q and the force F of the moving body will decrease. This explains the Kaufmann experiment of 1901. Similarly, the analysis showed that the Bertozi experiment in 1964 also did not prove that over the speed of light c was impassable.

Secondly, it is well known that photons travel at the speed of light ($v=c$) which is fine, and the speed of photons (c) is not obtained by acceleration, but is inherent. Moreover, there may be FTL particles in nature whose speed is not obtained by means of acceleration. That is, a particle moving at subluminal speed ($v<c$) may not be able to accelerate to superluminal speed, but a Feinberg tachyon may have an imaginary rest mass:

$$m_0 = j\mu \tag{18}$$

where μ is a real number. In this case, even $v>c$ ($\beta>1$), does not appear imaginary energy. Nature is very complex, and science has long used neutrinos as an example of how Tachyon might have existed.

Let $\beta = v/c$, (v is the moving body velocity), it can be seen that there is a factor $(1-\beta^2)^{-1/2}$ in SR that works everywhere, which is called the light speed limit factor by the author, and it is this factor that creates the light barrier. In 2001, Cao Shenglin pointed out that there was no need to assume tachyon with imaginary rest mass as Feinberg did. In his theory, the speed limit factor $(1-\beta^2)^{-1/2}$ becomes $(1-\beta^2+\beta^4)^{-1/4}$, and there is no need to deny FTL motion. The crux of the matter, he argues, is that Lorentz transformation (LT) only works in subluminal speed systems. Einstein's past use of the local characteristics of LT to deny the possibility of FTL motion was neither sufficient nor necessary. Now we write the equations of moving mass and energy derived by Cao Shenglin:

$$m = \frac{m_0}{\sqrt[4]{1-2\beta^2+\beta^4}} \tag{19}$$

$$E = \frac{m_0 c^2}{\sqrt[4]{1-2\beta^2+\beta^4}} \tag{20}$$

Since $1-2\beta^2+\beta^4 = (1-\beta^2)^2 = (\beta^2-1)^2$, when $\beta<1$, the formula consistent with SR theory is immediately obtained. If it is β greater than 1, it can be written:

$$m = \frac{m_0}{\sqrt{\beta^2-1}} \tag{21}$$

$$E = \frac{m_0 c^2}{\sqrt{\beta^2-1}} \tag{22}$$

The denominator is still a real number. So the theory does not require FTL particles (or any physical matter) to have virtual mass. It can be seen from Equation (22) that, when v increasing (β increasing), the energy E decreases; As v decreases (β decreases), the energy E increases. This is exactly what A. Sommerfeld expected of FTL particles in the early 20th century — that FTL particles would speed up as their energy was reduced and slow down as their energy increased.

So is it possible for nature to have natural FTL particles? The most likely candidates are neutrinos. These are tiny but ubiquitous particles that have no electric charge, come in three types: e particle, μ particle and τ particle. It has physical properties somewhat similar to photons. It was long thought to be a FTL particle. ... In 2015, Huang Zhixun published a paper "On Different Explanations of Phenomena Following Supernova Explosion in 1987". On February 23, 1987, a supernova explosion occurred. 7.7 hours before the arrival of the first photons from 1987A, the first neutrino wave was detected by detectors under The Browne Peak in Italy, which contained five events. Three hours before the first photons arrived, the second neutrinos arrived, eleven events were received by the Kamioka II detector in Japan, eight events by the IMB detector in Ohio, USA, and five events by the Baksan detector in the Soviet Union. How could this happen? It has never been properly explained. There are only three possibilities: ① photons travel at the speed of light, neutrinos at superluminal speeds; ② Neutrino velocity is c, photon's velocity is subluminal; ③ The neutrino's velocity is superluminal and the photon's velocity is subluminal. From the analysis of the situation at that time, something may have happened ① or ③. We think it might be ①!

From 2008 to 2011, the European Nuclear Research Center (CERN) organized multinational scientists to conduct experimental research on the speed of neutrinos. Neutrinos are emitted from CERN on the border between Switzerland and France, and received at the underground laboratory in Gran Sasso, Italy. The distance between the two places is 730km, and the distance measurement error is less than 10cm. The neutrino time of flight measurement error is less than 10ns. This was obviously a sophisticated experiment, yielding faster than light results, $v-c/c = 2.48 \times 10^{-5}$. Then, the 174 scientists who took part in the experiment signed off on the papers to be published. But it turned out that the equipment was faulty, and the results were wrong—a major three year study (the OPERA Project) was destroyed (a loose cable connection), which is unthinkable even today... Anyway, this is over after a flurry of activity in 2011-2012: But some physicists still insist that neutrinos are tachyons.

We Should Pay Attention to the Comparative Study of Breaking the Sound Barrier and Breaking the Light Barrier

In aeronautical engineering, the ratio of the speed v of an airplane to the speed c of sound in air (c also expressed the sonic speed for comparison purposes) is called a Mach number and the prescribed symbol is $M_a = v/c$. The so-called sound barrier breakthrough refers to the aircraft to achieve supersonic flight ($M_a > 1$). On October 14, 1947, the American X-1 rocket engine aircraft reached speed was $v = 1078$km/h, corresponding to $M_a = 1.105$. On

February 28, 1954, the American F-104 fighter jet flew at twice the speed of sound ($M_a=2$).

The speed of light in vacuum $c=299792458$m/s, it is 8.8×10^5 times of the speed of sound (340m/s). With such a big difference, and the fact that the speed of light in vacuum c is one of the fundamental constants of physics (the speed of sound is not), they seems to be no comparison between the two realms put together. But the history of wave mechanics tells us otherwise. In 1759, L. Euler obtained the 2D wave equation for the first time, which is the analysis of the vibration of rectangular eardrum. Let $f(x,y,z,t)$ represents the membrane displacement, c is a constant determined by the membrane material and tension, he obtained

$$\frac{\partial^2 f}{\partial x^2}+\frac{\partial^2 f}{\partial y^2}=\frac{1}{c^2}\frac{\partial^2 f}{\partial t^2} \tag{23}$$

In his paper "On the Propagation of Sound", a further analysis resulted in the 3D wave equation:

$$\nabla^2 f = a^2 \frac{\partial^2 f}{\partial t^2} \tag{24}$$

where f is the vibration (mechanical vibration or acoustic vibration) variable. Thus, wave equations were developed from the very beginning across mechanics and acoustics, the boundary between which is ambiguous to mathematicians. Due to the electromagnetic nature of light, the relationship between acoustics and optics can be understood as that between acoustics and electromagnetism. The wave equation obtained from Maxwell equations is

$$\nabla^2 \psi = \frac{1}{v^2}\frac{\partial^2 \psi}{\partial t^2} \tag{25}$$

where ψ is the wave function, $v=(\varepsilon\mu)^{-1/2}$, and ε, μ are the macro parameter of wave propagation medium. The consistency of equation (25) and equation (24) indicates that there is a unified rule in the fluctuation process.

As we know, the electrostatic field is a non-curl field, and the potential function satisfies the Laplace equation in the region where the bulk charge density ρ is zero. In aerodynamics, two basic functions are used to study fluid motion, namely potential function φ and flow function ψ. When the flow velocity is low, the flow density in the plane flow is regarded as constant, and the two-dimensional flow is described by Laplace equation:

$$\frac{\partial^2 \varphi}{\partial x^2}+\frac{\partial^2 \varphi}{\partial y^2}=0 \tag{26}$$

$$\frac{\partial^2 \psi}{\partial x^2}+\frac{\partial^2 \psi}{\partial y^2}=0 \tag{27}$$

These are incompressible non-curl flow equations, and they are linear differential equations of order two. If the airflow speed increases to a certain extent, ρ should be regarded as a variable; The basic equation for a compressible fluid in plane non-curl flow is

$$\left(1-\frac{v_x^2}{c^2}\right)\frac{\partial^2\varphi}{\partial x^2} - 2\frac{v_x v_y}{c^2}\frac{\partial^2\varphi}{\partial x \partial y} + \left(1-\frac{v_y^2}{c^2}\right)\frac{\partial^2\varphi}{\partial y^2} = 0 \tag{28}$$

$$\left(1-\frac{v_x^2}{c^2}\right)\frac{\partial^2\psi}{\partial x^2} - 2\frac{v_x v_y}{c^2}\frac{\partial^2\psi}{\partial x \partial y} + \left(1-\frac{v_y^2}{c^2}\right)\frac{\partial^2\psi}{\partial y^2} = 0 \tag{29}$$

where c is the speed of sound; Obviously, if $c \to \infty$, the equation degenerates into a simpler Laplace equation, which is the case of incompressible fluid. We note that although the factor $\left(1-\frac{v^2}{c^2}\right)$ appear in equations, it does not appear that "the speed of sound c cannot be exceeded". So why in optics, when c is the speed of light, then this factor that make the speed of light c "can't be exceeded"? !

There are many solutions for the compressible flow of ideal fluid, one of which is the perturbation linearization method. Referring to the situation of direct uniform flow, the flow velocity of incoming flow is specified as v_∞, the sound velocity is c_∞; and Mach number is M_∞; Then the potential equation can be obtained after processing and linearization under 2D flow condition:

$$\left(1-M_\infty^2\right)\frac{\partial^2\varphi}{\partial x^2} + \frac{\partial^2\varphi}{\partial y^2} = \frac{\partial^2\varphi}{\partial t^2} \tag{30}$$

In the process of linearization, the limit of $\left(1-M_\infty^2\right)$ cannot be too large, it is not supersonic flow. It can't be transonic flow. We notice that in the subsonic flow field, $M_\infty < 1$, $\left(1-M_\infty^2\right) > 0$, the equation is elliptic; Its property is basically the same as that of the Laplace equation of incompressible flow. However, for the supersonic flow field, $M_\infty > 1$, $\left(1-M_\infty^2\right) < 0$, the equation becomes hyperbolic, and the situation has great changes. In short, the equations of motion describing subsonic and supersonic velocities are of different types. However, the equations describing transonic flow are mixed and nonlinear, so it is very difficult to find the analytical solution. Thus came computational fluid dynamics, which is very similar to computational electromagnetism and uses the same methods (e.g., finite element method, finite difference method).

The so-called sound barrier refers to the fact that the speed of the aircraft has been hovering at the level of subsonic ($M_a < 1$) for a long time, and the attempt to fly at the speed of sound ($M_a = 1$) has encountered real difficulties. Early aircraft were slow enough to handle aerodynamic problems as incompressible fluids. When $M_a \geqslant 0.4$, the compressibility effect becomes obvious gradually, and the air density in front of the head increases sharply when approaching the speed of sound ($M_a \to 1$). When $M_a=1$, the disturbance in the fluid does not propagate relative to the aircraft, but instead concentrates to form a wave surface: when the nose meets the air in front of it, it compresses strongly, and the density increases to form an invisible wall (shock wave). The resistance caused is called wave resistance. It consumes about 75% of the engine's power, making it difficult. At this time, "near-sonic aerodynamics" and "supersonic aerodynamics" need to be developed. There were theoretical researches on transonic flow in the 1920s and 1930s, but the decisive progress was made in the 1940s. In 1945, American scientists put forward the swept-back wing theory, to overcome the impact of shock waves is to increase the speed of the aircraft to near the speed of sound. The effort to overcome the sound barrier was a collaborative effort of scientists, engineers and designers. From theoretical research to successful supersonic flight, it took only about 20 years for the scientific and aeronautical communities to work together to solve the problem.

In aerodynamics, the wave equation of the velocity potential of compressible fluid, linearized, can be expressed as

$$(1-\beta^2)\frac{\partial^2 \varphi}{\partial x^2} + \frac{\partial^2 \varphi}{\partial y^2} = \frac{\partial^2 \varphi}{\partial t^2} \qquad (31)$$

Here we use the symbol β instead of the symbol M_a, in order to compare relativity with aerodynamics. In essence, the basic operation of wave mechanics is the identification and solution of differential equations. Qian Xuesen (1911-2009), together with T.von Karman (1881-1963), first proposed the concept of supersonic flow in the 1930s, providing a theoretical basis for aircraft to overcome the thermal barrier and sound barrier. Their theory is applied to the design of high subsonic aircraft. In fact, the small perturbation theory is advanced to nonlinearity in the subsonic region. Although it cannot be used in the calculation of supersonic problems, it avoids singularities, the infinite mass density does not appear at $v=c$.

Now let's look at the singularity problem. The above mentioned "when $M_a=1$, the air density increases sharply to form a shock wave", does not say "when $M_a=1$, the air density increases to infinity". Yang Xintie pointed out that only subsonic flow was studied in the early stage. According to the small disturbance theory, for the flow in the shrink-pipe, if the mass density at relative rest is set as ρ_0, the mass density at relative speed β will increase as:

$$\rho = \frac{\rho_0}{\sqrt{1-\beta^2}} \qquad (32)$$

The above formula is identical with the mass-velocity formula of special relativity (SR). If the above equation is followed completely, the imaginary mass density will be calculated at supersonic speed. Later developments, however, proved that the infinite mass density was only mathematically infinite. Song Jian, the former chief engineer of China's Aerospace Industry Ministry, pointed out that the density of the gas in a supersonic aircraft only increases six times as it passes through the sound barrier (there is no such thing as infinite value). It is under the premise of nonlinear processing, the supersonic experimental study and the related theoretical options and processing are optimized, resulting in the success of supersonic motion (flight). The singularity is no longer mentioned here. Then we conclude the following three equations:

Subsonic velocity ($v<c$, $\beta<1$, $M_a<1$)——

$$\rho = \frac{\rho_0}{\sqrt{1-\beta^2}} \tag{32}$$

Supersonic ($v>c$, $\beta>1$, $M_a>1$)——

$$\rho = \frac{\rho_0}{\sqrt{\beta^2-1}} \tag{33}$$

Sound speed ($v=c$, $\beta=1$, $M_a=1$)——

$$\rho = k\rho_0 \tag{34}$$

where β is a Mach number, k is the coefficient, and ρ_0 is the density at relative rest. Now that the whole thing is perfectly explained, faster-than-light motion research must go the same way.

However, the rule of formula (30) can only be realized through technical improvement. In the late 19th century, the development of steam turbines required the highest possible flow of air. Conventional narrowing of pipe sections in the hope of obtaining supersonic flow has failed. Swedish engineer Carl Laval (1845-1913) used shrinking section and then expanding section, that is, adding a section of expanding tube with gradually expanding section behind the shrinking nozzle, and found that as long as the pressure is high enough, there is supersonic flow in the expanding tube. This proves that infinity only exists in mathematical formulas. In the 1940s, scientists and engineers conducted wind tunnel experiments and built supersonic aircraft inspired by the Laval tube.

FTL Particles Should be Searched with Improved Techniques at High Energy Accelerators

Relativists will say that the technical practice of accelerators has long shown that increasing energy is an effective means, or even the only way, to accelerate the flight of particles (electrons or protons). And the accelerated particles can actually only reach very close value c, such as $0.99999\,c$; Given this, how can Einstein's 1905 argument be opposed?... In this regard, the author puts forward the following views; First of all, "particles that have not been obtained with current accelerators by $v=c$ or $v>c$" is not the same as "there are no FTL particles in the universe". According to the electromagnetic field and electromagnetic wave principle design of the accelerator, in which the speed of flying charged particles can only be infinitely close to c but not reach c, is very natural, because the electromagnetic wave intrinsic velocity is c. That doesn't explain anything. Secondly, we do not deny that increasing electromagnetic energy can accelerate electrons, but this is not the same thing as proving the SR mass-velocity equation and the entire SR energy relationship. There is absolutely no experimental evidence for neutral particles (such as neutrons and atoms), so it makes no sense to suggest a general limit on speed. Moreover, the bigger problem is that Einstein only sees the electron as a general moving body of mass and speed, and does not consider the electron as a special moving body that carries an electric charge. Therefore, there is a lack of an electrodynamic theory which takes into account the effect of moving charge. Chinese scholars did the analysis and got significantly different results from Einstein.

It must be noted that we have the support of accelerator experts in trying to find FTL particles with high energy particle accelerators. For example, professor Pei Yuanji said in his article "Discussion on Faster than Light Experimental Scheme" in 2017:

"Up to now, charged particle dynamics have been based on the speed of light as a limit, on the dynamics of special relativity. Although the accelerators built so far have found no contradiction to this basic theory, the theoretical basis for all the parameters used to test the motion of particles is relativity, so any contradiction is hard to find. In order to find out if there is a contradiction, I propose an experimental method that may find some doubt, and if so, it can be further studied.

In the experimental device, the electron gun is an electron gun (such as photocathode microwave electron gun, external cathode independently tuned microwave electron gun, etc.) which can produce several MeV energy and ps(10^{-12}s) beam length. Acceleration tube 1 and acceleration tube 2 are conventional acceleration structures (their phase velocities are close to 1 and equal to 1 respectively). They accelerate the electron beam to the relative energy of the electron beam $\gamma=100$, that is, the velocity of the electron beam reaches $0.99995\,c$. Acceleration tube 3 is specially designed to make the phase velocity of wave greater than the speed of light. The fluorescent targets of magnet analysis 1, 2 and behind are devices for beam energy measurement with energy resolution better than 0.1%. The beam bin is a device for absorbing an electron beam. K1 is a device that provides microwave power for conventional accelerators, and its pulse power is about 50MW. K2 is a device that provides microwave power for faster-light phase accelerators, and its output power is 25MW. IAΦ is the element used to adjust the phase and power of the microwave power entering the accelerator tube 3."

Later, Professor Pei gave a detailed explanation of "energy simulation of electron beam in accelerator tube 3" and "test method", with several calculation diagrams attached. The goal is to find special phenomena that can be explained by electrons traveling faster than light.

To Achieve Faster than Light Astronautic Travel Whether Human Beings can Use Wormholes or Warp Drive Propulsion

In March 2019, the author received "Discussion on Superluminal Electron Accelerators" from Professor Yang Xintie. This is a research proposal jointly signed by many experts. In the "Project Basis", the reason proposed is similar to Pei Yuanji, that is, "it is impossible to discover the contradictions of SR itself with instruments designed according to SR". The report proposes an exploratory experimental scheme in which the last of the three accelerators (No.3) is replaced by a specially designed superluminal accelerometer in which the phase velocity of the wave is greater than the speed of light. Here the electrons are expected to accelerate in the FTL direction (this region of energy does not increase but decrease).

Professor Yang is an expert in aerodynamics, and he believes that these problems have also been encountered in the development of continuum media mechanics. According to the small perturbation approximation theory, the point of sound velocity is also infinite. The subsonic equation calculates the supersonic speed, but it also produces imaginary values. But none of the mechanics thought that spacetime should be described... The algorithm for the compressible flow of an ideal fluid originally contains a space transform, but aeromechanics call it a compression transform, which is essentially the same.

In order to transform the existing accelerator to search for special electron, the author thinks that the experience of Laval tube used in supersonic aircraft, design should be learned. Coincidentally, the author is an expert in the study of waveguide theory in microwave technology. My monograph "An Introduction to the Theory of Waveguide Below Cutoff" has won the National Award of Excellent Scientific Books. The special function in the cutoff waveguide (field intensity drops exponentially from the starting point), and the conic waveguide below cutoff mentioned in the book, may be suitable for accelerator transformation; So close up this paragraph for your reference.

The Unique Insights of Space Experts Should be Taken into Account

Theoretical physicists are respected, but they also have weaknesses. For example, replacing physics with mathematics, or even failing a formula, will make the whole project impossible. China is successful in space development, and it has trained many space experts, some of whom are concerned about the future of FTL space travel. Their comments were instructive, and here are two examples.

The first example is Academician Prof. Song Jian. He is an expert in engineering cybernetics and aerospace technology. He used to be vice minister and chief engineer of the Ministry of Aerospace Industry of China. He told an academic conference in 2004:

"Today's rocket engines are mainly chemical fuel engines with speeds of $v \cong 10^{-5} c$. Future deuterium fusion engines may have speeds as high as $(0.05{\sim}0.1)c$. If the spacecraft were traveling close to or even faster than the speed of light, it would take only a few years to travel to and from the nearest star... Einstein, in his 1905 paper, imagined a light barrier, saying that faster than light speed was impossible; But this is a guess, not a scientific law, because there is no experimental basis. From the technical point of view of optical barriers, with optical or radar round-trip signal time half to ranging, this is not to see the speed $v \geqslant c$ target, it is impossible to judge the situation there. Therefore, ignorance of FTL conditions is no excuse for its non-existence... In addition, it is found that the engineering practice of autonomous navigation conflicts with SR dynamics even in the case of $v \ll c$. For example, the dependence of engine thrust on its inertial velocity has never been observed."

Song Jian added:

"The light barrier problem in space is reminiscent of the sound barrier problem in aerospace engineering in the 20th century. Before the advent of supersonic aircraft, it was thought that the shock waves formed by aircraft approaching the speed of sound were impenetrable. But later theoretical analysis and wind tunnel experiments showed that the gas density at the head less than six times. The aviation community immediately began designing and building new planes, which flew at supersonic speeds in 1947... Does the light barrier have a similar future?"

Another example is Academician Prof. Lin Jin, a researcher at the China Academy of Launch Vehicle Technology, who is an expert on satellite navigation and inertial navigation. At the same academic conference, Lin Jin presented a new theoretical model on autonomous inertial navigation, which is used to analyze and deal with the time definition, measurement mechanism and faster than light motion of inertial navigation. He argued that a moving particle could measure its own position, velocity and acceleration relative to a given inertial frame as a function of the intrinsic time of its own motion clock. In principle, there is no need to exchange information with the outside world, and there is no problem with the speed of signal transmission. Autonomous inertial navigation is based on the nature of the gravitational field, even if the world has no electromagnetic field, no light, pure inertial system still work, as usual, independent positioning, speed measurement. So why 3×10^8m/s is the speed limit?! In short, the time definition of the inertial navigation spacecraft is the proper time of the spacecraft motion clock. As long as new power sources are developed in the future, there is no limit to the speed of the spacecraft. ...Lin also believes that photons should be restored to the ordinary status of other microscopic particles, even if they have a rest mass and their speed is not a limit speed.

Song Jian and Lin Jin, two space experts, are my friends, and I am also an aerospace fan. They have a thought-provoking way of looking at things; To be honest, these are words that theoretical physicists can't say!

Attention Should be Paid to the Study of Whether Information can be Transmitted at Faster than Light Speed

It is odd that messages, signals, which are not physical objects, are also on the SR's "no faster-than-light" list. NASA's Pioneer 10, launched in 1972, flew out of the solar system in January 2003, but stayed in contact with it for 11 hours, failing to complete command and communication in time. Relativity says not only that objects cannot travel faster than light, but that signals cannot travel faster than light. However, there is no such limitation in quantum theory. In August 2008, *Nature* published the experimental results of Swiss scientists, proving that the propagation speed of quantum entanglement is super luminal, that is, $v_G = 10^4 c \sim 10^7 c$; This is an important development and increases our confidence. The discussion of FTL communication is detailed in the literature.

ELECTROMAGNETIC PROPULSION (EM DRIVE) IS THE MORE PRACTICAL METHOD

It is interesting to compare Warp drive with electromagnetic propulsion (EM drive) and to see the trend of development. EM drive is a system based on electromagnetic energy drive, i.e. the conversion of electromagnetic energy into thrust without the need for rocket fuel. According to classical physics, this should not be possible because the law of conservation of momentum is violated. The law states that if a system is acted upon by an external force, then the total momentum of the system remains constant; this is why conventional rockets require fuel. The foreign press has been saying that although researchers in the US, the UK and China have argued for electromagnetic drive, they dispute the results because no one has a firm grasp of how it works. At the heart of the project is a microwave resonant cavity, which can also be seen as a closed cone-shaped waveguide (Figure 1). The non-uniform distribution of the electromagnetic field causes a thrust force that tends to accelerate the motion in the opposite direction. It has come through its early days of obscurity and is now attracting a great deal of attention from the scientific, technological, space and intelligence communities of various countries.

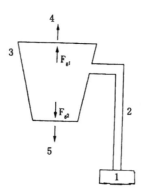

Fig.1 Schematic representation of the EM drive
(1 microwave source, 2 waveguide, 3 cavity, 4 direction of thrust, 5 direction of acceleration)

The UK Science Trends website reported on 2 August 2014 that "Incredible engine will change space travel forever", which was translated by the domestic newspaper *Reference News* as "Microwave engine may change space travel". The article says that senior British engineer R. Shawyer may soon be attracting a lot of attention. When he first built the engine now known as the EM drive, no one took it seriously. But that changed in 2012, when a group of Chinese scientists also built this engine, and succeeded. The EM drive is simple and light. Its thrust is generated by "microwaves bouncing back in a closed container". The engine design is so unusual that it would not work according to conventional mechanics. However, the engine can be powered by solar energy to generate microwaves. In addition, it does not require any form of propellant, so it can be used until the hardware stops working. In August 2013, a team from NASA built a slightly less powerful engine, but one derived from a similar concept.

There was a period of confused thinking about the EM drive. First of all, where does its force come from? And what kind of force is it (for example, is it the Newton force, or is it the Casimir force?) Secondly, does it violate the principle of conservation of momentum to say that it can fly, but then not eject matter backwards? Researchers at a Chinese university proved they had done useful work in a 2011 paper, but in 2013 the same authors arranged for a microwave radiation outlet from which they said a microwave beam could be ejected into the atmosphere or outer space. This is a conceptual retreat into the category of chemically fuelled rockets, as if this is the only way to not violate conservation of momentum. Now, we say it is wrong. For a chemically fuelled rocket, let the rocket and fuel (propellant) masses be m_1 and m_2 and their corresponding velocities be v_1 and v_2, then conservation of momentum requires:

$$m_1 \triangle v_1 = m_2 \triangle v_2 \tag{35}$$

If $\triangle v_2 \neq 0$ (accelerated ejection after combustion), then $\triangle v_1 \neq 0$ (rocket gains acceleration). So now why is this EM drive not ejecting material outwards but possibly having accelerated motion? Obviously, as long as there is force, the RF cavity does not need to be ejected outwards. In fact, the laws of conservation of momentum and energy are obeyed. ...However the above description may not be very satisfactory? !

At the 50th Joint Propulsion Technology Conference in July 2014, NASA scientists reported that test results showed that the RF resonant cavity thruster design, which is unique as an electronic propulsion device, produced thrust that could not be attributed to any classical electromagnetic phenomenon. Scientists note that NASA's tests were conducted on 8 August 2013, using a resonant cavity excited at 935 MHz to generate (30-50) mN of thrust on a low thrust torsion pendulum placed in a stainless steel vacuum chamber designed and built by NASA. Brady et al. suggest that one possible explanation for this type of propulsion is that it is caused by the interaction of quantum vacuum virtual plasma, which provides this anomalous thrust production from the RF test device. In this understanding, the purpose of the experimental activity at NASA was "to investigate the feasibility of using classical magnetic plasma dynamics to obtain propulsive momentum generated by quantum vacuum virtual plasma conversion". The microwave engine (called EM drive by Shawyer) under study was called the "quantum vacuum virtual plasma thruster" by Brady et al. These views are for reference. Whether the EM drive is relevant to quantum theory is a big question!

In conclusion, EM drive technology is now widely accepted. The acceleration and reaction forces generated in the opposite direction of the propulsive force obey Newtonian mechanics. In 2015, Zhixun Huang says that thrusts of 10mN/kW to 1000mN/kW can already be generated. The basic device is a cone-shaped closed resonant cavity, using the TE_{01} model. Due to the non-uniform field distribution in the cavity, the electromagnetic ensemble force $F_z \neq 0$ provides the thrust for the autonomous accelerated motion of the cavity. It carries no fuel and the microwave energy used in the thrusters is converted from solar energy, making it suitable for space flight. 88mN force was generated by Shawyer in June 2006 with 700W power and 96mN force was generated in May 2007 with 300W power. This shows that the (125-320)mN/kW level was reached early on. The force is small, but with force comes acceleration and the process of constant acceleration promises to eventually achieve very high speed.

It is important to note that the Chinese Academy of Space Technology (CAST) has been focusing on the development of this field since 2010, and has started to achieve results since 2012 with the independent development of several principle prototypes. After 10 years of effort, the team led by researcher Chen Yue has developed two innovative thrusters, one of which is a trapezoidal resonant cavity type (partially filled with dielectric inside); the other is a hemispherical thruster. They have both been granted patents recognised by the Chinese government. So, Chinese research is leading the way!

But we must not be blindly optimistic: for the EM drive to be practical, the thrust has to increase significantly (e.g. to several hundred N/W or more) and the acceleration has to be sufficient (e.g. $0.5m/s^2$ or more) to make it possible for it to become an interstellar probe. In any case, the space community can do real research and development on the EM drive; it is much better than the unrealistic talk of the Warp drive.

CONCLUSION

The *European Physics Journal C* recently reported on the research of faster than light astronautic travel led by American physicist H. White using the Warp drive theory proposed by M. Alcubierre in 1994. It is said that warp bubbles of nanometer size have been created by numerical simulation. As We all know, the theory of special relativity (SR) said that the FTL motion is impossible, but the Warp drive idea which based on the theory of general relativity (GR) holes that faster than light motion can be achieved. This is a spearstone, which warrants further study.

Although I have written this paper, my academic views remain unchanged and I still disagree with GR's "space time integration" and "spacetime curvature". For example, one of the diagrams drawn in many current books to illustrate the wormhole principle shows only "extremely curved space", not "extremely curved spacetime". The latter cannot be drawn because there is no physical image of the so-called spacetime. Even if GR were the correct theory, wormholes and Warp drive propulsion would not be possible. This article points out that EM drive is the project that the national space communities can actually engage in R&D on!

In the 1990s, NASA proposed the Breakthrough Propulsion Physics Program, a revolutionary change in propulsion design principles. Alcubierre's Warp drive concept is also being looked at by NASA. In addition, the US Department of Defense Advanced Research Projects Agency (DARPA) has commissioned NASA to carry out research with the following objectives: to achieve propellant-less propulsion; to require very high speeds; to switch on completely new physics principles; etc. For example:

—Utilizing the quantum vacuum, or quantum fluctuations;

—Designing space propulsion using the BB effect (Biefeld-Brown effect) is the use of high voltage electric fields;

—Utilizing the plasma drive;

In short, all kinds of physical principles can be considered for research and developments.

COMPETING INTERESTS

Author has declared that no competing interests exist.

REFERENCE

[1] Einstein A. Electrodynamics of moving bodies [J]. Ann. D. Phys., 1905, 17(7): 891–895.

[2] Einstein A. The field equations for gravitation. Sitzungsberichte der Deutschen Akademie der Wissenschaften[J]. Klasse f'ur Mathematik, Physik und Technik, 1915: 844–847.

[3] Morris M, Thorne K. Yurtsever U. Wormholes, time machines and the weak energy condition[J]. Phys. Rev Lett, 1988, 61(13): 1446–1449.

[4] Ford L, Roman T. Negative energy, wormhole and warp drive[J]. Sci. Amer., 2000, 282: 40–53.

[5] Alcubierre M. The warp drive: hyper–fast travel within general relativity[J]. Class Quant Grav, 1994, 11(5): 73–77.

[6] Huang Z X. Is Einstein's special theory of relativity correct[J]. Journal of Communication University of China (Natural Science), 2021, 28(5): 71–82.

[7] Huang Z X. The study and discussion of general relativity[J]. Journal of Communication University of China (Natural Science), 2022, 29(1): 64–80.

[8] Tang F. Finding warp bubbles in the real world? Don't worry. Faster than light travel is too early[N]. Science and Technology Daily, 2022–02.

[9] Flamm L. Beiträge zur Einsteinschen gravitations theorie[J]. Phys Zeit, 1916, 17: 448.

[10] Thorne K. Black holes and time warps[M]. New York: Norton & Company, 1994.

[11] Casimir H. On the attraction between two perfect conducting plates[J]. Proc Ned Akad Wet, 1948, 51: 793–797.

[12] Huang Z X. Two kinds of vacuum in Casimir effect[J]. Current Jour. of App. Sci. and Tech., 2021, 40(35): 61–77.

[13] Ford L, Roman T. Restrictions on negative energy density in flat spacetime[J]. TUTP–96–2, 1996, (2 oct): 1–17.

[14] Van Den Broeck C. A warp drive with more reasonable total energy[J]. Class. Quantum Gravity, 1999, 16: 3973.

[15] Natario J. Warp drive with zero expansion[J]. Class. Quantum Gravity, 2002, 19: 1157.

[16] Lobo F, Visser M. Fundamental limitations on warp drive spacetimes[J]. Class. Quantum Gravity, 2004, 21: 5871.

[17] Lee J, Cleaver G. Effects of external radiation on an Alcubierre warp bubble[J]. Phys. Essays, 2016, 29: 201.

[18] Santos–Pereira O, et al. Dust contents solutions for the Alcubierre warp drive spacetime[J]. Europ. Phys. Jour. C, 2020, 80: 786.

[19] Mattingly B, et al. Curvature invariants for the Alcubierre and Natario warp drives[J]. Universe, 2021, 7: 21.

[20] Santa–Pereira O, et al. Fluid dynamics in the warp drive spacetime geometry[J]. European Phys. Jour. C, 2021, 81: 133.

[21] Yang X T. Discussion on the theoretical basis of superluminal phenomena[J]. Jour. of Beijing Institute of Petrochemical Tech., 2002, 10(4): 27–32.

[22] Feinberg G. Possibility of faster than light particles[J]. Phys. Rev, 1967, 159: 1089–1105.

[23] Hawking S. A brief history of time[M]. New York: Bantom Books, 1988.

[24] Hu N. General relativity and gravitational field theory[M]. Beijing: Science Press, 2000.

[25] Lamoreaux S. Demonsration of the Casimir force in the 0.6 to 6μm range[J]. Phys. Rev. Lett., 1997, 58: 5–8.

[26] Wang L J, Kuzmich A, Dogariu A. Gain–assisted superluminal light propagation[J]. Nature, 2000, 406: 277–279.

[27] Huang C G, Zhang Y Z. Poynting vector, energy density, and energy velocity in an anomalous dispersion medium[J]. Physical Review A, 2001, 65(1): 015802.

[28] Zhang Y Z. New progress in the study of superluminal phenomena in anomalous dispersion media[J]. Physics, 2001, 30(8): 456–460.

[29] Veselago V. The electrodynamics of substances with simultaneously negative values of permittivity and permeability[J]. Sov. Phys. Usp., 1968, 10(4): 509–514.

[30] Smith D, et al. Composite medium simultaneously negative permeability and permittivity[J]. Phys Rev Lett, 2000, 84(18): 4184–4187.

[31] Smith D, Kroll N. Negative refractive index in left-handed materials[J]. Phys Rev Lett., 2000, 85(14): 2933–2936.

[32] Steinberg A, Kuwiat P, Chiao R. Measurement of the single photon tunneling time[J]. Phys. Rev. Lett, 1993, 71(5): 708–711.

[33] Huang Z X. Faster than light research — the intersection of relativity, quantum mechanics, electronics and information theory[M]. Beijing: Science Press, 1999.

[34] Huang Z X. New progress in faster than light research[M]. Beijing: National Defense Industry Press, 2002.

[35] Huang Z X. Theory and experiment in faster than light research[M]. Beijing: Science Press, 2005.

[36] Huang Z X. Faster than light research and electronics exploration[M]. Beijing: National Defense Industry Press, 2008.

[37] Huang Z X. Wave science and superluminal physics[M]. Beijing: National Defense Industry Press, 2014.

[38] Huang Z X. Study on the Superluminal Light Physics[M]. Beijing: National Defense Industry Press, 2017.

[39] Ma Q P. On the self-consistency of relativistic logic[M]. Shanghai: Shanghai Sci. and Tech. Literature Publishing House, 2004.

[40] Cao S L. Relativism and cosmology in Finsler spacetime[M]. Beijing: Beijing Normal University Press, 2001.

[41] Huang Z X. On different explanations of phenomena following supernova explosion in 1987[J]. Frontier Science, 2015, 9(2): 39–53.

[42] Adam T, Agafonova N, Aleksandrov A, et al. Measurement of the neutrino velocity with the OPERA detector in the CNGS beam[J]. Journal of High Energy Physics, 2012, 2012(10): 1–37.

[43] Ai X B. A suggestion based on the OPERA experimental apparatus[J]. Phys. Scripta, 2012, 85: 045055 1–4.

[44] Ehrlich R. Tachyonic neutrinos and the neutrino masses[J]. Astroparticle Phys., 2013, 41: 1–6.

[45] Ehrlich R. Six observations consistent with the electron neutrino being a tachyon with mass $m^2=-0.11\pm0.016eV^2$[J]. Astroparticle Phys., 2015, 60: 11–17.

[46] Huang Z X. The possibility of mass particles moving faster than the speed of light[J]. Journal of Communication University of China (Natural Scicnce), 2015, 22(3): 1–15.

[47] Song J. Spaceflight—the pull of spaceflight to basic science[M]. Beijing: Higher Education Press, 2007.

[48] Liu X G. Self-shielding effect of charge motion[J]. Journal of Chongqing University (Special Edition), 2005, 27: 26–28.

[49] Pei Y J. Discussion on superluminal experimental scheme[J]. Frontier Science, 2017, 11(2): 22–24.

[50] Huang Z X. An introduction to the theory of waveguide below cutoff (2nd edition)[M].

Beijing: China Metrology Press, 1991.
[51] Salart D, et al, Testing the speed of spoky action at a distance[J]. Nature, 2008, 454: 861–864.
[52] Huang Z X. Faster than light communication based on quantum nonlocality[J]. Frontier Science, 2016, 10(1): 57–78.
[53] Shawyer R. Microwave propulsion-progress in the EMdrive programme[C]. 59th International Astronautical Conference. IAC-2008. Glasgow, UK. 2008.
[54] Yang J. Net thrust measurement of propellantless microwave thrusters[J]. Acta Physica Sinica, 2012, 61(11): 110301.
[55] Shawyer R. Second generation EmDrive propulsion applied to SSTO launcher and interstellar probe[J]. Acta Astronautica, 2015, 116: 166–174.
[56] Tajmar M. Direct thrust measurements of an EM drive and evaluation of possible side effects[C]. 51st AIAA/SAE/ASEE Joint Propulsion Conference, Propulsion and Energy Forum. Orlando, Florida. 2015.
[57] McCulloch M. Can the Emdrive be explained by quantised inertia[J]. Progress in Physics, 2015, 11: 78–80.
[58] White H, et al. Measurement of impulsive thrust from a closed radio frequency cavity in vacuum[J]. Jour. of Propulsion and Power, 2016, (Nov): 1–12.
[59] McCulloch M E. Testing quantised inertia on the emdrive[J]. Europhysics Letters, 2015, 111(6): 60005.
[60] Yang J, et al. Thrust prediction of a workless microwave thruster at different powers[J]. Jour. of Phys., 2011, 60(12): 124101 1–7.
[61] Yang J, et al. Prediction and experimental measurement of the electromagnetic thrust generated by a microwave thruster system[J]. Chin Phys B, 2013, 22(5): 050301 1–9.
[62] Huang Z X. Microwave propulsion electromagnetic engine for space technology[J]. Jour. of Communication University of China (Natural Science), 2015, 22(4): 1–10.
[63] White H G, March P. Advanced Propulsion Physics: Harnessing the Quantum Vacuum[J]. Nucl. Emerg. Technol. Sp, 2012: 1–2.
[64] Caligium L, Musha T. Quantm vacuum energy, gravity manipulation and the force generated by the interaction between high-potential electric fields and zero-point-field[J]. Intern. Jour. of Astrophysics and Space Science, 2014, 2(1): 1–9.
[65] Fong K Y, Li H K, Zhao R, et al. Phonon heat transfer across a vacuum through quantum fluctuations[J]. Nature, 2019, 576(7786): 243–247.
[66] Musha T. Explanation of dynamical Biefeld-Brown effect from the standpoint of ZPE field[J]. Jour. of British Interplanetary Society, 2008, 61(9): 379–384.

The Completeness of Classical Electromagnetic Theory*

INTRODUCTION

The contributions of the English physicist J. Maxwell (1831-1879) in electromagnetic theory are comparable to those of I. Newton in classical mechanics. Newton established the first theoretical system of natural science in human history, and Maxwell incisively summarized the essence of electromagnetism in the 19th century and made innovations. Both of them are great, and the construction of industrial civilization and information society owes to them.

Faraday's scientific ideas are known for their concept of "lines of force," which include electric and magnetic lines of force. Maxwell studied Faraday's book *Experimental Studies in Electricity* and then tried to express Faraday's ideas mathematically. In 1856, Maxwell published his first electromagnetics paper "On Faraday's Force Line", in which he proposed six relations, namely: $\oint \mathbf{A} \cdot dl = \Phi$, $\mathbf{B} = \mu \mathbf{H}$, $\oint \mathbf{H} \cdot dl = \sum I$, $\mathbf{J} = \sigma \mathbf{E}$, $W = \oint \mathbf{J} \cdot \mathbf{A} dl$, $\mathbf{E} = -d\mathbf{A}/dt$. In 1861-1862, Maxwell published another electromagnetics paper, "On Physical Lines of Force", in which he proposed a new concept, "displacement current", which was calculated by $\partial \mathbf{D}/\partial t$. In 1865, Maxwell published his most important paper "A Dynamic Theory of Electromagnetic Field", in which he not only proposed the term "electromagnetic field" for the first time, but also presented a set of electromagnetic field equations composed of 20 scalar equations. More than 10 years later, the German physicist H. Hertz summarized four vector equations, namely Maxwell equations (ME). In his 1865 paper, Maxwell derived the electromagnetic wave equation (MWE) from ME.

ME is widely used in many aspects of information engineering because the theory is logically rigorous and self-consistent. In the 20th century and beyond, many electronic computer software programs have been developed to extend its applications, and many engineering problems have indeed been solved. ... However, ME cannot be treated as absolute, because it is a phenomenological theory based on classical physics and classical mathematics, and its shortcomings are gradually recognized by people. Experts point out that its theory is not strict enough; Slightly out of touch with modern mathematics and modern physics; Ignore the particle essence; The theory of completeness of the field is

---

* The paper was originally published in *European Journal of Applied Sciences*, 2022, 10 (4), and has since been revised with new information.

difficult to solve. These problems have come to the fore today. People are asking, are there any questions that traditional ME solutions can't solve? Is there a better system of ME?

To prove that Maxwell's equations still need to be re-examined today, here are a few more examples. First of all, ME and the electromagnet: wave equation derived from it are vector equations, that are difficult to solve strictly unless repeated approximations are made. Classical electromagnetic theory assumes rigor only when dealing with systems of scalar partial differential equations.

Another example, in 1964 physics master P. Dirac said: "We intend to study the possibility that Maxwell equation are not accurate; When the disdance is very close to the charge that generates the field, perhaps we may have to modify Maxwell field theory to change it for nonlinear electrodynamics." ... Indeed, although there is considerable literature on nonlinear Schrödinger equation (NLSE), there are even monographs entitled Nonlinear Quantum Mechanics, or Nonlinear Optics; We have not heard of anyone working on nonlinear Maxwell equations (NLME).

For another, the history of physics tells us that Maxwell field theory cannot explain the photoelectric effect, in which photons knock electrons off from a metal surface. It was from this point of view that Einstein proposed the photon hypothesis, which was used to explain the photoelectric effect perfectly, for which he was awarded the Nobel Prize in Physics of 1921. Then, why does the Schrödinger equation explain the microscopic world perfectly by taking into account both wave and particle properties, but the Maxwell field theory doesn't consider particles at all right? Of course, at that time in 1865, even the understanding of atoms was poor, so we can excuse Maxwell himself. However, after A. Proca put forward the theory of considering the rest mass of photon (i.e. $m_0 \neq 0$) in 1936, the scientific community still pays little attention to Proca equations even up to now. Does that make sense?!

The purpose of this paper is to study the possible defects of Maxwell equations, and to discuss the methods of it's solution.

GENERAL RELATIONS OF MAXWELL EQUATIONS

ME has different forms in different systems of units; The most common is the MKSA system of units, called the standard form ME:

$$\nabla \cdot \mathbf{D} = \rho \tag{1}$$

$$\nabla \cdot \mathbf{B} = 0 \tag{2}$$

$$\nabla \times \mathbf{H} = \mathbf{J} + \frac{\partial \mathbf{D}}{\partial t} \tag{3}$$

$$\nabla \times \mathbf{E} = -\frac{\partial \mathbf{B}}{\partial t} \tag{4}$$

where **D** is the electric induction intensity, **B** is the magnetic induction intensity, **E** is the electric field intensity, **H** is the magnetic field intensity, **J** is the current density, the above items all are vectors. The other thing: ρ is the body charge density, which is a scalar quantity. Formula (3) is known as Maxwell's First Electromagnetic Law, which we refer to as Law I for short. Formula (4) is known as Maxwell's Second Electromagnetic Law, which we call Law II for short. Law I and Law II are the essence of ME. However, equations (1) and (2) cannot be called Maxwell's law for the following reasons. Let's look at formula (1) first, since the following relation can be established under certain conditions:

$$\mathbf{D} = \varepsilon \mathbf{E} \tag{5}$$

where ε is permittivity; Then we have:

$$\nabla \cdot \mathbf{E} = \frac{\rho}{\varepsilon} \tag{1a}$$

The above equation shows that the body charge density is the source of electric field intensity. The experimental basis of this equation is that electric charges can exist independently in nature and that the forces between them obey Coulomb's law. The relationship between electric field and charge established in equation (1a) is also known as the Gauss theorem for electricity. So, strictly speaking, formula (1) is not Maxwell's original contribution, although it is an integral part of Maxwell field theory.

Another reason is that, under certain conditions, equation (1) can be derived from Equation (3), which damages the independence of equation (1). Formula (3) can now be written

$$\nabla \times \mathbf{H} = \mathbf{J} + \varepsilon \frac{\partial \mathbf{E}}{\partial t} \tag{3a}$$

Here is the total current density vector: **J**

$$\mathbf{J} = \sigma \mathbf{E} + \mathbf{J}' \tag{6}$$

where $\sigma \mathbf{E}$ is the conduction current density (σ is the conductivity), and $\mathbf{J}'$ represents the current density vector generated by non-current sources.

We take the divergence of left and right of formula (3):

$$\nabla \cdot \nabla \times \mathbf{H} = \nabla \cdot \mathbf{J} + \nabla \cdot \frac{\partial \mathbf{D}}{\partial t}$$

According to vector algebra, the left side of the above equation is zero, so:

$$\nabla \cdot \mathbf{J} + \frac{\partial}{\partial t}(\nabla \cdot \mathbf{D}) = 0$$

According the continuous equation:

$$\nabla \cdot \mathbf{J} + \frac{\partial \rho}{\partial t} = 0 \tag{7}$$

the ρ is the body charge density; so we have:

$$\frac{\partial}{\partial t}(\nabla \cdot \mathbf{D} - \rho) = 0$$

we can write

$$\nabla \cdot \mathbf{D} - \rho = f(x, y, z)$$

where f is a function of any position independent of time; Consider that $\nabla \cdot \mathbf{D} = 0$ and $\rho = 0$ before field generation, i.e. $f(x, y, z) = 0$; Therefore, $\nabla \cdot \mathbf{D} = \rho$ i.e. equation (1). Equation (1) can be considered a corollary of Law I, but that two conditions are fulfilled. The first is that the continuity equation hold, and the second is that the field is generated in finite time.

Similarly, we can prove that, under certain conditions, equation (2) can be derived from Law II. However, the situation is more complicated when it comes to the nature of the magnetic field. Equation $\nabla \cdot \mathbf{B} = 0$ is based on the experimental fact that the so-called magnetic charge has never been found; The N and S poles of a magnetic body always come in pairs, and the field lines are always closed (a situation Faraday noted early on). In 1931, P. Dirac proposed that only one magnetic polarity of magnetic elementary particles should exist, called magnetic monopole. However, the long-term search of experimental physicists did not find magnetic monopoles, only in 1982, B. Cabrera announced that he might find magnetic monopoles with the help of low-temperature superconductivity technology; However, the scientific community did not recognize the results.

However, magnetic charge is a useful concept in theoretical operation. Starting from the Coulomb law of magnetism, magnetic field intensity can be defined as follows:

$$\mathbf{H} = \frac{\mathbf{F}}{q_m} \qquad (8)$$

where $\mathbf{F}$ is the magnetic force (vector), q_m is the magnetic charge. In this view, magnetic fields are represented by $\mathbf{H}$, not $\mathbf{B}$. Different from the magnetic charge view is the current view, which is based on the Biot-Savart law and uses magnetic induction intensity (vector) $\mathbf{B}$ to characterize the magnetic field. However, for the same physical object, the results are essentially the same.

CONSTITUTIVE RELATION IN ELECTROMAGNETIC FIELD THEORY

In 1952, professor C.Böttcher of Leyden University in the Netherlands published his monograph *Theory of Electric Polarisation*. This book has a total of 492 pages. It first discusses some concepts and problems in electrostatics, including vectors $\mathbf{E}$ and $\mathbf{D}$, electric polarization vector $\mathbf{P}$, and the permissibility ε. Then it discusses dipole reflection field, polarization and energy, polarized dielectric material in static field or low frequency alternating field, non-polarized dielectric material in static field or low frequency alternating field, polarization under optical frequency, high frequency polarization, dielectric loss, solid polarization; and so on. As the book shows, when electromagnetic fields and matter interact, the theory is rich and complex.

For gases, most liquids, and many solids, the vectors. **E**, **D** and **P** are in the same direction. In this case formula (5) can be written, i.e. $\mathbf{D} = \varepsilon \mathbf{E}$. In a static field of moderate intensity, ε depends on the chemical composition and density of the material, and is independent of **E**. This is also called the dielectric constant. The above relation is true not only for static fields but also for low frequency alternating fields. For high frequencies (e.g., all the way to the optical frequency), this will result in a phase difference between **P** and **E**, and thus between **D** and **E**. In the above statement, **P** is the density of electric polarisation, referred to as electric polarization, it is a vector, **P** is the value at the space point p, its physical significance is detailed later.

The definition of electric polarization comes from the following formula:

$$\mathbf{\mathit{m}}_{ind} = \mathbf{P} \cdot \Delta v$$

$\mathbf{\mathit{m}}_{ind}$ is the induced dipole moment, v is the volume; Therefore,

$$\mathbf{P} = \frac{\mathbf{\mathit{m}}_{ind}}{\Delta v}$$

So the induced dipole moment per unit volume is the electric polarization.

In the theory of dielectric polarization, there are the so-called constitutive equation:

$$\mathbf{D} = \varepsilon \mathbf{E} = \varepsilon_0 \mathbf{E} + \mathbf{P} \tag{9}$$

where ε_0 is the vacuum dielectric constant; From the above formula:

$$\mathbf{P} = (\varepsilon - \varepsilon_0) \mathbf{E} \tag{9a}$$

This means that the existence of **P**, wants the condition of $\mathbf{E} \neq 0$.

It must be noted that the derivation of electric polarization density (also called electric polarization intensity) is sometimes linked to the argument from the concept of magnetic charge. The following is the constitutive equation of magnetic field:

$$\mathbf{B} = \mu \mathbf{H} = \mu_0 \mathbf{H} + \mathbf{M} \tag{10}$$

The vector **M** is the magnetization density (also called magnetization intensity) vector, and its position corresponds to **P**. Some researchers (such as Zhu Lancheng) have skillfully used **P** and **M** to help solve the derivation of ME when the medium is not stationary.

The constitutive relation in electromagnetic theory refers to the relationship between electric polarization and magnetization, relate to electric field intensity and magnetic field intensity in the medium. Every kind of media, as long as the constitutive relations describing these media are established and substituted into ME, correct results can be obtained.

NECESSITY OF INTRODUCING VECTOR WAVE FUNCTION (HANSEN FUNCTION)

The development of electromagnetic theory has proved that the following vector differential equations are generally applicable:

$$\nabla^2 c - \varepsilon\mu \frac{\partial^2 c}{\partial t^2} - \mu\sigma \frac{\partial c}{\partial t} = 0 \tag{11}$$

The vector function c can be $\mathbf{E}$, $\mathbf{H}$, or vector potential $\mathbf{A}$, or Hertz vector Π, while the above equation is called the vector wave equation of electromagnetic wave, which applies to the region without charge source. If the space has no conduction current ($\sigma=0$), then

$$\nabla^2 c - \varepsilon\mu \frac{\partial^2 c}{\partial t^2} = 0 \tag{12}$$

For monochromatic waves $e^{j\omega t}$, the vector Helmholtz equation holds:

$$(\nabla^2 + k^2)C = 0 \tag{13}$$

In this equation, $k = \omega\sqrt{\varepsilon\mu}$. According to the definition of the operator Laplacian, the above equation is

$$\nabla(\nabla \cdot C) - \nabla \times (\nabla \times C) + k^2 C = 0 \tag{13a}$$

For the rectangular coordinate system, 3 unit vectors are constant vectors, so they can be decomposed into 3 components, each of which C_i ($i=x,y,z$) satisfies the following scalar Helmholtz equation:

$$(\nabla^2 + k^2)C_i = 0 \tag{14}$$

Thus creating the premise for the use of the separated variables method. However, the above component method is often unable to be implemented, because the three unit vectors in the curvilinear coordinate system are not (or not all) constant vectors, so the above method can only get the coupling equation containing three components. In other words, when any component of ∇^2 is separated as a variable, the Laplacian of the product of three functions of each component is related to the 3-coordinate, that is, they are coupled together, so it cannot be carried out.

In the 1930s, W. Hansen proposed the idea of directly solving the vector wave equation, whose vector wave function is called Hansen function. However, the Hansen function directly satisfies the vector wave equation, as follows. Let function $\psi(1)$ be the eigen solution of the following equation:

$$(\nabla^2 + k^2)\psi = 0 \tag{14a}$$

Boundary conditions $\psi|_s = 0$ (boundary value problems of the first kind), or $i_n \cdot \nabla\psi|_s = 0$ (boundary value problems of the second kind); Let:

$$\mathbf{L} = \nabla \psi \tag{15}$$

$$\mathbf{M} = \nabla \times (\psi \mathbf{a}) \tag{16}$$

$$\mathbf{N} = \frac{1}{k} \nabla \times \mathbf{M} \tag{17}$$

where $\mathbf{a}$ is the constant vector.

$\mathbf{L}$ is a non-curl but with divergence field, because $\nabla \times \mathbf{L} = 0, \nabla \cdot \mathbf{L} = \nabla^2 \psi$. It can also be proved $\nabla \cdot \mathbf{M} = 0$, $\nabla \cdot \mathbf{N} = 0$, $\nabla \times \mathbf{M} \neq 0$, $\nabla \times \mathbf{N} \neq 0$; So $\mathbf{M}$ and $\mathbf{N}$ are with curl but passive field. In addition, it can be proved that $\mathbf{L} \cdot \mathbf{M} = 0$, $\mathbf{L} \cdot \mathbf{N} = 0$, $\mathbf{M} \cdot \mathbf{N} \neq 0$; in order to make $\mathbf{M} \cdot \mathbf{N} = 0$, select $\mathbf{a}$ use for the special vector. In fact, if we choose the unit vector $\mathbf{a}$ on the axis in cylindrical coordinates and the unit vector $\mathbf{1}$ in spherical coordinates, we can make $\mathbf{M} \cdot \mathbf{N} = 0$. Therefore, it is called $\mathbf{a}$ leading vector. Thus, $\mathbf{L}$, $\mathbf{M}$, $\mathbf{N}$ are perpendicular to each other.

Because $\nabla^2 \nabla \psi = \nabla \nabla^2 \psi = \nabla(-k^2 \psi)$, we obtain

$$\nabla(\nabla^2 \psi + k^2 \psi) = (\nabla^2 + k^2)\mathbf{L} = 0 \tag{18}$$

In addition, it can be proved:

$$(\nabla^2 + k^2)\mathbf{M} = 0 \tag{19}$$

$$(\nabla^2 + k^2)\mathbf{N} = 0 \tag{20}$$

It can be seen that when ψ is satisfied the scalar Helmholtz wave equation, the defined functional $\mathbf{L}$, $\mathbf{M}$, $\mathbf{N}$ can satisfy the vector wave equation.

Since there may be infinitely many eigen functions satisfying Equation (14a) and boundary conditions, i.e

$$(\nabla^2 + k^2)\psi_i = 0 \quad (i = l, m, n) \tag{14b}$$

the value of i is: $i = 0, 1, 2, \ldots$; So $\left(\sum_n a_n \mathbf{L}_n + \sum_m b_m \mathbf{M}_m + \sum_l c_l \mathbf{N}_l\right)$ also satisfies the vector wave equation.

As such, it can be used to represent $\mathbf{E}$, $\mathbf{H}$, $\mathbf{A}$, etc. When the electromagnetic field is thus represented, it is known that the field must contain both curl and non-curl parts. If only the curl field is known, then only expanded $\mathbf{M}$, $\mathbf{N}$.

For example, a rectangular cylindrical waveguide has traveling wave along the z axis, a standing wave on the section, and an electric field is written as:

$$\mathbf{E} = \mathbf{E}_t \, e^{-j\beta z}$$

Subscript t represents transverse time; Therefore,

$$\mathbf{E}_t = \sum \left(a_n \nabla \psi_n + b_n \nabla \psi_n \times \mathbf{k}_t + C_n \mathbf{N}_n\right)$$

where $\mathbf{k}_t$ is the transverse wave number, and ψ is ψ_t; while

$$\mathbf{N}_n = \frac{1}{k_t} \nabla \times \mathbf{M}_n$$

Since ψ_t is satisfied the transverse scalar Helmholtz equation

$$(\nabla^2 + k_t^2)\psi_t = 0$$

and $k_t^2 = k^2 - \beta^2$; now

$$\mathbf{N}_n = \frac{1}{k_t}\nabla \times \nabla \times (\psi \mathbf{k}) = \frac{1}{k_t}\left[\nabla(\nabla\psi_t \cdot \mathbf{k}) - \mathbf{k}\nabla^2\psi_t\right]$$

Because $\nabla^2\psi_t$ is in the cross section, the preceding term is zero, so

$$\mathbf{N}_n = \frac{1}{k_t}\left[-\mathbf{k}\nabla^2\psi_t\right]$$

But $\nabla^2\psi_t = -k_t^2\psi$, so

$$\mathbf{N}_n = k_t\nabla^2\psi_t\mathbf{k} \qquad (21)$$

So $\mathbf{N}_n$ is in the z direction; $\mathbf{L}_n$, $\mathbf{M}_n$ are in cross section, they are perpendicular to each other.

For example, TE wave, the electric field is in the cross section, not contain component $\mathbf{N}_n$ And because in the sourceless region, it should be curl field, so it can be expressed by $\mathbf{M}_n$. A magnetic field must have a longitudinal field $\mathbf{N}_n$, described by $\mathbf{L}_n$. The electric field is expressed as

$$\mathbf{E}_t = \sum_n a_n \mathbf{M}_n = \sum_n a_n (\nabla\psi_t \times \mathbf{k})$$

the different n means different modes; now in keeping with custom, replace n with i:

$$\mathbf{E}_{ti} = \sum_i a_i (\nabla\psi_i \times \mathbf{k})$$

let $a_i = 1$, the normalized electric field is

$$\mathbf{e}_{ti} = \nabla\psi_{ti} \times \mathbf{k} \qquad (22)$$

The normalized transverse magnetic field is

$$\mathbf{h}_{ti} = \nabla\psi_{ti} \qquad (23)$$

The normalized longitudinal magnetic field is

$$\mathbf{h}_{zi} = \frac{k_t^2}{j\omega\mu}\psi_{ti} \qquad (24)$$

These expressions can be used in engineering calculation of some structures, such as GTEM chamber of microwave electromagnetic compatibility equipment.

The completeness of electromagnetic fields is now briefly discussed (without reference to C.T.Tai's theory). According to Helmholtz's theorem, the field $\mathbf{F}$ is decomposed into the sum of a non-curl field and a curl field:

$$\mathbf{F} = \mathbf{F}_l + \mathbf{F}_r = \nabla\Phi + \nabla \times \mathbf{A} \qquad (25)$$

Φ and $\mathbf{A}$ are scalar potential and vector potential respectively; $\nabla\Phi$ is a divergence field without curly, and $\nabla\Phi \times i_z = \nabla \times (\Phi i_z)$ is a curly passive field; The set of the two solutions constitutes the set of the solutions of the complete field, which can be proved as follows:

$$\iiint \nabla\Phi \cdot (\nabla \times \mathbf{A}) dv = \iiint (\nabla \times \mathbf{A}) \cdot \nabla\Phi dV$$
$$= \iiint_S \Phi(\nabla \times \mathbf{A}) \cdot i_n dS - \oiint \Phi \nabla \cdot (\nabla \times \mathbf{A}) dv$$

the first term on the right side of the equation is zero when means that on the boundary $\Phi|_s = 0$ or when $\mathbf{A} = \nabla \times (\Phi i_z)$, $(\nabla \times \mathbf{A}) \cdot i_n = \left.\frac{\partial \Phi}{\partial n}\right|_{s=0}$, and the second term is always zero, so we have

$$\iiint \nabla\Phi \cdot (\nabla \times \mathbf{A}) dV = 0$$

This shows that the vector function space in the given space can be uniquely decomposed into two complete and disjoint subspaces, i.e. fields

$$\{\mathbf{F}\} = \{\mathbf{F}_r\} \cup \{\mathbf{F}_l\} \text{ or } \{\mathbf{F}_r\} \cap \{\mathbf{F}_l\} \tag{26}$$

The $\mathbf{F}$ could be an electric field or a magnetic field; And $\Phi|_{s=0}$ and $\left.\frac{\partial \Phi}{\partial n}\right|_{s=0}$ are boundary value problems of the first and second kinds.

DISCUSSION ON SOLVING VECTOR ELECTROMAGNETIC FIELD EQUATIONS AND VECTOR ELECTROMAGNETIC WAVE EQUATIONS

Many people now think they have solved many engineering problems using Maxwell electromagnetic theory. This is not wrong, but current practices may not be strict; Because we are dealing with vector partial differential equations and vector wave equation, it is not easy to solve them rigorously.

The vector function $\mathbf{M}, \mathbf{N}, \mathbf{L}$ is the Hansen functions, so what does it mean? As we know, classical electromagnetic theory only shows rigor when dealing with scalar partial differential equations, but Maxwell equations are vector partial differential equations, and people have been lacking solution method. When the vector wave equation is converted into the vector Helmholtz equation under monochromatic wave condition, it is found that in the general curviline coordinate system, only the coupling equation of components can be obtained, and the variables cannot be separated. From 1935 to 1937, W. Hansen put forward the suggestion of directly solving vector wave equation in a group of papers on antenna radiation; He constructed an independent vector function solution ($\mathbf{M}, \mathbf{N}$) for the vector wave equation, so the vector wave function is also called Hansen function. In fact, he uses the eigen function of scalar Helmholtz equation as the original basis, and constructs a new orthogonal basis, which directly satisfies the vector Helmholtz equation and corresponding boundary conditions. This kind of functional ($\mathbf{M}, \mathbf{N}$) satisfies the vector wave equation directly, which makes the direct solution work begin. In 1941, J. Stratton introduced Hansen's idea, gave the solution, discussed the form of understanding, and added the vector function solution $\mathbf{L}$. At that time and thereafter, functions were not considered to be qualitatively different from and $\mathbf{L}, \mathbf{M}, \mathbf{N}$.

In 1971, C. T. Tai pointed out that the diadic Green function could be used to solve the boundary value problem of Maxwell equations directly, and gave the expression of perfect symmetry. But then it was pointed out that the two sides of the formula were not identical and the solution was not complete. Specifically, although the dyadic Green function obtained by Tai satisfies coordinate symmetry, it does not contain function $\mathbf{L}$. In 1973, Tai himself modified the new expression by adding a singular term containing the function $\mathbf{L}$ (representing the source field), which caused controversy. The derivation given by W.M.Song in his monograph *Dyadic Green Function and Operator Theory of Electromagnetic Field* in 1991 has no singular term. Because the work is mathematically difficult and complex, and there is no way to prove experimentally which one is right, academic consensus is hard to come by. This is the "electromagnetic field completeness problem" at issue. In 1998, X. Y. Ren pointed out in his doctoral thesis that the controversy about the Green function of the electric dyadic rectangular cavity stems from different understanding of the function $\mathbf{L}$— people have not realized that $\mathbf{L}$ and $\mathbf{M}, \mathbf{N}$ are wave functions with different physical meanings (the equations $\mathbf{L}$ satisfying are not the solution of the passive electromagnetic field, but the other two are); He thought the function $\mathbf{L}$ was just a mathematical tool. In the same year, Ren and Song proposed that it might be correct to exclude functions $\mathbf{L}$— although the eigen function system of electromagnetic field without function $\mathbf{L}$ seems incomplete, this might be the characteristics of electromagnetic field itself, that is, electromagnetic field is inherently incomplete in Euclid space?! In a physical sense, one view holds that the non-singular term (the main part of the Tai solution) is the required physical field, and the singular term represents the non-physical field or "pseudo-mode". Song believes that electromagnetic field actually includes electromagnetic field (light quantum field) and field describing the interaction between charged particles (virtual photon field), that is, particle property is no longer ignored, which is similar to the author's academic view and research work in recent years... In a word, people have expressed opinions on the existing theory from different angles.

In fact, Born-Infeld's electrodynamics modified Maxwell's theory on the basis of a different action integral-one consistent with Maxwell's action in weak fields, but not in strong fields. Born-Infeld's theory belonged to quantum field theory (or quantum electrodynamics), whose success Dirac described in 1964 as "very limited and constantly troubled". One must consider the quantization of electromagnetic fields and waves. In classical theory, the motion equation of electromagnetic field vector is Maxwell equation. In quantum theory, field vectors are treated as operators and governed by Maxwell equations. The state of a physical system is represented by a state vector. In the description of quantized systems, the Heisenberg image is the basis vector formed when the eigenvector is regarded as a state vector that does not change with time. The apparent state vector of Schrödinger image is a time function, and its motion equation is defined by Schrödinger equation. ...While the quantization of electromagnetic waves is not a new problem, quantum electrodynamics has long been established. But the complete quantization is too complicated in theory, so the semiclassical method is often used.

Electronics scientists have found that Schrödinger equation is now often involved in macroscopic electromagnetic problems (metal-walled waveguides, dielectric waveguides, optical fibers, etc.). Engineers are not familiar with this equation, but it is what largely overcomes and resolves the weaknesses of classical electromagnetic theory (based on Maxwell's equations).

The days of relying solely on Maxwell's equations are over, even with the theoretical tools electronics scientists use to deal with engineering problems. People also rely on Schrödinger equation, Klein-Gordon equation, Dirac equation to deal with complex electromagnetic systems. For example, for microwave cyclotron electronic devices, the dynamics theory relies on Maxwell equations and Vlasov equations, which are related to each other and constitute a self-consistent field problem.

IMPROVEMENS OF TRADITIONAL ELECTROMAGNETIC THEORY

The electromagnetic wave was discovered at the end of the 19th century, and its propagation in space presents energy and momentum, which H. Poincarè deduced and expounded in his paper in 1900. At the same time (1897) electrons were discovered, and in 1924 de Broglie proposed the concept of matter waves based on electrons. However, these matter waves are only probability waves, quite different from electromagnetic waves. "Light is a type of electromagnetic wave" was proposed by J. Maxwell in 1865, and "light is both a wave and a particle" seemed logical when A. Einstein proposed the light quantum hypothesis in 1905. However, quantum mechanics (QM) emerged in the 1920s shows that electromagnetic wave is actually the probability wave of photon, and the macroscopic umber of photons realizes the probability wave as the energy and momentum distribution changing with time. This is different from the case of electrons—photons are bosons, lots of them in a certain electromagnetic pattern; but electrons are Fermion, subject to Pauli exclusion principle, and there can only be one electron in a quantum state... Such complications are reflected in the discussion of wave-particle duality.

Can the development of wave science only follow the path of QM? For example, in the electromagnetic wave at the macro level, can it be theoretically improved or even leap-forward by deepening Maxwell's equations and further referring to modern mathematical methods? That's what scientists are thinking.

In 2003, Song pointed out in his book *Basic Equations of Electromagnetic Waves* that Maxwell's equations (and later the simplified form organized by Hertz) could not be solved, and only some special solutions could be found for scalar wave equations. In order to change the mathematical foundation from classical mathematics to modern mathematics, the basic equations of electromagnetic wave represented by two scalar functions can be derived from Maxwell equations by using "vector function space" and "vector partial differential operator theory". It is actually a system of pure curly fields. These two scalar functions are called "state functions", which reflect the characteristics of the electromagnetic wave group, not the state functions of a single light quanta. The equations are in the form of (in the domain):

$$\nabla^2 \phi_m + k^2 \phi_m = -\rho_m \tag{27}$$

$$\nabla^2 \phi_n + k^2 \phi_n = -\rho_n \tag{28}$$

For simple harmonics $e^{j\omega t}$, we have

$$\rho_m = j\omega\mu_0 \mathbf{i}_z \cdot \nabla \times \mathbf{J}$$

$$\rho_n = j\omega\mu_0 \mathbf{i}_z \cdot k^{-1}(\nabla \times \nabla \times \mathbf{J})$$

On the boundary:

$$\mathbf{i}_n \times \{\nabla \times \phi_m \mathbf{i}_z + k^{-1}(\nabla \times \nabla \times \phi_n \mathbf{i}_z)\} = 0$$

It can be proved that the solution of these equations is equivalent to the solution of the curly field operator equation in the curly field space.

In fact, wave theory is not perfect even at the macro level. The motion picture of matter without trajectory, without acceleration, continuously distributed in ever-increasing volume, has not been understood until now. But it is also a fact of matter in motion. At the heart of wave theory is the theory of wave function space, which is not quite the same as Newton's concept of space and time. In the mathematical model of continuous function space, only in the process of establishing the "element" (wave function or basis function) in this space, can the relation directly occur with the coordinate of the element in Euclid space. All subsequent operations are performed not on the elements of Euclidean space (coordinate points) but on the elements of wave function space (basis functions), not according to the rules of Euclidean space. Although Newton's classical mathematical theory also solved the concept of continuity, but in Euclid space not only allow functions and their derivatives to have discontinuity, or in general, Euclid space operations often appear all kinds of discontinuity caused by singularity, and wave function space is impossible to have any singularity. Because a point in Euclidean space is no longer a quantity with direct meaning in the wave function space, singularity becomes meaningless for spatial points.

THE ELECTROMAGNETIC FIELD EQUATIONS OF THE PHOTONS WHICH REST MASS ARE NOT ZERO

In college, Maxwell's theory of electromagnetism is generally unquestioned and widely applied. But because Maxwell's theory could not explain the photoelectric effect, Einstein came up with the photon hypothesis and succeeded. But we can't understand photons without Maxwell's theory — so that creates a kind of logical loop or paradox. Photon fields (free electromagnetic fields) are described by Maxwell wave equation, and the theory of photons emerged because Maxwell's theory failed miserably in the face of the photoelectric effect.

Another example is that waves in nature (wheat in a field, waves on water) are only an appearance of the motion of matter, not matter itself. Is it a matter for electromagnetic waves? If not, the photon corresponding to the electromagnetic wave is matter or not? In 1923, the particle collision experiment of A. Compton proved that photon and electron are all material entities with positive real dynamic mass. It is also proved that the conservation of momentum and energy is correct in a single collision event of microscopic particles. In the wave-particle duality description, the wave corresponding to the electron is "matter wave" or "probability wave". So what waves do photons correspond to? It is usually considered a classical electromagnetic wave, but this wave is not a probability wave. So can we get rid of probability waves for photons? However, the concept is still used in two-photon entanglement analysis. Such contradictions and problems plague physicists; It is only natural that they should move from classical physics to quantum physics.

The quantum mechanical wave equation (Schrödinger equation, SE) emerged in 1926 gave us great enlightenment:

$$\frac{\hbar^2}{2m}\nabla^2\Psi + j\hbar\frac{\partial\Psi}{\partial t} - U\Psi = 0 \tag{29}$$

where $\Psi = \Psi(\mathbf{r}, t)$ is the wave function, $\mathbf{r}$ is position vector, $\hbar = h/2\pi$ is the normalized Planck constant, m is particle mass, $U = U(\mathbf{r}, t)$ is the potential energy of particles in the force field. This formula contains two aspects of the wave-particle duality, one is $\left[\frac{\hbar^2}{2m}\nabla^2\right]$ representing the kinetic energy Operator of particle, the other $\left[j\hbar\frac{\partial\Psi}{\partial t}\right]$ is to reflect the wave process. As well known, SE is to be very successful in describing the movement of electrons. Due to the importance of SE in optical fiber analysis, it is also successful in describing photon motion.

However, the aforementioned classical wave equation of electromagnetic wave, only reflects character of the wave, but not the particle, and there is no particle mass m in the equation. Proca electromagnetic field equations and Proca electromagnetic wave equation, which will be described below, make up the original defect in this problem.

Particle physics generally assumes that the Lorentz-Einstein mass velocity formula is true:

$$m = \frac{m_0}{\sqrt{1-v^2/c^2}} \tag{30}$$

where v is the particle velocity, c is the speed of light, and m_0 is the rest mass at $v=0$. Physics textbooks never say that the upper form does not apply to photons, so one might as well try. However, for photons, the two relations are both valid ($m_0=0, v=c$), so its motion mass $m=0/0$ is an indeterminate form; It doesn't make any sense that the mass of the photon m is arbitrary. The problem can only be caused by the following three aspects: ① The mass velocity formula is wrong; ② The photon rest mass is not zero; ③ Photons do not travel at the speed of light. It is clear that any of these three are incompatible with special relativity (SR). However, it has long been men suspected that photons may have a very small rest mass, and research has followed this up.

Now take the discussion further; From the above equation, can be obtained

$$v = c\sqrt{1 - m_0^2/m^2} \tag{30a}$$

According to Einstein's photon hypothesis, light radiation is a collection of large numbers of photons (each of which carries energy hf), and the mass of a single photon is

$$m = \frac{hf}{c^2} \tag{31}$$

where h is Planck constant; Now we can prove:

$$v = c\sqrt{1 - m_0^2 c^2/h^2 f^2} \tag{32}$$

The above equation indicates that the particle velocity depends on the rest mass m_0 and frequency f; If f specified, then v is determined; So there are three possible scenarios for photons:

① $m_0 \neq 0$, but it is a real number; In this case, the particle is moving at sublight speed, i.e. $v<c$.

② $m_0 = 0$, then $v=c$, particles move at the speed of light.

③ $m_0 \neq 0$, but it is imaginary ($m_0 = j\mu$); Then $v>c$, particles move faster than the speed of light.

Conventional electromagnetic theory chooses case ②.

In particle physics, let a particle have energy E_0 when it is at rest, and kinetic energy E_k when it moves. Then the total energy is $E = E_0 + E_k$, so $E_k = E - E_0$; If particle momentum $p = mv$, it can be proved when admitting Lorentz-Einstein mass velocity equation:

$$E_k = \sqrt{p^2 c^2 + m_0^2 c^4} - m_0 c^2 \tag{33}$$

So available

$$\frac{v}{c} = \sqrt{1 - \left(\frac{m_0 c^2}{m_0 c^2 + E_k}\right)^2} \tag{34}$$

This is generally accepted formula for the ratio of v to c in physics. If $E_k = 0$, $v/c = 0$; As E_k increases, v/c will increase; As E_k equals infinity, v/c reaches its maximum value ($v/c = 1$). Such an analysis, on the one hand, is used to show that particles can only travel at the maximum speed c, and sometimes to show that SR does not require photons to be zero rest mass. Even if the photon's rest mass is not zero, it seems no matter, except that the velocity depends on kinetic energy E_k.

Another view that needs to be considered is that SR's second postulate (the principle of invariance of the speed of light) determines that there will be no photon stationary system, so the photon rest mass $m_0 = 0$. If this is true, then it would be inappropriate to say that SR does not require the hypothesis that the photon rest mass is zero. In short, it is an indisputable fact that existing physical theories are full of contradictions.

Since the paradox arises, the subject can be studied experimentally. There have been many experimental results over the years. For example, de Broglie used binary star observation method in 1940, $m_0 \leq 8 \times 10^{-40}$g; in 1969, G. Feinberg used pulsed star light for observation, $m_0 \leq 10^{-44}$g; in 1975, L. Davies et al. observed Jupiter's magnetic field, and the result was $m_0 \leq 7 \times 10^{-49}$g, and so on. Many other researchers used the test of Coulomb's law to calculate the rest mass of photon, and the results were $m_0 \leq 3.4 \times 10^{-44}$g, 3×10^{-46}g, 1.6×10^{-47}g. Other people started with Ampere's law and got results of 2×10^{-47}g, 8×10^{-48}g, 4×10^{-48}g, etc. In the 21st century, scientists are still designing experiments to measure the rest mass of photons. For example, In 2003, *Phys. Rev. Lett.* published an article by J. Luo et al., a Chinese scholar, which reported that the detection result of their precision torsion balance method was $1.2 \leq 10^{-48}$g. Later Luo gave the following measurement data: $m_0 \leq 2 \times 10^{-50}$g obtained by static torsion balance experiment in Lakes in 1998, Luo et al. obtained $m_0 \leq 1.2 \times 10^{-51}$g by dynamic torsion balance modulation experiment in 2003. In 2006, Tu et al. obtained $m_0 \leq 1.5 \times 10^{-52}$g by using improved dynamic torsional balance modulation experiment. ... All these data are the upper limit of photon rest mass; But Luo said, "one day we will be able to observe the rest mass of photons, not the upper limit." ... By the way, in 2005, the author visited the photon rest mass measurement system in the cave built by Luo's team in Huazhong University of Science and Technology, and was deeply impressed. A few years later, Professor Luo was promoted to the Academician of Chinese Academy of Sciences.

In 1936, A. Proca deduced the equations which were not exactly the same as Maxwell's equations on the assumption that the rest mass of photons $m_0 \neq 0$. However, Proca's electromagnetic field equations are not the complete negation of Maxwell's equations, but the former is more comprehensive than the latter. In other words, the appearance of Proca equations reveals the approximation of Maxwell equations. According to Proca, mathematical variational principles and physical quantum electrodynamics (QED) thinking lead to the following equations:

$$\nabla \cdot \mathbf{D} = \rho - \kappa^2 \varepsilon \Phi \tag{35}$$

$$\nabla \cdot \mathbf{B} = 0 \tag{36}$$

$$\nabla \times \mathbf{H} = \mathbf{J} + \frac{\partial \mathbf{D}}{\partial t} - \frac{\kappa^2}{\mu} \mathbf{A} \tag{37}$$

$$\nabla \times \mathbf{E} = -\frac{\partial \mathbf{B}}{\partial t} \tag{38}$$

It can be seen that, compared with Maxwell equations, only two formulas in Proca equations are changed, and the other two are unchanged. In the above formula, $\mathbf{A}$ is the magnetic vector potential, Φ is the electric scale potential. K is a coefficient:

$$K = \frac{m_0 c}{\hbar} \tag{39}$$

Therefore, K is a coefficient proportional to the m_0, we call it Proca constant. If the photon has no resting mass ($m_0 = 0$), the familiar Maxwell equations are immediately obtained.

Proca equations are not only different in form, but also bring about changes in physical thought. For example, we know that the invariance of electromagnetic field equations under gauge transformation is called gauge invariance, and this transformation forms the local gauge group U(1), which means that the matrix representing the transformation is one-dimensional, i.e. the invariance of the field equations under the U(1). In the Lagrange theory of Maxwell field equations, a variable such as Lagrange density of electromagnetic field can be variated to obtain Maxwell equations. If the U(1) canonical invariance is abandoned, the Lagrange quantity needs to be modified by adding an associated term m_0, and the derivation gives the Proca equations. In this case, vector potential $\mathbf{A}$ and scalar Φ

appear directly in the equations, the gauge transformation loses its meaning and the gauge invariance is destroyed.

It can be proved that in the case of Proca equations, the phase velocity and group velocity of electromagnetic wave are

$$v_p = \frac{c}{\sqrt{1-\left(\frac{\kappa c}{\omega}\right)^2}} \tag{40}$$

$$v_g = c\sqrt{1-\left(\frac{\kappa c}{\omega}\right)^2} \tag{41}$$

make

$$\omega_c = \kappa c = \frac{m_0 c^2}{\hbar} \tag{42}$$

ω_c is called cutoff angular frequency, the meaning of which will be explained later; Therefore, to

$$v_p = \frac{c}{\sqrt{1-\left(\frac{\omega_c}{\omega}\right)^2}} \tag{40a}$$

$$v_g = c\sqrt{1-\left(\frac{\omega_c}{\omega}\right)^2} \tag{41a}$$

Let $p = \omega/\omega_c$, $q = p/\sqrt{p^2-1}$, we obtain:

$$v_p = qc \tag{40b}$$

$$v_g = \frac{c}{q} \tag{41b}$$

Therefore, a

$$v_p v_g = c^2 \tag{43}$$

so when Proca equations is valid, the product of phase velocity and group velocity is a constant value c^2.

It can be seen that at the cutoff point ($\omega = \omega_c$, $p=1$), $v_p = \infty$, $v_g = 0$; As the frequency increases, the wave speed rapidly approaches the value c. For example, when $p=10$, we calculate $q=1.005$, $1/q=0.995$, which is only 0.5% different from 1. In fact, the value of p is much larger than 10 (e.g. $p=10^6 \sim 10^{10}$), then v_p and v_g are nearest by c. Theoretically, because v_p and v_g are related to ω, under the vacuum condition, the dispersion effect of electromagnetic wave velocity is presented. Only when $\omega \to \infty$, the phase velocity and group velocity in vacuum are consistent (c). Obviously, the principle of invariance speed of light in vacuum has lost its meaning. ... So we have now shown that the theory that the rest mass of a photon is not zero is a physical theory incompatible with special relativity (SR).

In addition, in the vacuum condition (i.e., in the completely free space propagation condition), Maxwell theory shows that $v_p = v_g = c$. And Proca says, $v_p > c$ (superluminal), $v_g < c$ (subluminal). Therefore, the cognition of the two different theoretical systems is very different. ... In short, what appeared to be a minor modification of Maxwell's equations turned into a major problem that shook one of the fundamental theories of physics, relativity. However, the Proca theory is consistent with quantum electrodynamics. It also seems to confirm my long-held view that quantum theory and relativity are fundamentally incompatible.

The Theory of Proca may be seen as a consequence of the introduction of QED into conventional electromagnetic theory, but it also introduces other anomalies. For example, the transverse wave properties of conventional electromagnetic waves are destroyed. As is known to all, **A** and Φ cannot be completely determined from a given **E** and **B** in traditional theory, so H.Lorentz introduced the following relation:

$$\nabla \cdot \mathbf{A} + \frac{1}{c} \frac{\partial \Phi}{\partial t} = 0 \quad (44)$$

This is the Lorentz guide, characterized by $\nabla \cdot \mathbf{A} \neq 0$. If the canonical transformation is not true, then

$$\mathbf{k} \cdot \mathbf{A} = 0 \quad (45)$$

In the formula, **k** is wave vector, so **A** is perpendicular to the wave propagation direction, that is, electromagnetic verses is transverse waves. In other words, the polarization of the photon is perpendicular of **k** (only horizontally polarized), and in three components of **A** there are only two independent polarization states.

If $m_0 \neq 0$, the gauge transformation is destroyed, and the number of independent polarization states of **A** is 3, so longitudinal wave appears. Light waves will vibrate longitudinally like sound waves, that is, there will be longitudinal photons. This further complicates our understanding of photons.

Proca theory also has influence on current electrostatic field theory. The plane wave solution of Proca equation in vacuum can be written

$$\Psi = \Psi_0 e^{j(\mathbf{k} \cdot \mathbf{r} - \omega t)} \quad (46)$$

where Ψ can be **E** or **H**; It can now be proved that:

$$k^2 = k_0^2 - \kappa^2 \quad (47)$$

$k_0 = \omega/c$; For the electrostatic field $\omega = 0$, so

$$k = -j\kappa \tag{48}$$

This gives rise to the evanescent wave state, where the field strength drops exponentially in accordance with the law e^{-kr}. As a result, the potential of point charge in the electrostatic field will decay exponentially with distance, so the "inverse square law" in Coulomb's law will be destroyed, and the force between two point charges will be

$$F \propto r^{-n} \quad (n>2) \tag{49}$$

Although n close to 2, the Coulomb law of electrostatic field needs to be modified.

Proca's theory is rejected by some and approved by others. In 1999, V. Majernik discussed the complex quaternion algebraic analysis method of classical electromagnetic field, which not only considered the Proca equation, but also considered the equation proposed by T. Ohamara in 1956. In addition, the influence of magnetic monopoles on the theory is considered in the theory. For example, S. Kruglov discussed generalized Maxwell equations and their solution methods in 2001. The generalized Maxwell equations derived include Proea equations. In 2004, S. Kruglov discussed the square root of Proca equation and obtained the field equation of spin 3/2. This paper deals with the problems of faster-than-light, negative energy and supergravity. These works not only answer the question of whether Maxwell's equations are imprecise, but also expand our understanding of photons.

Table 1 shows the comparison of the two theoretical systems. The "essential evaluation" is the author's personal view, only for reference.

Table 1 Comparison of the two theoretical systems

	Maxwell electromagnetic theory	Proca theory of heavy photons
The rest mass of a photon	$m_0 = 0$	$m_0 \neq 0$
Number of independent polarization states of the vector potential	2	3
The characteristics of the wave	Transverse wave	Transverse wave, longitudinal wave
Gauge transformation	Gauge fields	Gauge invariance failure
The speed of light in vacuum	$c = (\varepsilon_0 \mu_0)^{-1/2}$	when $\omega \to \infty$, $c = (\varepsilon_0 \mu_0)^{-1/2}$
Principle of invariance of light speed	Yes	No
Attitude towards Coulomb law of static field	Admit	Some admit
Essential evaluation	Make a great contribution to; But it goes some paradox; Discrepancy with physical reality	More comprehensive, more consistent with reality

DERIVATION OF PROCA ELECTROMAGNETIC WAVE EQUATION BY THIS ARTICLE

How to understand the relation between photon and electromagnetic wave? The question seems simple, but in fact it is not easy to answer. The author's cautious attitude is expressed as follows: the wave corresponding to the photon is generally regarded as electromagnetic wave; But if a photon is considered to be a microscopic particle, it should have probability wave properties. However, there is no equation of photon probability wave at present. It is difficult to define a wave function for photons.

So how to think about the wave function and wave equation of photons? The author's answer is — there is a view that the wave function of the free state photon is the wave function of the plane electromagnetic wave. Accordingly, Maxwell electromagnetic wave equation is the wave equation of free state photon. ...But this is a simplistic view and does not provide the dynamics that give the physical image of the photon. The photon wave equation still needs to be studied.

We also believe that the "photon free rest mass hypothesis" leads to the lack of theoretical consistency. For massive photons, Maxwell equations can be replaced by Proca equations proposed in 1936. We derive a new wave equation for electromagnetic waves and photons, called the Proca wave equation (PWE). In PWE, there is a term containing particle mass parameter (m), which is consistent with Schrödinger wave equation and Dirac wave equation. This allows the theoretical relationship to improve, and the massive photon to be separated from the point particle.

The author's above understanding comes from three aspects; First, as mentioned earlier, SR's idea that photons have no rest static mass does not support itself. Second, a growing number of scientists, not the only Chinese physicist J. Luo, accept the idea that photons have a rest mass. For example, American physicist R. Lakes, who has been doing theoretical and experimental research on the rest mass of photons, once stated with certainty that "The photon is massive!" Finally, scientific theories are always rigorous and perfect: If you have a theory that works for a photon with a rest mass, and you want to know what happens if the photon has no static mass, then you just take the exact equation, $m_0 = 0$ and immediately get the answer you want.

For convenience, we call the Maxwell wave equation MWE and the Proca wave equation PWE. Just as MWE is easy to derive from Maxwell equations, PWE is easy to derive from Proca equations. But somehow Proca didn't do the job. In 2019, Z.X. Huang deduced PWE and made up for this defect. However, there was a careless mistake that resulted in the asymmetry of the results. Professor Lingjun Wang pointed this out, and we corrected it in our 2021 book. The following derivation, of course, follows the correct narrative.

The equations composed of equations (35), (36), (37) and (38) have been given previously. Take the curl of both sides of Equation (38):

$$\nabla \times \nabla \times \mathbf{E} = -\nabla \times \frac{\partial \mathbf{B}}{\partial t} = -\mu \frac{\partial}{\partial t}(\nabla \times \mathbf{H})$$

Substitute equation (37) in, and get

$$\nabla \times \nabla \times \mathbf{E} = -\mu \frac{\partial}{\partial t}\left[\rho \mathbf{v} + \varepsilon \frac{\partial \mathbf{E}}{\partial t} - \frac{\kappa^2}{\mu}\mathbf{A}\right]$$

That is

$$\nabla \nabla \cdot \mathbf{E} - \nabla^2 \mathbf{E} = -\varepsilon \mu \frac{\partial^2 \mathbf{E}}{\partial t^2} - \mu \frac{\partial}{\partial t}(\rho \mathbf{v}) + \kappa^2 \frac{\partial \mathbf{A}}{\partial t}$$

Arrange it and get

$$\nabla^2 \mathbf{E} - \varepsilon \mu \frac{\partial^2 \mathbf{E}}{\partial t^2} = \nabla \nabla \cdot \mathbf{E} + \mu \frac{\partial}{\partial t}(\rho \mathbf{v}) - \kappa^2 \frac{\partial \mathbf{A}}{\partial t}$$

Can be obtained from formula (35):

$$\nabla \cdot \mathbf{E} = \frac{\rho}{\varepsilon} - \kappa^2 \Phi \tag{35a}$$

After plug in

$$\nabla^2 \mathbf{E} - \varepsilon \mu \frac{\partial^2 \mathbf{E}}{\partial t^2} = \nabla\left(\frac{\rho}{\varepsilon}\right) + \mu \frac{\partial}{\partial t}(\rho \mathbf{v}) - \kappa^2 \frac{\partial \mathbf{A}}{\partial t} - \kappa^2 \nabla \Phi$$

For free space without charge, $\rho = 0$, so

$$\nabla^2 \mathbf{E} - \varepsilon \mu \frac{\partial^2 \mathbf{E}}{\partial t^2} + \kappa^2\left(\nabla \Phi + \frac{\partial \mathbf{A}}{\partial t}\right) = 0$$

Due to Lorentz guide, can be obtained $\mathbf{E} = -\left(\nabla \Phi + \frac{\partial \mathbf{A}}{\partial t}\right)$, so

$$\nabla^2 \mathbf{E} - \varepsilon \mu \frac{\partial^2 \mathbf{E}}{\partial t^2} - \kappa^2 \mathbf{E} = 0 \tag{50}$$

This is PWE as a vector of electric field intensity $\mathbf{E}$, but there is one more term on the left side of the equation than in the classical electromagnetic wave equation (MWE).

The Proca electromagnetic wave equation expressed by the vector of magnetic field intensity $\mathbf{H}$ is derived. Starting from Equation (37), take the curl of both sides of the formula:

$$\nabla \times \nabla \times \mathbf{H} = \nabla \times \mathbf{J} + \nabla \times \frac{\partial \mathbf{D}}{\partial t} - \frac{\kappa^2}{\mu}(\nabla \times \mathbf{A})$$

However,

$$\nabla \times \frac{\partial \mathbf{D}}{\partial t} = \varepsilon \frac{\partial}{\partial t}(\nabla \times \mathbf{E}) = -\varepsilon \mu \frac{\partial^2 \mathbf{H}}{\partial t^2}$$

Therefore, a

$$\nabla \times \nabla \times \mathbf{H} = \nabla \times \mathbf{J} - \varepsilon\mu \frac{\partial^2 \mathbf{H}}{\partial t^2} - \frac{\kappa^2}{\mu}(\nabla \times \mathbf{A})$$

That is

$$\nabla\nabla \cdot \mathbf{H} - \nabla^2 \mathbf{H} = \nabla \times (\rho \mathbf{v}) - \varepsilon\mu \frac{\partial^2 \mathbf{H}}{\partial t^2} - \frac{\kappa^2}{\mu}\mathbf{B}$$

Considering equation (36), $\nabla \cdot \mathbf{H} = 0$, so there is in free space ($\rho = 0$)

$$\nabla^2 \mathbf{H} - \varepsilon\mu \frac{\partial^2 \mathbf{H}}{\partial t^2} - \kappa^2 \mathbf{H} = 0 \tag{51}$$

This is PWE as a vector of magnetic field intensity $\mathbf{H}$; Equations (50) and (51) are the main results of this paper. They are symmetric! If $m_0 = 0$, the Proca wave equation reverts to Maxwell wave equation.

It must be noted that the application of PWE is an important and interesting topic that is not discussed in this article.

COMMENT OF RELATIONS BETWEEN THE MAXWELL EQUATIONS AND RELATIVITY

This is a big topic, and we try to analyze it objectively and seriously from the perspective of Maxwell equations (ME) and relativity (SR and GR). First look at the two theoretical systems themselves, after all whether correct, reliable and complete? Can it stand up to some tough questioning?

The classical electromagnetic theory system can be traced back to its origin. On the one hand, it comes from the formal logic system established by ancient Greek philosophers (which is prominently reflected in Euclid geometry). On the other hand, it also comes from the outstanding idea of "discovering the causal relationship of things through systematic experiments" in the European renaissance. The development of classical electromagnetic theory system is based on classical mathematics composed of calculus, differential equation theory, vector algebra, etc. At the end of the 20th century, after the classical electromagnetic theory encountered difficulties in solving Maxwell equations, scientists put forward a method to solve Maxwell's self-consistent solution by adding an intermediate axiom into Euclid space, namely a mathematical method of vector partial differential operator space. Another foundation was the experimental framework of natural science, built by Kepler and Galileo Galilei, which gave classical electromagnetic theory a foothold as firmly as Newtonian mechanics. ...However, the author thinks that there is a third aspect, that is, when we attach importance to the field and wave, we should take particle into consideration. For example, the research of photon rest mass and Proca theoretical system are paid attention again. In this case, the philosophical significance of the whole classical electromagnetic theory can be fully demonstrated.

As one of the methods of philosophical thinking, dialectics teaches us that we must pay attention to the interrelationships between different phenomena in nature. Maxwell wave equation leads to a deep understanding of the nature of light — light is a kind of electromagnetic wave; The discovery quickly brought electromagnetism closer to optics. However, the speed of light in vacuum $c = 1/\sqrt{\varepsilon_0 \mu_0}$, this relation gives vacuum an electromagnetic property. Of course, vacuum also has other properties (polarization, phase transitions, etc.) that have led to the surprising realization that it can be thought of as a medium that is too complex to be explained without quantum theory.

Maxwell once said, "You should know that physical similarity exists, that is, a local legal similarity between two disciplines." Thus he used the analogy of the streamline of an incompressible fluid to the Faraday force line. In his view, the $\mathbf{E}, \mathbf{H}$ in electromagnetism corresponds to the force in a fluid, and the D, B corresponds to the flow quantity of a fluid.

Similarly, contemporary aerodynamicists have developed some formulae for relativity that substitutes the speed of sound for the speed of light. In 2006, Yang pointed out that scientists use fluid mechanics to derive electromagnetic field equations, hoping to provide more material derivation for the improvement of electromagnetic field equations. It can be proved that fluid also has some similar expressions to Maxwell's equations, which he called "continuum media mechanics equations", in which the ratio of material motion $\beta = v/c$ comes into play, but in this case the speed of sound c rather than the speed of light. Indeed, advances in aerodynamics overcame the shock waves of the $\beta = 1$ singularity and enabled aircraft to fly at supersonic speeds in 1947. Therefore, it is theoretically possible and reasonable to cause superluminal motion transformation. Then, Yang put forward a proposal to search for tachyons on the improved high energy physics accelerator. Later efforts were made to enlist the support of accelerator experts. And so, starting from the topic of the completeness of classical electromagnetic field theory, it will lead to the intersection and fusion of many different disciplines, which can not be described in a single article.

Now let's look at another set of theories, relativistic mechanics. Relativity has long been given biblical status in physics. Any new theory inconsistent with relativity immediately disqualifies it from inclusion in human knowledge. On the one hand, papers that differ from relativity are often rejected by physics journals. On the other hand, some physicists criticized, why are your papers not published in physics journals? ... The long history of deifying Einstein and his theory hasn't stopped scientists (physicists, astronomers, aerospace experts, electronicians, etc.) from criticizing the theory of relativity for its logical inconsistencies (even absurdities). Since I have already published two long articles, here are some other experts' opinions.

Qing-Ping Ma, a professor at The University of Nottingham in England, has published two books criticising SR: *Questions About the Self-Consistency of Relativistic Logic* (462 pages) in Chinese, and *The Theory of Relativity* (503 pages) in English. The two books, which total 965 pages, are filled with scathing judgments of SR, detailing exactly where Einstein went wrong. He points out, for example, "that if SR refused to communicate to verify the twin clock, relativity would be a complete quack pseudoscience; Since this SR required clock slowing can never be verified, your own clock stays the same and the other's clock can only be guessed (how do you observe the clock in a speeding spaceship?) ... If you fly it back, he says, you change the conditions for constant speed (though experiments show that the lifetime of muons is affected only by speed, not by acceleration). Therefore, the attitude towards relativity should adopt the experimental scientific method following Galileo. Einstein's claim of relative character has never been proved."

Another example is astrophysicist and particle physicist Ti-Pei Li's criticism of GR. Einstein's problem, he says, was to treat the GR equation, which describes the curvature of the gravitational field in detail, as an expression of the laws of gravity, leading to a geometric description of gravitational phenomena instead of the search for laws of gravity. The bending of physical manifold is attributed to the bending of spacetime, and the field equation without gravity law of moving mass is called gravity equation. This is how GR was created. Later, Penrose et al. proved that GR must lead to the existence of black hole and big bang singularities. Yet this is both nonphysical and anti-rational. ... Li said that the inversion of causality and the lack of logic self-consistency in GR theory are in fact the special "gravitational hegemony". It is amazing that something like this has been hailed as the crowning achievement of Western science for more than a century.

In literature [36], the author points out that there are serious conflicts between special relativity (SR) and quantum mechanics (QM). In [37] we tabulated the great debate on SR and quantum mechanics. It can be seen that the local description in relativity is incompatible with the wave of particles in QM and with the allowed transformation of particles in QM. In particle physics, non-relativistic QM is a logically self-consistent single-particle theory. However, the premise of so-called relativistic QM is not logically self-consistent, and it is difficult to be a single-particle equation of motion like Schrödinger equation (SE).

It can be said that from SR to GR, Einstein's fundamental mistake was the incorrect view of time and space. For example, he believed that time and space could not exist independently, but must be combined into spacetime. And forced the curved space of every planet to be called curved spacetime. As Li said, mass generates gravitational potential according to linear law, while GR uses Riemann geometry to express gravitational law, thus making Einstein fall into the trap woven by nonlinear tensor analysis. Einstein did not understand the laws of gravity of moving body and could not work out the gravitational field, so he transferred the complexity of the gravitational field to the spacetime context and left the physical difficulties to the mathematicians. Now, we know the universe is a very uniform and flat system of inertia with absolute time, namely Galileo space. However, GR treats the universe with a curved spacetime trap, and enforces an anti-rational origin of the Big Bang and gravitational waves, causing ideological confusion in astronomy and physics.

So how does 26-year-old Einstein's famous paper, "Electrodynamics of Moving Bodies," relate to classical electromagnetic theory (ME)? A fundamental question is whether ME is subject to Galileo transformation (GT) invariance or Lorentz transformation (LT) invariance. As early as 2014, Xiao-Chun Mei published a paper entitled "Roof that There is no LT Invariance in Microscopic Particle Interaction Theory" in a foreign physics journal in 2022, Mei Xiaochun published another paper, pointing out a previously unnoticed problem in Einstein's paper, namely the invented relativistic transformation of electromagnetic fields (we call it RT for short), whose formula is:

$$E'_x(x') = E_x(x)$$
$$E'_y(x') = \frac{E_y(x) - vB_z(x)/c}{\sqrt{1 - v^2/c^2}}$$
$$E'_z(x') = \frac{E_z(x) + vB_y(x)/c}{\sqrt{1 - v^2/c^2}}$$
$$B'_x(x') = B(x)$$
$$B'_y(x') = \frac{B_y(x) + vE_z(x)/c}{\sqrt{1 - v^2/c^2}}$$
$$B'_z(x') = \frac{B_z(x) - vE_y(x)/c}{\sqrt{1 - v^2/c^2}} \tag{52}$$

There is no proof that these formulas are true, but Einstein forced them down to show that he was in agreement with Maxwell's theory of 40 years earlier. So now how do we prove that these formulas don't work? Mei assumed that there was a particle with an electric charge q moving in a uniform straight line in the vacuum. The electromagnetic field generated by Lorentz transformation can be compared with that obtained by RT, and the results are different! ... It is also proved that classical electromagnetic theory has no relative character from other aspects. According to literature [41], for example, many formulas in quantum mechanics and quantum field theory do not have LT transformation invariance. Therefore, the relativity principle does not hold in either the macro world or the micro world.

In addition, in early 2022, Guo-Fu Ji presented a paper entitled "On Maxwell Equations Covariant Under Galileo Transformation". In this paper, it is proved that the wave equation is covariant according to Galileo transformation formula in different inertial systems by analyzing and deducing the electromagnetic wave equation. This article can be used as a reference.

Despite the analysis of the above experts and scholars, it is a big problem whether ME and relativity are contradictory. Maxwell founded ME 40 years before the year 1905. He certainly had no idea that Einstein would come along, much less that every theory of physics has been subject to the requirements of relativity since the 20th century. But can we say that ME is subject to Galileo transformation invariance but not LT invariance? ... I asked professor Ling-Jun Wang of Tennessee State University, an internationally renowned physicist, about this. His pithy answer, in an email dated June 7, 2022:

"Maxwell's equations conform to the Lorentz transformation is based on the assume that relativity is correct. Special relativity proposes the transformation formula of electric field and magnetic field in different coordinate systems. This transformation formula is based on the transformation formula of force in different coordinate systems in relativity, and the electric field intensity is only the force per unit charge. The same goes for magnetic field strength. Therefore, if relativity is not correct, the relativistic field strength transformation is also not true, and therefore ME does not conform to LT. People hope to use ME's Lorentz covariance to prove the correctness of relativity, and use Maxwell's great achievements in electromagnetic field theory to "endorse" relativity, this playing a logical cycle trick: ME conforms to LT according to the relativistic field intensity transformation formula, and in turn, ME conforms to LT to prove the correctness of relativity. Many of relativity's logical contradictions and basic premises (including the principle that the speed of light does not change) had proved it impossible, and its logical cycle had collapsed.

"Even if we concede that ME obeys LT according to the relativistic field transformation formula, Lorentz covariance cannot be regarded as a universal physical law. An equation conforming to a certain covariance is only a mathematical feature of the equation, so it cannot be regarded as an physical law, requiring all physical theories to conform to Lorentz covariability.

"In fact, none of these mathematical physical equations are subject to Lorentz covariability, perhaps except wave equation. In the wave equation, only the electromagnetic wave equation in vacuum obeys, but not the electromagnetic wave equation in media. The wave equation of acoustic wave does not obey Lorentz covariance. Therefore, Lorentz covariance is only a mathematical property of a particular equation among many equations describing various phenomena in nature, and there is no reason to regard it as a universal law of all physical theories, as an important law of nature.

"Lorentz covariance is not universally observed even in relativity, not in general relativity. This is very important in research of theory."

COMMENT OF RELATIONS BETWEEN THE MAXWELL EQUATIONS AND RELATIVITY THEORY (CONTINUED)

Now we introduce the extended Maxwell equations (in after referred to as WE) proposed by Chinese scientist Zhong-Lin Wang not long ago. And relationship between standard ME and SR is further discussed.

It was reported that "On January 13,2022, Chinese Academy of Sciences, Beijing Institute of Nano Energy and Systems at a major research progress on the news conference, released by the institute, member of chief scientist Wang recently and the major scientific research achievements, the Maxwell's equations based on the static electromagnetic field theory to sports medium, successful development of Maxwell's equations. The theoretical basis of electrodynamics of moving media is determined by using the scope. As soon as the news came out, there was a lot of discussion in the physics community and a lot of skepticism."…

In an article, Wang said:

"In Maxwell's equations, there is an electric displacement vector $\mathbf{D}$, where $\mathbf{D} = \varepsilon_0 \mathbf{E} + \mathbf{P}$, where $\mathbf{P}$ represents the density of the polarization field. It must be pointed out that the condition of ME formula is that the volume, surface and spatial distribution of the medium are fixed with no change in time.

"In general, the medium will be polarized in the presence of an electric field $\mathbf{E}$. For isotropic dielectric, $\mathbf{P}$ is expressed as $\mathbf{P} = (\varepsilon - \varepsilon_0)\mathbf{E}$, which is the result of dielectric polarization induced by electric field. If $\mathbf{E} = 0$, $\mathbf{P}$ disappears, then $\mathbf{D} = \varepsilon \mathbf{E}$, which means that if there is no electric field, there is no displacement current, or if there is no external electric field, there is no polarization. This is the general case of electromagnetic waves for which all previous theories and applications have been developed. The nanogenerator is made of a dielectric that generates a strain-sensitive electrostatic charge on the surface, an electrode with a free charge distribution $\mathbf{P}$, and an interconnecting wire across an external load that carries a free-flowing current ($\mathbf{J}$). When a mechanical disturbance is applied to a medium (e.g.TENG), the distribution and/or configuration of the static charge and the shape of the medium will change with time, resulting in a change in the dielectric polarization field. Therefore, an additional polarization term $\mathbf{P}$ must be introduced into the displacement vector $\mathbf{D}$ to account for the dielectric polarization charge."

In another paper, Wang said that his Expanded Maxwell's equations (of moving charged stationary media) is as follows:

$$\nabla \cdot \mathbf{D}' = \rho_f - \nabla \cdot \mathbf{P}_s \qquad (53)$$

$$\nabla \cdot \mathbf{B} = 0 \qquad (54)$$

$$\nabla \times \mathbf{E} = -\left(\frac{\partial}{\partial t} + v \cdot \nabla\right)\mathbf{B} \tag{55}$$

$$\nabla \times \mathbf{H} = \mathbf{J}_f + \left(\frac{\partial}{\partial t} + v \cdot \nabla\right)(\mathbf{P}_s + \mathbf{D}') \tag{56}$$

The proposal of WE has aroused heated discussion in Chinese academic circles. One theory, the electrodynamics of moving body, was solved by Einstein in 1905: the motion of electromagnetic fields is relativistic LT, not GT, must be satisfied, and Landau's work can be read first for Wang Zhonglin's mistake. In short, you can't be right if you break the principle of relativity. Others say that they know there is something wrong with WE "at first glance", because WE can be obtained by GT from ME, and WE destroys the relativistic nature of electromagnetic fields, and so on.

These arguments are plausible, but they are one-sided. Therefore, it is still necessary to further explore the relationship between ME and relativity (mainly SR). Not long ago, it must be noted, a physicist published a scathing critique of relativity on the Internet, but also expressed views about the covariance of physical laws. He said:

—"All the laws of Newtonian mechanics can remain unchanged under the Galileo transformation (GT), i.e., there is Galileo covariance. But Maxwell equations (ME) do not have Galileo covariance and satisfy Lorentz transformation (LT)."

—"Newton mechanics and Galileo relativity principle fit well, electromagnetism and Galileo relativity principle do not match."

—"ME is not covariant under GT."

—"The particle system described by GT is more demanding than the charge system described by LT."

—LT is suitable only for electromagnetic systems, Einstein is naturally transplanted into mechanical systems.

Therefore, the scholar opposed relativity, but advocated "ME and LT covariant." Therefore, the following question becomes a fundamental one—ME subject to Galileo transformation (GT) invariance or Lorentz transformation (LT) invariance?

We believe that many experts and scholars are ignorant of the history of physics and view the problem in isolation, which leads to misunderstandings. Professor Qing-Ping Ma is well aware of this. He says:

"The question of whether Maxwell equations (ME) are subject to Galileo transformation invariance becomes clearer after understanding the historical background of ME, Lorentz transformation (LT) and special relativity. Maxwell proposed his electromagnetic theory based on the fact that electromagnetic interactions require a medium, which is commonly known as the ether. In Maxwell's own theory, like any physical phenomenon that requires a medium (such as mechanical waves), ME obeys Galileo transformation invariance, but the medium is a superior frame of reference for the wave."

Since the electromagnetic interaction requires a medium, it is necessary to understand the relationship between the medium and the observer's frame of reference, that is, the relationship between the moving object and the medium. From the theoretical and experimental phenomenon, people have three basic views:

① Moving objects have no effect on the medium. Experiments with water-filled telescopes and many others seem to support that the motion has no effect on the medium.
② Partial dragging medium of moving object. Fresnal proposed the formula for partial drag, and the Fizeau flow experiment seemed to support a moving partially drag medium.
③ The moving body completely drags the medium. Stokes proposed the model of complete drag. Later, the model of complete drag of massive body and the model of local dominant gravitational field corresponding to light medium were proposed.

Scholars who held any of the three viewpoints still believed that ME was subject to Galileo transformation invariance in the middle of the 19th century, as long as the relation between the moving body and the medium was taken into account. If the moving object has no effect on the medium, then GT can be used to determine the velocity of the moving object relative to the electromagnetic/optical medium through optical experiments. The earth's velocity with respect to electromagnetic/optical media can be determined by optical experiments on earth. One can even further deduce that if the motion has no effect on the medium, then the electromagnetic/optical medium should be absolutely stationary in space, so that the earth's absolute velocity can be determined by optical experiments on earth. Michelson's 1881 and Michelson and Morley's 1887 experiments were designed to measure the phases of the earth. For the velocity of the electromagnetic/optical medium to determine which of the above points is positive, the underlying theoretical basis is of course that M obeys Galileo transformation invariance. Michelson's 1881 physical conclusion was that the Earth was completely exposed to the light medium and therefore the earth's velocity with respect to the electromagnetic/optical medium could not be measured. Michelson and Morley's 1887 experiment confirmed Michelson's 1881 experimental conclusion.

Scholars who claim that ①, including Fitgerald and Lorentz, do not accept the experimental conclusion that the Earth completely drags the optical medium. In a text message published in *Science* in 1889, Fitgerald suggested that Michelson and Morley's 1887 results might be caused by the resolution of the magnitude of the length of the object in the direction of motion. Lorentz then proposed LT based on length contraction and the "temporal brain suit hypothesis" to reconcile the idea that "moving objects have no effect on the optical medium" (i.e. the ether is absolutely static in space) with Michelson and Morley's 1887 experiment (according to GT). The earth completely drags the light medium. Therefore, the reason why ME obeys Lorentz covariance but not Galileo covariance lies in Lorentz et al.'s insistence that the aether is absolutely static in space on the one hand, and on the other hand, they have to accept the experimental results of optics and electromagnetism when the ether is completely dragged by the earth. LT is the result of these two aspects. Abandon the view of the etheric light medium is stationary in space, according to the result that the light medium is completely g by the earth and other massive objects, ME is completely subject to GT invariance, but not at all subject to LT invariance. Both the working principle of GPS and the variation of the speed of light in the generalized and binary theory suggest the existence of light media, which correspond to locally dominant gravitational fields.

Einstein's approach to the problem was to inherit the mathematical form of Lorentz's ether theory and discard its physical basis (light medium). Thus, in SR, ME is subject to Lorentz covariance as in Lorentz ether theory. Einstein took the results of Lorentz's ether theory (length contraction and time dilation), which caused ME to submit to LT and kept the speed of light constant as the cause and premise, abandoned the light medium and proposed the principle of the invariance of the speed of light and the viewpoint of "ME consistent with LT". The invariance of the speed of light and the relativity principle derived LT, and then derived from LT, length contraction and time dilation. The existence of optical media in Lorontz's ether theory provides a physical basis for velocity effects and avoids paradoxes/paradoxes. SR abandons the medium of light, leading to many paradoxes.

The above analysis of Professor Ma is very profound, from which we can see that Einstein actually deviated from Lorentz and carried out a whole set of logical cycles based on the wrong cognition. In the past, the author pointed out in the paper: from 1892 to 1904, H. Lorentz assumed that the length of the moving body was shortened and the time was delayed in order to explain the Michelson-Morley experiment. In 1905 and 1952, Einstein gave the derivation of length shortening, but these relativistic length shortening existed in logic. Lorentz's theory stated that the length of objects at rest in the ether had this relationship with the length of objects moving relative to the ether. However, there are many paradoxes in the special theory of relativity (SR) that the mutual view of physical phenomena causes the shortening of length, because SR is based on relative motion and causes a paradox in principle.

In Lorentz's theory, the delay in time is caused by the absolute motion of the moving body. Clocks with high absolute velocity slow relative to clocks at rest: this is Lorentz time delay in ether theory. But in SR, replacing the absolute velocity with the relative velocity of the moving body, the situation is completely different. Einstein explained length shortening and time delay by replacing the relationship between the observer and the ether with the relative motion of the reference frames of different observers. Many paradoxes, most notably Langevin's twin paradox in 1911, question the self-consistency of SR.

The consistency of the relativity principle of one of the laws of physics from the perspective of any inertial system was first introduced by H.Poincarè, which was embodied by the Lorentz transformation (LT). However, the relativity thought published by H. Lorentz in 1904 was derived under the existence of the ether. In 1905 Einstein published a famous paper, in which there was a postulate—the invariance of the speed of light. The argument that there was no need for an ether, that is, a preferred reference system, always included the question of whether Einstein's special theory of relativity (SR) or the modified Lorentz theory (MOL) better described the nature. The main difference between the two is that SR believes that all inertial frames are equal and equivalent, while MOL believes that there is a preferred reference frame. Over the years, many studies have shown that SR is logically untreatable, and there is a lack of definitive experimental confirmation." I want to go back to before Einstein, Pomcare and Lorentz," John Bell, a scientist at CERN (Europe's Nuclear Research Centre), said in 1985. It is worth noting that SR cannot explain the recent research results—gravitational propagation superluminal and quantum entangled propagation superluminal, while MOL can explain it.

It can be seen that some physicists who are used to wielding the big stick of relativity do not understand the history of the development of physics and do not understand the theory of relativity. As for Z. L. Wang's equations, this paper is not to guarantee their correctness, but to say that all new physical theories should not be dismissed with the accusation of transmutation LT covariance at every turn. As Professor Lin-Jun Wang says, LT covariability is not a physical law of iron. To use it to measure ME (or any other equation, such as WE), you have to be sure that relativity is correct. But either SR or GR is so flawed that how can ME (as Einstein and others hope) be consistent with relativity?!

In 2021, I published a paper entitled "Is Einstein's Special Relativity Correct?" The article states that the first postulates of SR (Special relativistic principle) have been sharply criticized by scientists (e.g. H. Bondi, Guo, And Tang). They point out that the relativity principle is not compatible with cosmology, and the large scale space-time is not Minkowski space-time at all, but the universal standard time and cosmic background space. In fact, the SR space-time view is not tenable at macro level, and it is fundamentally wrong to admit only the relativity of motion but not its absoluteness. In addition, some scholars (such as Professor Ru Yong Wang) specially designed experiments to prove the false principle of SR. In fact, Professor Wang has been writing articles for many years that the two principles of SR (two postulates) are wrong and have never been really verified by experiments.

Einstein once wrote: "Oh, Newton, please forgive me!" It was an arrogant, disingenuous remark, showing that he actually felt he had Newton underfoot. ...But he did not dare to do so with Maxwell, because electromagnetism was not his specialty and

Maxwell's theory had a solid experimental foundation. In 1905, Einstein, a 26-year-old who had never experimented with electromagnetism, despised Maxwell. Although he wrote his thesis with a good title "On the Electrodynamics of Moving Bodies ", the content is not appropriate to the title. Therefore, I do not accept the statement of a physicist ("Z. L. Wang's problem was solved by Einstein 117 years ago"). In addition, the constitutive equation of dielectric electromagnetic field does not have LT invariance; that is to say, ME does not satisfy LT invariance. In addition, relativity should not be used as an excuse to attack all new theories.

CONCLUSION

Maxwell equations (ME) as the core of the classical electromagnetic theory is a long-term, has maintained a strong vitality. It is one of the three pillars of physics, the other two being the Newton system of classical mechanics and the quantum theory system with quantum mechanics (QM) as its core. Classical electromagnetic theory is complete in general, but there are some problems pointed out in this paper. The in-depth discussion of these questions is not to negate classical electromagnetic theory. Since it is born on the basis of the observation and measurement of numerous electromagnetic phenomena, it is a vivid embodiment of the physical reality of nature, and has been tested by long-term practice A further exploration will improve it, make it more profound and complete, and make this scientific pearl more bright and attractive!

In this paper, the direct solution of ME and vector electromagnetic wave equations is discussed by means of in-depth mathematical analysis. It is necessary to further deepen this subject and I believe it will arouse people's interest. This paper points out that the deepening of mathematics and the modernization of physical concepts complement each other. In this paper, the improvement direction of ME particle property is discussed in a considerable amount of space, and the electromagnetic Proca wave equation is derived, in order to arouse the academic circle's new understanding and attention, and reflect the insistence on the concept of "wave-particle duality".

Finally, this paper discusses the relationship between ME and relativity, and points out that during the recent discussion, some people use "not satisfy LT invariance" to deny the proposed expand the ME theory, the reason is not sufficient, however, due to we not in-depth study, we don't have the right of judgment for WE, but think it should consider and study, this is the active academic, encouraging innovation is the only way.

Acknowledgments: Thanks to Prof. Ling-Jun Wang, Prof. Xiao-Chun Mei, Prof. Qing-Ping Ma and Prof. Xin-Tie Yang for their support and helpful discussions.

REFERENCES

[1] Huang Z X. Faster-than-light research — the intersection of relativity, quantum mechanic, electronics and information Theory[M]. Beijing: Science Press, 1999.

[2] Pang X F. Nonlinear quantum mechanics[M]. Beijing: Publishing House of Electronics Industry, 2009.

[3] Sheng J L. Fundamentals and applications of nonlinear optics[M]. Beijing: Capital Normal University Press, 2019.

[4] Tai C T. Dyadic Green's function in electromagnetic theory[M]. Pennsylvania: Intext Educational Publishers, 1971.

[5] Song W M. The operator theory of dyadic Green's functions and electromagnetic fields[M]. Beijing: University of Science and Technology of China Press, 1991.

[6] Song W M. Basic equations of electromagnetic waves[M]. Beijing: Science Press, 2003.

[7] Zhang Y Z. Experimental basis of special relativity[M]. Beijing: Science Press, 1979 (first edition), 1994(reprint).

[8] 黄志洵. 关于电磁波特性的一组新方程 [J]. 中国传媒大学学报, 2019, 26(5): 1–6.

[9] 黄志洵. 从声激波到光激波 [J]. 中国传媒大学学报, 2020, 27(4): 1–15.

[10] 黄志洵. 爱因斯坦的狭义相对论是正确的吗? [J]. 中国传媒大学学报, 2021, 28(5): 71–82.

[11] 黄志洵. 对广义相对论的研究和讨论 [J]. 中国传媒大学学报（自然科学版）, 2022, 29(1): 64–80.

[12] Ma Q P. The question of relativity's logic consistency[M]. Shanghai: Shanghai Science and Technology Press, 2004.

[13] Tang S S. From special relativity to standard time-space theory[M]. Changsha: Hunan Science and Technology Press, 2007.

[14] 黄志洵. 非线性 Schrödinger 方程及量子非局域性 [J]. 前沿科学, 2016, 10(02): 50–62.

[15] 黄志洵. 光子是什么 [J]. 前沿科学, 2016, 10(03): 75–96.

[16] 黄志洵. 运动体尺缩时延研究进展 [J]. 前沿科学, 2017, 11(03): 33–49.

THE APPENDIX

About the Early Study on Maxwell's Equations (ME)

The great edifice of physics is actually built on three foundations: Newton's Classical Mechanics, Maxwell's Classical Electrodynamics, as well as the multi-established Quantum Mechanics and Quantum Electrodynamics. As for the origin of classical electromagnetic field theory, literature [32] has a good summary; it believes that the formation of Maxwell's equations comes from four aspects:

—Electrostatics: Coulomb→Poisson→Gauss→Maxwell;
—Theory of Current: Galvani→Volta→Ohm→Maxwell;
—Electrodynamics: Ampere→Newman→Weber→Maxwell;
Austead→Biot-Savart→Maxwell;
—Electromagnetic induction and field theory: Faraday→Kelvin→Maxwell;

It all comes to Maxwell. Maxwell's creative years began in 1865 (age 34), when he digested and summarized the work of his predecessors and added his own innovative idea (displacement current) to create the cornerstone of the physical edifice, the ME.

The equations of electromagnetic field proposed by Maxwell in 1865 were all in differential form, consisting of 20 scalar equations. According to him at the time, ME included the following concepts: amount of charge, electric potential, conduction current, electrical displacement, full current, magnetic force (magnetic field strength), electromagnetic momentum, etc. Among them, the so-called electromagnetic momentum is later vector potential $\mathbf{A}$.

It is well known that H. Hertz confirmed Maxwell's prediction with his experimental discovery of electromagnetic waves, but Hertz's theoretical contribution to ME is unknown. It is therefore necessary to give a brief account of Hertz and its history, which helps to understand Maxwell's time. The German physicist H. Hertz (1857-1894) lived a shorter life than Maxwell, but he was unforgettable as a meteor flying through the sky of history. Hertz was born in Hamburg and graduated from the University of Berlin at the age of 23. He was a student of the famous physicist H. Helmholtz (1821-1894). After graduation, he became an assistant professor with Professor Helmholtz. It was this practitioner who advised him to do theoretical and experimental research on Maxwell's theory. Moreover, in 1870 Helmholtz proposed the concept of media polarization.

One of the big questions in European scientists at the time was whether Maxwell's theory was correct. Hertz had been thinking and studying this question from 1879 to 1883. In 1885, Hertz became a professor at the Karlsruhe Polytechnic School, which was so well equipped that he was able to prove experimentally the existence of electromagnetic waves in 1887. The experiment has been carried out many times in different ways, with success. Hertz was found to have published two papers in 1888, both of which stated that the wavelength of the electromagnetic wave he produced was 2.8 meters. The paper published in 1889 mentioned the wavelength of 4.8m, 2.8m; Another article in the same year said the wavelength was 4~5m, 66cm, 58cm, etc. The papers were published in *Annalen der Physik*, an authoritative German journal.

He also did a lot of research in theory. In 1889, Hertz published the paper "on Maxwell's Fundamental Equations of Electromagnetism". In 1888 and 1890, he published papers on how to improve ME. It was not just Hertz, of course, but also H. Lorentz and other famous physicists. On the basis of respecting Maxwell's original invention right, it is refined, summarized and integrated to form the final combination of four vector differential equations, which is the standard form of ME which is very popular later generations [see Equation (1), (2), (3) and (4) in this paper].

However, in different units, the form of ME is not exactly the same. For example, in the Gauss system of units, ME is of the form

$$\nabla \cdot \mathbf{D} = 4\pi\rho$$

$$\nabla \cdot \mathbf{B} = 0$$

$$\nabla \times \mathbf{H} = \frac{4\pi}{c}\mathbf{J} + \frac{1}{c}\frac{\partial \mathbf{D}}{\partial t}$$

$$\nabla \times \mathbf{E} = -\frac{1}{c}\frac{\partial \mathbf{B}}{\partial t}$$

For the form of ME in other units, see [12].

On the Non-Relativistic Space-Time View and the Covariation of Maxwell's Equations*

INTRODUCTION

In 1905, A. Einstein published his first paper on Relativity, the basic document of his theory of Special Relativity (SR). The starting point of SR is two postulates (also known as two principles), namely the special relativistic principle and the principle that the speed of light is constant in one direction. The special relativistic principle states that the laws of physics are the same in all inertial systems, and this holds true not only for the laws of Mechanics, but also for the laws of Electromagnetism. Therefore, according to SR, the law of electromagnetism should have the same form in all inertial frames moving in a straight line with uniform velocity relative to each other. It is known that the property of physical laws that remain unchanged in form under transformation of inertial reference frames is called covariation. The space-time transformation between different inertial frames required by the relativistic principle of SR must be the Lorentz transformation (LT). This covariation can be called SR covariation or Lorentz covariation, and it has dominated the rules of physics for a long time. Some non-relativistic views of time and space boldly abandon the principle of Special Relativity because the development of theoretical analysis and experimental facts requires it. It is possible that physics can be changed to a new orbit by assuming that the equations that express physical theories and laws remain unchanged under other transformations, an example of which is the generalized Galilei transform (GGT). These things will be described in this article. We might as well explore whether physics can emerge without SR.

This paper is one of several scientific papers that reject the term "Relativistic Electromagnetics". In taking this view, we shall assume that electromagnetic laws do not obey the principle of Special Relativity; In other words, the absolute reference frame (i.e., the ether) is assumed to exist; In any inertial frame moving uniformly with respect to the absolute reference frame, the form of the electromagnetic field law will be different from that in the absolute reference frame. But the problem can be dealt with appropriately.

Einstein always pursued a physical theory that explained everything. Early SR had a fundamental requirement that the laws of physics should be invariant to LT. And then he decided that this was not enough, that the law should be invariant to any coordinate

---

* The paper was originally published in *Current Journal of Applied Science and Technology*, 2022, 41 (45), and has since been revised with new information.

transformation. He argued that a "generalized covariation" could be derived from the principle of relativity. But these two are different, the former is a statement of the uniformity of space and time, which is reflected in the fact that there is a transformation group; This homogeneity with LT can be stated not only in Galilei coordinates but also in terms of generalized covariation. In fact, following the phrase "law invariance" hides Einstein's conceptual confusion. He argues that General Relativity (GR) differs from SR in that SR uses LT groups and GR uses other groups. Such statements are difficult to understand. He confused the meaning of the word "covariation".

Another example of Einstein's conceptual confusion is his resolute abandonment of ether in SR and his talk of the "ether of General Relativity" in GR. Why is that? He later decided that he could not allow "nothing void," so he tried to pick up the ether that had been thrown away. In this way, Einstein did not do science to explore the laws of nature, but to make nature subject to the needs of his research work — he thought the ether was superfluous in his SR work, so there is no ether in the universe; When he started GR, he decided that "nothingness" would not work without ether, so the universe could be filled with ether again.

This paper discusses the development of physical theory by discussing the covariation of Maxwell's equations (ME) and comparing SR with the non-relativistic view of time and space. This paper points out that non-locality in quantum mechanics (QM) represents a non-relativistic view of time and space, which is completely opposite to SR. The Schrödinger equation (SE) in QM is derived from Newton mechanics. Although it does not have covariation with LT, it does not prevent it from becoming the most basic and important quantum equation of motion, and its application is not limited to "low speed". This paper puts forward and discusses the "new ether theory". At the same time, on the basis of its criticism of SR and experimental facts, we advocated a new discipline, "Superluminal Light Physics", and pointed out the outstanding contributions of Chinese scientists.

COMMENTS ON "RELATIVISTIC ELECTROMAGNETISM"

The theory of Relativity has long been regarded as an absolute truth, a sacred object to be observed by all disciplines. The following terms have become common in the literature: Relativistic Mechanics, Relativistic Quantum Mechanics, Relativistic Electrodynamics, Relativistic Electromagnetism, etc. What has not yet emerged is Relativistic Optics, Relativistic Heat Theory, Relativistic Biology. Now let's take out Relativistic Electromagnetics to see is this concept can holds?

The development of electromagnetism began in Europe and is associated with the names of many great masters. For example, in electrostatics. there were C. Coulomb (1736-1806), S. Poisson (1781-1840), G.Green (1793-1841), K.Gauss (1777-1855) and others. In electrodynamics, there were H. Oersted (1777-1851), A. Ampere (1775-1836), W. Weber (1804-1891), J. Biot (1774-1862), F. Savart (1791-1841), H. Lorentz (1853-1928) and others. In the field of electromagnetic induction and field, there were M. Faraday (1791-1867), J. Henry (1797-1878), J. Maxwell (1831-1879), H. Hentz (1857-1894) and others. These physicists, each with outstanding theoretical and experimental contributions, together performed a grand symphony that lasted for 200 years, and was finally summed up by Maxwell-Hertz in mathematics, achieving the great achievements of the electromagnetic theory mansion... Table 1 shows several cases of Maxwell's equations (ME) induced by us.

Table 1 Several cases of Maxwell's equations (ME)

Serial number	Common form	Static field	Steady static field
1	$\nabla \cdot \mathbf{D} = \rho$	$\nabla \cdot \mathbf{D} = \rho$	$\nabla \cdot \mathbf{D} = \rho$
2	$\nabla \cdot \mathbf{B} = 0$	$\nabla \cdot \mathbf{B} = 0$	$\nabla \cdot \mathbf{B} = 0$
3	$\nabla \times \mathbf{H} = \mathbf{J} + \frac{\partial \mathbf{D}}{\partial t}$	$\nabla \times \mathbf{H} = 0$	$\nabla \times \mathbf{H} = \mathbf{J}$
4	$\nabla \times \mathbf{E} = -\frac{\partial \mathbf{B}}{\partial t}$	$\nabla \times \mathbf{E} = 0$	$\nabla \times \mathbf{E} = 0$

Serial number	quasi static field	Free space	Free space, single frequency
1	$\nabla \cdot \mathbf{D} = \rho$	$\nabla \cdot \mathbf{E} = 0$	$\nabla \cdot \mathbf{E} = 0$
2	$\nabla \cdot \mathbf{B} = 0$	$\nabla \cdot \mathbf{H} = 0$	$\nabla \cdot \mathbf{H} = 0$
3	$\nabla \times \mathbf{H} = \mathbf{J}$	$\nabla \times \mathbf{H} = \varepsilon_0 \frac{\partial \mathbf{E}}{\partial t}$	$\nabla \times \mathbf{H} = j\omega\varepsilon_0 \mathbf{E}$
4	$\nabla \times \mathbf{E} = -\frac{\partial \mathbf{B}}{\partial t}$	$\nabla \times \mathbf{E} = -\mu_0 \frac{\partial \mathbf{H}}{\partial t}$	$\nabla \times \mathbf{E} = -j\omega\mu_0 \mathbf{H}$

Obviously, none of these developments have anything to do with Einstein. In fact, for some of the most important results in electromagnetism, Einstein was not even born (Einstein was born in 1879). So why did "Relativistic Electromagnetism" emerge after Einstein became famous?... As an example, Chapter 9 of reference [2] is entitled "Relativistic Electromagnetic Fields," which departs a section (§2) to "the covariation of electromagnetic field equations and relativistic time-space transformations." How can we call it Relativistic Electromagnetism, since the core content of Electromagnetism has not chanted, and now only the space-time transformation and covariation have been added?! The key points of "relativistic transformation of electromagnetic field" mentioned in literature [2] are as follows: First, since the special relativistic principle of SR requires all physical laws to have the same form in different inertial systems, ME must also have covariation. Secondly, the relativistic transformation formulas of charge density ρ and current density $\mathbf{J}$ can be derived, and then the equations of ($\nabla \cdot \mathbf{B}$) and ($\nabla \times \mathbf{E}$) in ME can be deduced to be covariant in the inertial system, and then the other two equations ($\nabla \cdot \mathbf{D}$) and ($\nabla \times \mathbf{H}$) can be proved to be covariant in the inertial system. The results are as follows: ME accords with the relativity principle; SR is self-consistent and harmonious.

What's wrong with this argument? First, the question of whether ME has LT covariation, some experts have pointed out that Einstein, in order to prove that the electromagnetic field is relativistic invariant, introduced a so-called relativistic transformation of the electromagnetic field itself, which, unlike the LT of the electromagnetic field, cannot be established. Secondly, the purpose of this self-circular argument is to use ME to enhance the prestige of SR, and at the same time, to express a kind of SR's "approval" of ME. But ME does not need Einstein's approval; it already exists and has proven itself right by being widely used. In short, these superficial arguments do not prove "relativistic electromagnetic fields" or "Relativistic Electromagnetism" was established.

Any electromagnetic phenomenon occurs in a certain time and space. The prerequisite for studying electromagnetism is to have a correct view of time and space. We can try to move away from relativity and start with some new assumptions:

① Maxwell and Lorentz believed that ether existed; It is not only the carrier of electromagnetic field, but also the medium on which electromagnetic wave propagates. We therefore assume the existence of an ether, that is an absolute frame of reference.

② The 21st century view of the ether is different from the 19th century; Today's "new ether" has three options: physical vacuum, microwave background radiation (CMB), and gravitational field. We think that ether is the ubiquitous physical vacuum, also a kind of matter, is the basic form of matter in the universe; The demonstration of electromagnetic phenomena itself contains the role played by this background.

③ The law of electromagnetic field does not obey the principle of Special Relativity. The ether is the absolute reference frame by which all matter moves. That is to say, ME is only valid in the absolute reference frame of the ether.

If one starts from the above premise, a problem arises: in any inertial frame moving uniformly with respect to the absolute reference frame, the electromagnetic field law will take a different form from that in the absolute reference frame. This is a complex theoretical problem that may involve many mathematical derivations. The Chinese physicist Shusheng

Tan did this work, resulting in a system of equations containing the effect of the term v/c^2. But it's not relativity's either. It's not relativistic electromagnetics. In fact, Professor Tan proposes and believes in the "Theory of Standard Space and Time", but not SR.

The starting point of the so-called "Standard Space and Time Theory" is also two principles: the absolute reference system principle and the loop speed of light invariant principle. Unlike SR, they do not involve the principle of special relativistic principle, nor the principle that the speed of light is constant in one direction. Prof. Tan started from his own two hypotheses (principles) and combined GGT's way of thinking, so it is his independent contribution.

Therefore, for the theory of space-time which does not accept the principle of special Relativity, there is a task to determine the transformation relationship between the electromagnetic field quantity in the absolute reference frame and the electromagnetic field quantity in the general inertial frame. There is also a definition for the standard form of ME—it refers to a system of electromagnetic fields in an absolute reference frame. Qing-Ping Ma, professor of the Nottingham University of UK, has made sharp criticism of SR; When I asked him to comment on the so-called relativistic electromagnetic, he said that he agreed with me that the laws of the electromagnetic field did not obey the special relativistic principle, and that the ether system was the superior reference frame for electromagnetic phenomena. In addition, in 1980, a professor at the University of Yunnan, Yong-Li Zhang has a point, he did not in order to prove the lack of Relativity of electromagnetism "in trouble", the electromagnetic induction phenomenon, for example, said in the magnet coil in and out of the experiment, "if the magnet still, space is only static magnetic field and electric field, and the movement of the coil will generate an electric current, in addition to additional assumptions, It's hard to understand." But Prof. Ma believes that this view is actually using his relativistic view to limit classical electromagnetic theory. Firstly, the classical electromagnetic theory is put into the framework of SR, and then the dynamic electromotive force of electromagnetic induction in the classical electromagnetic phenomenon is difficult to understand. In the classical electromagnetic theory, Faraday electromagnetic induction theory considers the change of magnetic flux in a conductor (especially a closed conductor loop), where the conductor can be regarded as the medium of the electromagnetic field. In the classical electromagnetic theory, the medium reference frame is the superior reference frame, and the electromagnetic induction theory should consider the change of magnetic flux in the conductor medium system. Whether the magnet is moving, the coil is not moving; or the magnet is not moving, the coil is moving; from the point of view of the conductor medium system, there are changes in magnetic flux. Therefore, in the classical electromagnetic theory, there is no such thing as the magnet is not moving, the coil is moving when the induced current is difficult to understand.

It can be considered that Maxwell extended Faraday's electromagnetic induction theory and Ampere's law to non-conducting media systems, especially non-physical space media systems (ether system). The ether can propagate " displacement currents" (changing electric fields), which generate magnetic fields and changing magnetic fields generate changing magnetic fields, resulting in the ether system. The ether system is a superior reference system compared with other reference systems when the internal situation of various physical objects (such as metal conductors and electrolyte solutions) is not considered. Faraday and Maxwell's electric field changes and magnetic field changes are fundamentally changes relative to the medium reference frame.

SR believes that electromagnetic waves electromagnetic fields do not need medium, and the change of electric field and magnetic field in Faraday and Maxwell theory is not the change of reference frame relative to medium, but the change of reference frame relative to motion (observer). Prof. Zhang thinks that the classical electromagnetic theory only considers the space ether system, and does not consider the other medium system. Therefore, he believed that the induction of current when the magnet is stationary and the coil is moving is difficult to understand according to classical electromagnetic theory. SR argues that electromagnetic phenomena are due to changes in the relative motion (observer) frame of reference, a view that leads to paradoxes. If the magnetic phenomenon is only the relative motion effect of the observer, two identical charges at rest on the ground will be repulsed by the Coulomb force, and the moving observer will find that they will also be attracted by the Ampere force, reducing the repulsive force. When the observer moves fast enough, the Ampere force will exceed the Coulomb force, and the moving observer will find that the two charges of the same species attract. An observer on the ground would always find the two charges repealing each other.

It can be seen that Relativity not only does not help electromagnetism, but also leads to wrong understanding and conceptual confusion.

Consider now the papers on electromagnetism written in the non-relativistic mind, and the new work that may be called "non-relativistic electromagnetism". Examples of the former are two papers by Prof. Zhixun Huang (one in Chinese and one in English); the paper points out. Although there is room for improvement, ME is undoubtedly a brilliant scientific pearl. It not only expresses a way of looking at nature, but also has fascinating scientific beauty. It has been used all over the world to solve many engineering problems. It is also argued that the covariation of ME must be justified by the correctness of Relativity theory. If the theory of Relativity does not hold, the following happens: the relativistic transformation of the field strength does not hold. In this way, ME does not comply with LT covariation. ... An example of the latter is the emergence of Expanded Maxwell Equations (EME); In July 2022, Zhonglin Wang, an academician of the Chinese Academy of Sciences, published a long paper, which was both a description of his achievements and an answer to some of the accusations. It is stated that the view of time and space based on GT should not be ignored, and that his own system of equations (called WE by the author) "does not maintain LT covariability". It was a brave statement.

THE SO-CALLED "LOW SPEED APPROXIMATION" MAY BE A DEROGATION OF NON-RELATIVISTIC EQUATIONS

In recent years, Chinese scientists have done a lot of research on the establishment of new electromagnetic theory, and have achieved fruitful results. For example, in order to solve the self-consistent solution of ME, Prof. Wen-Miao Song introduced a new mathematical method, that is, on the basis of Euclid space, he added an intermediate term — the mathematical norm method of vector partial differential operator. Another example is that Zhonglin Wang extended ME to the case of moving media, expanding the application scope of the theory. This WE came under a lot of attack for not maintaining LT covariability.

Now let's write the titles of Prof. Wang's two most representative papers:

I. Dynamic Maxwell Equations for Engineering Electromagnetics and Their Solutions (published on November 16, 2015);

II. Maxwell's Equations for a Mechano-Driven Varying-Speed Motion Media System Under Slow Motion and Nonrelativistic Approximations (published on July 8, 2022).

Obviously, paper II is more noteworthy, as Wang will present his final thoughts after spontaneous online discussions in the first half of 2022. The title of thesis II actually says: "My system of equations may contradict or not conform to the requirements of Relativity; But it is an approximation at low speed, so it should be allowed." We are not clear why Prof. Wang said so: is it after careful analysis and calculation to determine that WE can only be used at low speed, or to avoid the criticism of the relativists, to fight for the survival space for WE?

Here, look at the history of how the Schrödinger equation (SE) has been belittled and later gradually corrected by relativists. Erwin Schrödinger (1887-1961) was one of the founders of Quantum Mechanics (QM); In early 1926, he published the first of a series of papers called "Quantisation as a problem of proper values". The idea was to consider simple (non-relativistic and unperturbed) microscopic systems, such as hydrogen atoms, in order to discover the true nature of quantum rules. At this time, he proposed the function Ψ, which is a single valued and continuously differentiable real function, and defined S by the Hamilton-Jacobi differential equation:

$$S = K \lg \Psi$$

He tried to replace the quantization condition by a variational problem with discrete eigenvalue spectra (corresponding to Balmer terms) and continuous eigenvalue spectra (corresponding to hyperbolic orbits). The following equation arises for the one-electron case

$$\nabla^2 \Psi + \frac{2m}{\hbar^2}\left(E + \frac{e^2}{r}\right)\Psi = 0 \tag{1}$$

The latter called the upper equation "a stationary nonrelativistic quantum wave equation," and Schrödinger himself called it "an Euler-type differential equation for variational problems", and said that it had a solution for every positive value. Schrödinger then derives a condition, Equation (15) in his text, from which the Bohr level corresponding to the Balmer terms in hydrogen atom is derived.

Hamilton equations and variational methods are commonly used in classical mechanics, and are now used by Schrödinger to construct new kinetic equations, whose results are consistent with the hypotheses and experimental facts about the hydrogen atom.

On June 23, 1926, Schrödinger submitted paper IV, which was published in *Ann. D. Pysik*, volume 81, number 4. This long article presented a time-dependent equation that marked the maturity of wave mechanics thinking and the birth of Quantum Mechanics. This combination of the analysis of waves and particles is excellent. If the general expressions of wave function and potential function are $\Psi(r,t)$ and $U(r,t)$, then there is a time-dependent Schrödinger wave equation:

$$j\hbar \frac{\partial \Psi}{\partial t} = \frac{\hbar^2}{2m}\nabla^2 \Psi + U\Psi \tag{2}$$

Let $\hat{H} = \frac{\hbar^2}{2m}\nabla^2 + U$, therefore have

$$j\hbar \frac{\partial \Psi}{\partial t} = \hat{H}\Psi \tag{2a}$$

This is a quadratic differential equation in time t and space coordinates, so it has no covariation under Lorentz transformation (LT) and does not meet the requirements of Relativity, so it is a non-relativistic equation. In other words, the time and space coordinates must have the same degree of differentiation in the equation that satisfy the requirements of Relativity.

On the other hand, if the wave function is stationary, that is, $\Psi(r)$; If the potential field is constant, i.e., $U(r)$, then the time-independent stationary Schrödinger equation can be written as follows.

$$E\Psi = \hat{H}\Psi$$

where E is the energy of the system; The above equation is similar to the wave equation of electromagnetic wave.

It can be proved that Schrödinger equation can be obtained by introducing de Broglie wave concept on the basis of Helmholtz scalar wave equation. The fact that SE is "accessible" to the Helmholtz equation will be seen more clearly later on in the application of SE to high-speed particles (for example, the photons).

"Why do we derive SE from Newtonian mechanics instead of relativistic mechanics?" Schrödinger once explained that although he was "a little embarrassed to be forced to abandon Relativity in the search for wave equations, the difficulty of introducing Relativity was increasing, even alarmingly so ". Anyway, he don't need Relativity!

Is the Schrödinger equation "valid for low-speed phenomena"? Some works in physics assert that SE is based on a number of strict approximations, one of which is that "all relevant velocities are assumed to be sufficiently small", so that the non-relativistic nature of the Schrödinger equation is not understood correctly. Those words were not in Schrödinger's original paper.

Newton's equation for particle kinetic energy is well known:

$$E_k = \frac{1}{2} mv^2 = \frac{1}{2} m_0 v^2 = \frac{p^2}{2m} \tag{3}$$

The above formulation shows that the mass does not change with velocity in Newton mechanics, and the moving mass m is no different from the rest mass m_0; But in SR, the kinetic energy is:

$$E_k = \sqrt{p^2 c^2 + m_0^2 c^4} - m_0 c^2 \tag{4}$$

In the above equation, momentum $p = mv$, and $m = m_0 \left(1 - \frac{v^2}{c^2}\right)^{-1/2}$. The results show that the E_k calculated value of Newton mechanics is larger than that of SR mechanics when the same value is taken. So, the two theories are fundamentally different.

As we all know, SE is useful and effective when dealing with the problem of photons passing through potential barriers, or when dealing with phenomena in optical fibers. Therefore, it is wrong to say that "the Schrodinger equation is derived under the assumption of low speed" (and can only be applied at low speed).

In short, it is not true to say that Newton mechanics and ME can only be applied at low speeds. The non-relativistic quantum wave equation (SE) has been successfully used to calculate the high-speed problem (see [14] for the work of the author on the calculation of optical fibers by SE).

AS ONE OF THE NON-RELATIVISTIC SPACE-TIME VIEWS, QUANTUM NON-LOCALITY FINALLY WINS

According to media reports, on October 4, 2022, the Royal Swedish Academy of Sciences announced that the 2022 Nobel Prize in physics has been awarded to French scientist Alain Asper, American scientist Johann Krauser and Austrian scientist Anton Zeilinger. For their contributions to "entangled photon experiments, verification of violations of Bell inequalities, and pioneering quantum information science." ... At the Nobel committee's press conference, the Nobel committee showed an image of a Chinese quantum satellite showing an experiment in intercontinental quantum communication between China and Europe. In fact, in 2017, Chinese and Austrian scientists successfully conducted the world's first quantum confidential intercontinental video call with the help of China's "Micius" quantum satellite. It is safe to assume that a Chinese physicist won't be far away from winning the Nobel physics prize.

On Oct 8, I wrote an article entitled "A Victory not too Late — Congratulations to Alain Aspect on Winning the 2022 Nobel Physics Prize", published in *Science Network*. The article said that quantum mechanics (QM), which was born in 1926, has been advancing rapidly and invincible for more than 90 years. It is regarded as one of the most important and beautiful achievements in the history of human thought, and its application scope is extremely broad. The essence of QM lies in its non-classical, microlity and non-locality, while quantum non-locality can be colloquially interpreted as Superluminality and can obtained quantum entangle states. In contrast to this is the classical, macroscopic and local nature of Relativity (mainly SR). The main contents of this local reality are as follows: believe in the classical physical reality, believe in local causality, and oppose probabilistic thinking; In SR, the light speed is considered to be the limit of the speed of moving bodies in the universe and the speed of information transmission. It does not accept the possibility of physical entanglement.

Einstein's EPR paper in 1935 was against Quantum Mechanics, and the essence of the disagreement was that SR and QM had different worldview, time and space views. The author believes that these two theoretical systems are not just "existing contradictions" as some people say, but fundamentally incompatible. Some of the content in the EPR paper is only foreshadowing (e.g., "that a physical theory must be not only correct but complete"; Another example is that "wave functions in Quantum Mechanics give an incomplete description of reality"). The fundamental thing is in the analysis of the interaction of "two-body systems" (systems consisting of two subsystems, see Fig.1), where subsystems I and II should be understood as microscopic particles. The states of the two subsystems are known until t=0, when they interact between t=0 and t=T, and when t>T they no longer interact (e.g. away from apart in different directions). Let be the quantum state of the system is, which can be expanded according to the eigenfunction system of the physical quantity (such as mechanical quantity) A of the measurement I, and also according to the eigenfunction system of the physical quantity B of the measurement I. According to QM, the wave packet will collapse during measurement and $\Psi(x_1, x_2)$ reduction after measurement, so that measurement of I will affect the state of II. However, since I and II are separated, such a strange influence at a distance is unlikely to occur. Since SR stipulates that the interactions in nature can only be realized at speeds lower than the speed of light, the spatially separated system should be local, but QM gives a non-local condition, so QM is not self-consistent and incomplete. These are the most important things in the EPR paper.

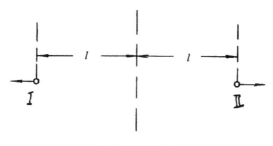

Fig. 1 EPR test

It follows that there is an invisible thread connecting SR and EPR; In other words, EPR thinking is put forward on the basis of SR. Secondly, we say that there is a sharp contradiction between SR and QM worldview, which is reflected in the issue of "local realism or non-local realism". The EPR paper was Einstein's maximum use of his intelligence at the age of 56 to give Quantum Mechanics the body blow he had hoped for. Einstein was shocked by the appearance of Heisenberg's uncertainty principle in 1927, but he thought the EPR paper could refute the principle and prove that QM was imperfect. The discussion of "two systems" (I and II) in EPR seems to indicate that "knowing both position and velocity" is feasible, because the velocity of I is the velocity of II. When the article was published, Bohr refuted it. Bohr means that the EPR paper's setting can be dismissed — uncertainty affects both I and II, and II is affected immediately when I is measured so that the result is consistent with Newton's law; This effect occurs immediately, even if I and II are far apart. But younger scientists (W. Heisenberg, for example) could not argue with Einstein the way Bohr did. This is not only because Einstein was their predecessor, but also because he was already a well-known figure in the world and enjoyed great prestige. The Russian academician V. Fok said: "It is particularly surprising that Einstein, who has done so much for quantum theory in its early development, has taken a negative attitude towards modern Quantum Mechanics. There is no direct force interaction between the two subsystems of the EPR mind, and one can also affect the other, which Einstein considered incomprehensible and thus incomplete." According to Fok, the interaction (influence) of Pauli's principle in QM is an example of a non-force. The interaction (influence) between two particles with a common wave function (EPR system) is another form of non-force interaction (influence) of QM. The existence of non-force interaction (influence) is beyond doubt, and it would be wrong to deny it.

The CERN's scientist J. Bell was a fan of Einstein's theory when he developed his inequality (Bell's inequality) in 1965; Bell's analysis builds on Bohm's theory of spin-dependent schemes (spin two-valued particle systems) and hidden variables. Assume that the spin component of the related particle has only two possible values, namely A(**a**, λ) or B(**b**, λ)=±1, where **a** and **b** are unit vectors. Under ideal correlation conditions, A(**a**, λ)=-B(**a**, λ) in any direction **a**. In addition, it is assumed that when the two particles are separated, the measurement result A(**a**, λ) of I is independent of the orientation of **b**, and the measurement result B(**b**, λ) of II is independent of the orientation of **a**. There are three assumptions above, namely, spin two state system, perfect correlation and locality condition. The following correlation functions are also defined:

$$P(\boldsymbol{a},\boldsymbol{b})\int \rho(\lambda)\mathbf{A}(\mathbf{a},\lambda)\mathbf{B}(\mathbf{b},\lambda)d\lambda$$

where, $\rho(\lambda)$ is the probability distribution function of λ. From this, Bell derived the following inequality:

$$|P(\boldsymbol{a},\boldsymbol{b})-P(\boldsymbol{a},\boldsymbol{c})| \leqslant 1+P(\boldsymbol{b},\boldsymbol{c}) \qquad (5)$$

This is in conflict with QM's prophecy. From 1981 to 1982, A.Apect led and completed a number of experiments in France, and proved that the results greatly violated Bell inequality and were very consistent with QM by high-precision experiments.

Although the initial experiments (e.g. R. Holt, 1973; G. Faraci, 1974) had obtained the result of "deviation from QM, consistent with Bell Inequality", it was not recognized by the scientific community due to its low accuracy and poor credibility. Another 10 experiments from 1972 to 1982 (including the experiment done by Jian-Xiong Wu in 1975 and the 3 experiments done by Aspect and others in 1981 to 1982, see Fig.2) were all "in violation of Bell inequality and consistent with QM", and very consistent with QM prediction. This is no accident, nor is it a surprise to QM experts. It is worth noting that Aspect experiment is dynamic rather than static, that is, the experimental device changes with time during particle flight; This was J.Bell's hope, because the local conditions would then be a direct result of Einstein causality (the inability of any signal to travel faster-than-light). The measured result, Bell parameter S=0.101±0.020, is very close to the QM calculation result (S=0.112), but far from the specified data of Bell inequality (-1 ⩽ S ⩽ 0). Not only that, but in 1998, the magazine *Phys. Rev. Lett.* In the experiment completed by G.Weihs et al., under the condition of space distance of 400m (Aspect experiment is only 15m), the experiment was conducted with two-photon wavelength of 702nm, and the result also violated the inequality and fully supported QM. Of these experiments the French physicist B. d' Espagnat commented that "there is almost certainly something wrong with local realism";

"The violation of the Bell Inequality can only be explained by abandoning the hypothesis of Einstein separability". He also argues that although J.Bell had three premises in deriving the inequality, the local reality hypothesis is the most basic. In the author's opinion, in fact, N.Bohr has already clarified the principle of "indivisibility", that is, in the quantum field, the two subsystems of the indivisibility failure system do not exist completely independently even if separated, and the measurement of one must affect the other.

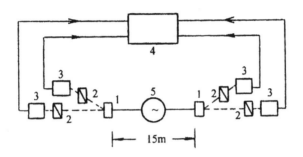

Fig. 2 Aspect experiment
(1: switch, 2: polarization sheets, 3: photo-electric tube, 4: monitor, 5: source of photons)

From the quantum entanglement (quantum entangled state) research progress can be seen, QM worldview has completely defeated the SR - EPR worldview. The distance of entanglement between two photons successfully developed from 15m in the earliest Aspect to 25km, and even 144km in 10 years ago. According to a report in the June 15, 2017 issue of *Science*, a team of Chinese scientists led by academician Jian-Wei Pan made a new achievement with a quantum satellite — achieving quantum entanglement at the thousand-kilometer level (the distance from Delingha Station in Qinghai province to Gaomeigu Station in Yunnan Province is 1203km). The result shocked the world. In a word, a series of experiments perfectly prove that wrong of SR space-time view is an indisputable fact.

In the mid-1960s, J. Bell of CERN published two papers proposing a hidden variable model compatible with QM, arguing that "no local variable theory can reproduce all the statistical predictions of Quantum Mechanics". Some inequalities for correlation functions when two particles spin projection along different directions of space and time are presented. Bell turned out to be a staunch supporter of Einstein and a believer in physical reality and locality. He believes that some hidden variables are responsible for the mysterious action at a distance in QM. In fact, it is possible to construct a theoretical inequality (which must be followed by particle observations) to confirm the QM incompleteness stated in the EPR paper. Bell's analysis builds on Bohm's spin-dependent scheme and hidden variable theory. We now dispense with the mathematical analysis and emphasize that Bell's inequality is not consistent with QM. Bell's theorem says that a hidden variable theory cannot reproduce all the predictions of QM. ... But just how that is must be determined experimentally. The breakthrough was due to the precise experiments of the French physicist Alain Aspect. Experiments led by spect show that the results violate Bell's inequality with high accuracy and are in good agreement with the predictions of Quantum Mechanics. The Bell inequality is not proved to be true by exact experiment, which means that the EPR paper is wrong and the QM is correct. John Bell opened the door to quantum informatics!

It is important to note that advances in experimental physics can change the views of some of the best theoretical physicists. Examples are P. Dirac and J. Bell; Although they maintained their faith in Einstein and Relativity in their early years, they changed a lot in their later years. In the second half of the 20th century, experimental physicists made two major discoveries; First, American microwave scientists A. Panzias and R. Wilson jointly discovered the microwave background radiation. This was in 1965, and the experiment was conducted in the centimeter-wave band, measuring noise temperatures of (2.5 to 4.5) K; Finally, the Cosmic Microwave Background (CMB) temperature was determined by the physical community, and the standard value was 2.7K. The CMB is isotropic in nature and has nothing to do with Earth's rotation or revolution. It is thought (debatable) to be an ember of the Big Bang; The other, more important, is considered an alternative to the "new ether". Anyway, the 1978 Nobel Physics Prize was awarded to Penzias and Wilson... Another thing happened in 1982, when the American journal PRL published an experiment led by a team of A.Aspect, which examined whether two photons emitted simultaneously by an energy level transition in a single atom followed Bell's inequality. It turned out that QM was correct, while Einstein's space-time and world views (represented by the EPR paper) were wrong. ... Both of these experiments sent shock waves through the physics world.

We think Ohanian's *Einstein's Mistakes* is a good book, but the author obviously doesn't know the details of P. Drac's and J. Bell's life work and falsely asserts that both men supported Einstein and the theory of Relativity. Yes, they were strong supporters in the early days; But as new experiments continued to emerge, such as these two and others, the late Dirac distanced himself from Relativity, saying "Lorentz was right and Einstein was wrong". As for Bell in his later years, he not only said, "go back to before Einstein (1905), namely Lorentz and Poincarè," but also condemned Relativity for bringing various difficulties to the development of quantum theory, and affirmed the fact possibility of superluminal phenomenon. Dirac died in 1984 and Bell died in 1990. The transformation of the two masters of physics was dramatic—E.Schödinger and P.Darc shared the Nobel Prize in 1933; In his speech of thanks, Schrödinger, who succeeded from Newton's mechanics, avoided the theory of Relativity. The young Dirac made the mistake of raving about his "derivation from Relativity". In fact, the equations for mass-velocity and mass-energy that he used as his starting point had been derived by Lorentz in 1904 and by Poincarè in 1900, both without Relativity. Late in his life, Dirac said, "There are insuperable difficulties in combining Relativity with Quantum Mechanics," a euphemism for saying that he had invented "Relativistic Quantum Mechanics." In fact, since Einstein went through life rejecting QM, it makes no sense that this RQM exists at all. As for J. Bell, he came up with the theory of hidden variables in 1965, and he gave inequalities, which was supposed to support Relativity; The results were contradicted by Aspect's precise experiments (which were later supported by multiple experiments). In addition, considering the impact of the CMB's discovery, Bell was finally announced Einstein's Relativity in 1985.

"All three scientists experimented with quantum entanglement," Reuters reported on Oct. 4, 2022, after the Nobel committee announced its decision for the physics prize. "In quantum entanglement experiments, two particles are connected to each other no matter how far apart they are. This bothered Einstein, who called it spooky action in distance".

But I have a different view. The Bell inequality has been widely tested since 1982 and has become an important means of identifying entanglement that can be described by discrete measurements. Such as measuring the spin direction of one quantum particle, and then determining whether that measurement correlates with the spin of another particle. If a system violates this inequality, then entanglement exists. In short, the Bell inequality became a signature method of checking whether it was obeyed. Both theory and experiment show that nonlocality is a fundamental feature of QM — the experimental results violate Bell's inequality and suggest that nonlocality exists. John Bell's name entered the history of science, and his inequality was hailed as "one of the greatest scientific discoveries in human history".

In short, Alain Aspect has become a figure in the history of physics, and his award is well deserved. John Bell would have been eligible, but he died young and the Nobel Prize can only be awarded to people who are still alive. In fact, whether the award itself is not important, the key is to establish a correct view of time and space, world outlook. Professor Wen-Miao Song, a famous Chinese expert on electromagnetic theory and a good friend of mine, once commented on the widespread belief in the theory of Relativity: "Truth cannot be obtained by faith and worship, only the nature is the standard by which we scientists test everything." He speaks very well!

From 1982 to 2022, that's exactly 40 years. The Nobel Committee is late, but not too late. We congratulate Alain Aspect!... From this matter, we can also see that the study of natural science is a difficult thing, of course, is also a happy thing. The right concept to establish, not in the short term can be effective. For example, it can be seen that the local description in relativity is not compatible with the particle fluctuation in QM, nor is it compatible with allowing particle transformation in QM. In particle physics, non-relativistic QM is a logically self-consistent single-particle theory, but the premise of the so-called "relativistic QM" is logically inconsistent. It is difficult to act as a single particle equation of motion like SE.

WHY IS POSSIBLE OF FASTER-THAN-LIGHT

Let us now turn to the early faster-than-light studies, when the goal was not to "overthrow the SR". However, in the process of research, the theory of Relativity is changed from belief to doubt, and finally may be deviated and abandoned. In addition, in the 21st century, there have been a series of decisive experiments that deny the principle of the invariance of light speed, some of which are organized and carried out by faster-than-light researchers. All this suggests that researchers have a key code in Relativity and are working to crack it.

SR proposed the "light speed limit" using simple logic, such as from the so-called "mass-velocity formula", the motion and state of $v > c$ can not exist. In 1904 Lorentz derived the following formula:

$$m = \frac{m_0}{\sqrt{1-\beta^2}} \qquad (6)$$

where, $\beta = v/c$, m is the mass of moving body, and m_0 is the rest mass when $v=0$. This was Lorentz's formulation for the motion of the electron, which he assumed to be a sphere of radius R at rest, with electron charge e uniformly distributed on the surface. Before In 1892, Lorentz had proposed "the ruler shortening in motion"; and in 1895 the contraction factor was defined as $\sqrt{1-\beta^2}$. So in 1904 he assumed that "electrons change shape when they move in a straight line," and that electrons shrink in size in the direction they move. This clearly indicates that the source of the denominator term ($\sqrt{1-\beta^2}$) in the mass-velocity formula is the ruler factor. If the physical assumption of "ruler shrinkage" is wrong, Lorentz's formula for mass-velocity is also wrong.

However, referring to Einstein's paper, he failed to derive the same result as the above equation. Later, however, SR incorporated Lorentz's mass-velocity formula, and Einstein also acknowledged Lorentz's invention right to this formula. Now SR extends this formula, derived for the electromagnetic mass of electrons, to any neutral particle or neutral matter, and the mass is general, which is very problematic. In 1909, Lewis and Tolman analyzed it as a two-ball collision; When the conservation of momentum and mass is assumed in the collision process, the formula of mass-velocity can be derived. But the premise of the derivation is not only "the conservation of total mass and total momentum of the two particles in the collision process", the velocity addition formula SR should be quoted. The author thinks this is "using the formula in his own theory to prove the theory is correct", it is a circular argument. Lewis and Tolman's treatment, therefore, does not prove that the mass of a neutral particle depends on the speed of motion. Although experiments on electrons have been proved to be consistent with Lorentz's mass-velocity formula (for example, Kaufman and Buchrer). But the experiments on neutral particles are still not available (neither confirmed nor falsified), because there is no technology to accelerate neutral particles efficiently.

However, SR ignored all this and asserted that Lorenz's mass-velocity formula holds for any motion of matter; It says that if the speed of motion is close to c, the mass will be very large, and if $v=c$ ($\beta=1$), the mass and energy will be infinite. Therefore, it is not possible to use accelerators to make particles reach and exceed the speed of light, nor is it possible for any object, such as a spaceship. ... Although many physicists disagreed with the "light speed limit" theory, few publicly raised objections to SR before the 1960s. In 1967, Professor G. Feinberg of Colombia University wrote a paper like the "ice-breaking journey". He pointed out that photons travel at the speed of light c, and that they are not artificially accelerated, but are naturally present in nature. Second, quantum theory suggests the possibility of faster-than-light speeds (later echoed in 1985 by J.Bell). Finally, one can try to circumvent the difficulties caused by SR in theory—if $m_0 = j\mu$, then:

$$m = \frac{j\mu}{\sqrt{1-\beta^2}} = \frac{\mu}{\sqrt{\beta^2-1}} \tag{7}$$

In this case, the real number m remains real even if $\beta>1$. He called such particles with imaginary resting masses "tachyon." We know that some physicists still insist that neutrinos are tachyons recently.

Radio astronomy has long developed a technique for combining radio telescopes around the world called Very Long Baseline Interferometry (VLBI), which is equivalent to building a radio telescope about the diameter of the Earth. VLBI observations of the universe have yielded rich results. For example, on quasars (objects that look like stars), observations have shown complex structures in some quasars and galactic nuclei. There may be two internal radio sources (light-years apart); And they're moving away from each other at tremendous speeds (faster-than-light speed). For example, quasar 3C345, observations since 1971 show that the two parts fly apart at eight times the speed of light ($v = 8c$). Observations of quasar 3C273 show that the separation velocity is $9.6c$. In addition, quasar 3C279 and radio galaxy 3C120 were also found to be separated from each other at superluminal speeds. This is completely unexpected to astronomers, and has profound implications. Because after ruling out some possible explanations, it was accepted that these objects might indeed be moving faster than light.

As early as 1986, Prof. Sheng-Lin Cao of astronomy department of Beijing Normal University made a research on the discovery obtained with VLBI technology on the world, and thought that the superluminal expansion of radio source is the evidence that the real superluminal motion can exist. Further research using mathematical methods and statistical fitting of Finsler geometry was included in a monograph published in 2001. In

2019, Prof. Cao pointed out that NASA's Hubble Telescope observed supergiant star bursts, which showed stars in the Milky Way expanding at superluminal speeds ($4.3c$), this is a remarkable superluminal phenomenon.

In 2005, Xian-Gang Liu, an associate professor at Beijing Normal University, pointed out that Einstein's 1905 paper had an achilles' heel in its analysis, which treated electrons as general agents of mass m and speed v, while electrons were special agents of electric charge. Clearly, a kinetic theory of moving charges is needed. Assuming two stationary point charges q_1, q_2, position r_1, r_2, and, using Coulomb's law, the electric field intensity equation can be derived. And let's say that q_1 we're at rest, q_2 moving with velocity v, and arrive r_2 at time t, where the force of q_1 action on q_2 is

$$F_{12} = kq_1q_2 \frac{1}{|r_1 - r_2|} \left|1 - \frac{vv}{c^2}\right| (r_2 - r_1) \tag{8}$$

This is a vector equation, where $k = 1/4\pi\varepsilon_0$. Taking $r_2 = r$, $q_2 = q$, it can be proved that:

$$F = kq\left(1 - \frac{vv}{c^2}\right)E$$

where E is the electric field intensity vector; In the 1-dimensional case when we only think about the value of vector, then

$$F = kq\left(1 - \frac{v^2}{c^2}\right)E \tag{9}$$

where v is the speed of charge movement. Now consider the motion of the electron, which increases from its initial velocity v_0 due to the acceleration of the electric field to v, and we can see that:

$$v = c\sqrt{1 + \left(\frac{v_0^2}{c^2} - 1\right)e^{-2w/mc^2}} \tag{10}$$

where m is the electron mass and w is the energy; The work can be calculated by integrating the force F, we obtain:

$$J = \frac{m}{2}(c^2 - v_0^2) \tag{11}$$

Since $v_0 \ll c$, we obtain

$$J = \frac{m}{2}c^2 \tag{11a}$$

So the work done by the electric field on the electron, even at the speed of light, energy is not infinite.

In 2010, Liu published a monograph titled *Research on Electrodynamics of Moving Bodies*, and there is an interesting metaphor in the book. He used the term "bat mechanics" to explain Einstein's error — bats use sound waves to navigate; If the speed of signal propagation and response were the speed of sound, the bats would believe this to be the highest speed in the universe, according to Einstein's SR moving body theory, where the equations for mass, momentum, and energy are unchanged.

We give the examples above to show that Einstein and Relativity were respected at first; Although it points out that SR theory has problems, it is not completely negative. Discarding SR happens gradually.

THE NEW DISCIPLINE "SUPERLUMINAL LIGHT PHYSICS" CAN BE ESTABLISHED

The term "Superluminal Light Physics" first appeared in an English paper by Chinese scholars around 2012. The author thinks from the domestic and foreign research situation, it seems that the condition is ripe to put it forward as a branch of discipline. In 2014, I published a monograph with the title *Wave Science and Superluminal Light Physics*, a bold step for Chinese scientists. In the "foreword" of the book, I wrote:

"The creation of any new discipline is not at the will of anyone or an academic institution, nor does it need to be approved by anyone. As more people studied it, its achievements and implications became apparent, the direction gained more attention, and the discipline was established. This is the case with Superluminal Light Physics, a term that was never (or rarely) mentioned before. We think that after decades of work by scientists from all over the world, this discipline has been born."

It must be pointed out that there is also a very special case in China, that is, space-farer's advocacy and promotion of FTL research, which is also rare in the world. Maybe NASA does it, too; However, in China, as early as 2004, the academic conference of "Frontier Issues on Astronautics and Light Barriers" was convened by the aerospace community. About 50 experts and scholars (including 9 academicians) attended the conference. This is something the world does not know. A key leader is academician Jian Song, a former deputy minister and chief engineer of the Ministry of Space Industry and later director of the State Science and Technology Commission. He pointed out that:

"In 1905 Einstein declared that faster-than-light speeds were impossible, later known as the 'light barrier'. But this is only hypothetical. Because of the difficulty of observing faster-than-light movements, where nothing can be seen it can only be guessed or hypothesized. Now we call it spaceflight within the solar system, and astronautic flight outside the solar system. It is expected that the first astronauts will fly out of the solar system and return safely in this century, and flying out of the solar system is the great dream of mankind. But there are many theoretical and technical problems to solve. We must go faster; faster than the speed of light, if possible."

This is a forward-looking statement, and it is Song's words that have made many people (including me) realize that doing FTL research is not just a personal interest, but may be part of the larger cause. This is also an important research direction of "Superluminal Light Physics". Song also points out that examining SR from 40 years of space technology practice shows that the engineering practice of autonomous navigation conflicts with SR dynamics even at speeds well below the speed of light, such as the dependence of engine thrust on its inertial velocity, which has never been seen.

In 1999, Zhixun Huang, a professor at the Communication University of China, published the first monograph on the problem of superluminal—*Research on Faster-than-Light: The Intersection of Relativity, Quantum Mechanics, Electronics and Information Theory*. One of the features of this book is that the author uses his knowledge of cutoff waveguides and evanescent state electromagnetic theory to analyze the existing experimental phenomenon of superluminal group velocities of microwaves passing through waveguides below-cutoff. Wenmiao Song, a research fellow at the Institute of Electronics, Chinese Academy of Sciences, said:

"Professor Zhixun Huang, a visiting fellow in our laboratory, has done a lot of research on the connection between the macroscopic law of attenuation waves in cutoff waveguides and the law of quantum mechanics. Through these studies, the attenuation electromagnetic wave is related to the motion state of the photon quantum in the quantum potential and the motion state of the electron when it penetrates the barrier. The wave's propagation constant is imaginary, and its momentum is imaginary from the point of view of QM. It is a very difficult problem to study the conversion law between the physical wave with virtual momentum and the general propagating wave. However, the virtual electromagnetic wave is an inseparable part of the whole electromagnetic wave propagation process. For example, the propagation of light wave in optical fiber is the coexistence of virtual electromagnetic wave and normal electromagnetic wave."

Since then, the author (i.e. me) has been working on the FTL problem for 20 years and has published many theoretical and experimental works.

In 2000, Dr. Li-Jun Wang, a young Chinese scientist, performed an experiment in the laboratory of the United States, which was published in the famous journal *Nature*. The experiment caused much controversy. Taking quantum optics, rather than classical physics, makes it unique. The experiment succeeded in passing a pulse of light at a negative group velocity ($v_g = -c/310$) through a cell of size 6cm, in which the cesium gas was excited by a sophisticated technique. The negative group velocity (NGV) means that not only will the light pulse travel faster than light as it passes through the vacuum, but it will leave the cell before it enters it. Some people think this is impossible. I wrote to Dr. Wang in 2001, and he replied:

"Our experiment achieved negative group velocity; In short, this only occurs in fluctuations and is not inconsistent with causality because group velocity is not information velocity. The laser pulse passing through the gas cell (medium) arrives earlier than that from the vacuum condition, which is an equivalent condition for the group velocity of light in the medium to be greater than c. When this advance is larger than the vacuum propagation time L/c, the group velocity of light is negative. In our experiment, this advance is about 20m, the corresponding vacuum propagation distance is 6cm, and the corresponding group velocity is about ($-c/310$)".

In short, Dr. Wang insists that he has performed a faster-than-light experiment, which does not violate the law of causality or SR. But not everyone else sees it that way. Interestingly, Liao Liu, a well-known expert on Relativity and a professor at Beijing Normal University, has a different opinion. In 2002, Prof. Liu wrote that "we should see the possibility of faster-than-light pulses in the experiments, which would constitute a shock to the theory of Relativity. Specifically, the occurrence of negative velocity transforms a delayed (conventional) light pulse into a leading light pulse, resulting in the outgoing pulse being ahead of the incoming pulse in time. This seems to violate the conventional temporal causality, that is, the effect is ahead of the cause in time". Prof. Liu believes that the time sequence limitation should not be regarded as absolute, but the law of causality should be expressed as "effect can not affect cause through any way". In this way, the objectivity of the law (men cannot change history) is maintained, and new experiments are explained. In addition, Liu suggested the concept of "advance wave" to explain the work of Dr. Wang et al.

Some people think that "negative velocity" says "the direction of motion is reversed", but the former concept is different. Borm and Wolf's book *Principles of Optics* says the phase speed is a scalar, Brillouin's book *Wave Propagation and Group Velocity* stated that negative group velocity (NGV) is "a velocity faster than an infinite velocity". Now many countries have done successful NGV experiments, which has become a unique landscape of FTL research.

Whatever the evaluation of the experiment, it is a sign that a new field has been opened up, characterized by the experimental demonstration that light pulses can travel faster-than-light in a negative group velocity (NGV) mode, a possibility established by A. Sammerfeld and L. Brillouin in their wave velocity theory. Wang's experiment also shows the necessity of introducing quantum theory, which in turn makes the experimental system more complicated. Nevertheless, some universities, such as Peking University and Jilin University, have carried out relevant experiments.

In 2014, I conducted a study with my doctoral student Rong Jiang. In theory, we point out that because the wave velocity (such as phase velocity v_p and group velocity v_g) is a scalar rather than a vector in wave mechanics, the negative group velocity (NGV) can not be understood as the opposite direction of motion, but the advance in time, so we call it "negative characteristic motion of electromagnetic wave". Second, we obtained NGV, or the advance propagation of microwave pulses, from (-0.13c) to (-1.85c) using a left-handed transmission line composed of complementary Ω-like structures .Our study uses classical physics rather than quantum optics, see Fig 3 and Fig.4.

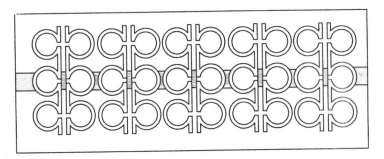

Fig. 3 The structure chart of the microstrip transmission line with complementary omega-like structures etched in the ground plane. The gray strip is the microstrip transmission line at the reverse side of printed circuit board.

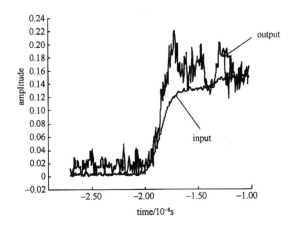

Fig. 4 Experimental results by using E5071C network analyser

That is the enlarged picture of step raise range. It's obviously shown that output pulse advance.

Since the beginning of the new century, two Chinese scientists have done valuable experimental research to against SR — Professor Ru-Yong Wang (St.Cloud State University, USA) and Academician Jin Lin (China Academy of Launch Vehicle Technology). Wang specially designed experiments to falsify two principles of SR (Einstein called "postulates"); For example, either the generalized Sagnac effect or the GPS system can be used to prove that the principle of constant speed of light is wrong. In particular,

Sagnac-type experiments were redone with modern technology, using moving optical fibers, hollow fibers, zigzagging fibers, and segmented fibers. Modern Sagnac experiments were performed at different speeds, proving that speed has an effect on the propagation of light in back-and-forth moving fibers, and the propagation time of light is different. "Our result," Prof. Wang said in 2005, "our result falsifying the principle of the light-speed constancy."... For many years, Prof. Wang has been thinking and experimenting on the two principles of falsifying SR in the spirit of unrelenting pursuit of truth. As for the principle of SR, Wang pointed out in 2006 that the most uncontroversial judgment experiments of SR are those that test the principle of Relativity. If you do an experiment in a closed system and find that the results are not the same for two states of uniform motion in a straight line, you falsify the relativistic principle of (and if the experiment uses the speed of light, you also falsify the principle of constant speed of light). He called it the "speedometer" project.

Lin is a renowned expert in satellite navigation and inertial navigation. He was praised in the scientific community for his original and novel insights and methods of redefining space and time based on rocket measurements. Different from Einstein's abstract discussion of time and space, he treated the concept of time and space with the thinking of space experts. Based on the experiments conducted in 2007-2008, Lin and his team published an important paper entitled "The Crucial Experiment for Checking Einstein's Postulate of the Constancy of the Light". The significance of Lin's experiment is as follows: ① unidirectional light speed measurement; ② To conduct experiments on a large distance of tens of thousands of kilometers, even the space powers (the United States, Russia) have not done; ③ It is proved that the speed of light traveling in different directions may be different; Thus, the "speed of light invariant principle" of SR is falsified. This shakes up one of the cornerstones of SR. we think it's a potential Nobel physics prize experiment (but Lin died in 2016).

In a word, Superluminal Physics as a new subject has a lot of rich and vivid content. And then there's the case of entangled states traveling faster than the speed of light, and that's included. This subject has attracted more and more attention.

We believe that FTL physics is an interdisciplinary and integrated discipline, including classical physics, quantum optics, particle physics, accelerator technology, electromagnetic field theory, microwave technology, etc. It is a model of crossover, penetration and synthesis. In addition, the development of aeronautical engineering, space technology, inertial navigation and satellite navigation progress, are closely related to the physics of FTL. It can be seen that this is a promising research direction, and we welcome the participation of experts from all aspects.

"LORENTZ RELATIVITY" IS SUPERIOR TO EINSTEIN RELATIVITY

When we study and discuss many problems in classical physics (such as space-time view, mass, covariation, FTL), we cannot do without the master Lorentz. The fact that the Dutch physicist Hendrik Lorentz (1853-1928) created the Relativity of Electro-magnetism in 1904 was known by people, but not by many others. Lorentz had been an expert in electromagnetic theory; He submitted his doctoral thesis on ME to Leyden University in 1873, only eight years after Maxwell published his brilliant work. Later he became an authority on electromagnetism at the time, and he did not start from Newton mechanics, but from electromagnetic theory, to study the relativistic problems. Lorentz had some prominent scientific ideas, such as:

— The idea of space-time transformation in motion, i.e. the Lorentz transform (LT);

— thoughts on the existence of the ether (that is, the absolute coordinate system);

—Thoughts on shortening the length of moving body and delaying the time of moving clock;

As these three are interrelated, the situation is more complicated.

Ether and ME were popular topics from the late 19th to the early 20th century. Lorentz, an expert in electron motion theory and electromagnetic theory, proposed that LT also came from thinking about both. If LT comes from the derivation of the electromagnetic field transformation relation, then is the ME naturally covariant with LT? We have to think independently about many problems. ... LT was born because Lorentz was trying to deal with the electrodynamics of moving bodies. He found that according to the Galilei transforms

$$x'=x, y'=y, z'=z-vt, t'=t$$

When the coordinate transformation between different reference frames was realized, the basic equations of electrodynamics changed obviously, which he thought was unreasonable. The point of GT is that the time in different reference frames is the same ($t'=t$), but LT has no such restriction.

Let inertial system A(coordinate x, y, z) with A' (coordinate x', y', z') of the axis parallel to each other, A' along the $z(z')$ direction for uniform motion (velocity v); An event (z, t) occurring in A corresponds to (z', t') of A'; Take the linear transformation:

$$z'=az+ht \qquad (12)$$

$$t'=bt+gz \qquad (13)$$

where, a, b, h, g are undetermined constants. At time t, $z=vt$, corresponds to o' position ($z'=0$), and can be substituted into equation (12) to obtain $h=-av$. Now the number of undetermined constants is down to three. Suppose that a pulse of light occurs when two reference frames coincide instantaneously, that is, a flash of light is emitted at o, o' (they are now a same point), and then the motion of a spherical wave front diffusing outward in both reference frames is observed. Assuming that the speed of light is the same in different reference frames, the equation of spherical wave front in different reference frames is respectively

$$x^2+y^2+z^2=(ct)^2 \tag{14}$$

$$x'^2+y'^2+z'^2=(ct')^2 \tag{15}$$

where c is the speed of light. If we substitute the transformation of GT into Equation (15), we can obtain

$$x^2+y^2+(z-vt)^2=(ct)^2 \tag{16}$$

Equation (16) is different from Equation (14). The reason for the problem is that the speed of light is different in different reference frames, or the speed of light is not constant. The reason for this situation is that the time in different reference frames is assumed to be equal in GT transformation. Substituting $x'=x$, $y'=y$ and equations (12) and (13) into Equation (15) and arranging; Then, by comparing with Equation (14), three equations composed of a, b, g, can be obtained. Solve the simultaneous equations and substitute a, b, g, and into equations (12) and (13), and specify $\beta = v/c$ to obtain the common expression of LT:

$$x'=x,\ y'=y,\ z'=\frac{z-vt}{\sqrt{1-\beta^2}},\ t'=\frac{t-\beta z/c}{\sqrt{1-\beta^2}} \tag{17}$$

Clearly, when $v \ll c$, LT reduces to GT.

We still going to take the inertial frame A, A' prime, but we don't want the axes to be parallel to each other. It is still assumed that the flash is emitted from o(o') point at $t=t'=0$, and the speed of light c in the two systems is the same, then the event satisfying Equation (14) must satisfy Equation (15). Now let's introduce the function S:

$$s^2 = x^2+y^2+z^2 - c^2t^2 \tag{18}$$

It can be proved by the second postulate of SR that $s'^2 = s^2$, that is, s^2 has invariance. There are now

$$x^2+y^2+z^2-c^2t^2 = x'^2+y'^2+z'^2-c^2t'^2 \tag{19}$$

The square root of the above equation (s) is the norm of a 4-dimensional vector, which remains the same in different systems. This analysis is called the 4-dimensional space-time continuum and is also called Minkowski space-time. Since $(jct)^2 = -c^2t^2$, jt is Minkowski imaginary time. Due to the introduction of virtual time, the 4D continuum invariant theory of space-time is similar to the 3D Euclid continuum invariant theory. However, in Euclidean geometry s^2 is always positive ($s^2>0$), so it must be real ($s>0$); Now, in Minkowski's space-time, s^2 it might be negative... Here we must point out that imaginary time has no physical meaning. We should also point out that the author does not agree with the "integration of time and space" at all, we say that time and space are independent.

Now Lorentz called t' local time or coordinate time. In 1904, Lorentz first considered the relation of time transformation, at which time the covariability of ME was guaranteed. But that can not explain the Michelson-Morley experiment, so "length contraction" was introduced.

The French mathematician Henri Poincarè (1854-1912) was an excellent and famous scientist. In 1904 he presented the idea of relativity in a lecture, and in 1905 he published a revision of Lorentz's paper. He pointed out that Lorentz's work actually provided a mathematical transformation group:

$$\begin{bmatrix} x' \\ y' \\ z' \\ ct' \end{bmatrix} = L, \quad L = \begin{bmatrix} x \\ y \\ z \\ ct \end{bmatrix} \begin{bmatrix} \gamma & 0 & 0 & \gamma\beta \\ 0 & 1 & 0 & 0 \\ 0 & 0 & 1 & 0 \\ \gamma\beta & 0 & 0 & \gamma \end{bmatrix} \qquad (20)$$

In the type $\gamma = \left[1 - \beta^2\right]^{-1/2}$.

Reference [1] makes no mention of the Michelson-Morley experiment, nor of Lorentz and Poincare, although these works were done before 1905. According to Einstein, he wrote his paper without knowing (or having read) the papers of Lorentz and Poincare. But this is highly unlikely. Professor Qing-Ping Ma pointed out that Einstein's 1905 paper might have plagiarized the ideas of Lorentz and Poincare. Einstein himself said, "The secret to creativity is to hide your sources." This sounds like a plagiarist's principle, says H. Ohanian in his 2008 book *Einstein's Mistakes*. In his 2002, the book *Einstein: An Incorrigible Plagiarist*, C. Bjerknes was even more scathing... However, Lorentz was a modest and gentleman. Instead of arguing for "priorities", he was generous and polite enough to say a few nice things about Einstein, but he never accepted SR. Einstein referred to Lorentz several times in the rest of his life as a public display of intimacy. The truth is, Lorentz kept a distance and never really had a close relationship with Einstein.

SR is based on two postulates (the special relativistic principle and the invariance of the speed of light), and the relation of space-time transformation depends on LT. Although we now know that LT is not absolutely necessary to establish a correct view of time and space; There are better transformation relation equations, such as generalized GT (i.e. GGT). But Einstein was eager to disavow Newton mechanics, and of course he wouldn't go back to GT (or anything like that).In this way, his fundamental need was to start from two postulates and develop a transformation of the space-time relationship like LT, published in 1904, in order to prove that Einstein, and no one else, had created a new theoretical system, SR, that differed from Newton. Thus, in his first paper, he not only made no mention of his earlier scientific work (Lorentz and Poincarè), but even pretended to be unaware of the Michelson-Morley experiment. In this state, one must come up with a derivation of the same result as LT's equation. Einstein actually knew the results of the MM experiment, as well as Lorentz's or Poincarè's papers, and all he had to do was piece together a "proof" based on "two principles." He handles theoretical relationships like magic, and the examples of his "trick" of switching concepts are endless. Thus, [1] is by no means an "unparalleled masterpiece" that will not stand up to expert scrutiny. Of course, when Einstein became famous, he was generous and attributed the space-time transformation entirely to Lorentz, instead of calling it Lorentz-Einstein transformation (LET). This is because LT is only part of SR (at least according to Einstein), and therefore it doesn't matter what the space-time transform is called.

Although LT and SR are formally identical, they are very different theories. This is because Lorentz theory takes absolute space-time and the existence of ether as its starting point, while SR abandons both and builds on the principle of relativistic and the constant speed of light in one way.

In fact, Lorentz's adherence to ether theory was based on Newton mechanics and GT, plus two assumptions (length contraction and time delay). In this way, the principle of Relativity will not hold and the covariability of ME to LT will be lost. However, LT was proposed to ensure ME covariation under the condition of the relativistic principle — which leads to a paradox.

Now the question is about the value of LT. It can be proved from LT that the speed of light is constant one way, but not from ether. So stick with ether or stick with LT? We think of course the former is more important. This corresponds to Prof. Ma's statement that "LT is not actually necessary".

If only the length is shortened while the time remains the same, there will be conceptual confusion. In 1904 Lorentz proposed "time dilation". Now he thinks: the absolute motion ruler is shorter, the absolute motion clock is slower. These are reflected in his paper.

In short, Lorentz deserves his reputation as a master of physics only in aspects unrelated to Relativity. But there are many things about Relativity that many people find confusing and controversial because they are mixed up with Relativity (Einstein's SR). The author discussed this with Professor Qing-Ping Ma. He said that Lorentz (and Fitzgerald respectively) proposed the contraction of the length of moving objects and the slowing of the moving clock (time dilation) in terms of contributions related to the theory of Relativity. Mass-velocity formula; Space time transformation; Lorentz theory of the ether. This LT later became the core formula for SR.

His lasting contribution, so to speak, was his formula for the slow-down of the movement clock and the mass-velocity formula. The current interpretation of both the motion clock slowing and the mass-velocity formula by Lorentz and SR may be wrong. The formula of moving clock slowing and mass-velocity reflects the effect of moving speed relative to the (electromagnetic) interacting medium on the interaction, that is, the electromagnetic interaction speed between objects moving relative to the (electromagnetic) interacting medium slows down, resulting in a slower clock (atomic clock with electromagnetic interaction as the mechanism); The electromagnetic interaction between objects in motion relative to the interacting medium becomes weaker, resulting in an apparent "increase in mass of motion" when in fact there is a decrease in force and acceleration. The great thing about Lorentz's contribution here is that even if we adopt the right interpretation, we may still have to use his formula for clock slowing; Then change the mass-speed formula to force-speed formula. This factor $\sqrt{1-v^2/c^2}$ will continue to be used in these formulas.

Lorentz's biggest weakness or mistake was that he insisted that the ether was absolutely stationary and could not be dragged due to the phenomenon of optical aberration, so he had to propose "motion length contraction" to explain the negative results of MM experiment. "Motion length contraction" leads to "slowing of motion clock" and "increasing of motion mass", laying the foundation of Relativity theory. However, the "length shortening" has never been confirmed by experiments, so Lorentz's "length shortening" is fundamentally wrong! This mistake came from his insistence that the ether was absolutely stationary and could not be dragged.

So, if Lorentz were alive today, would he approve of us doing FTL research? ... We figured he'd agree that FTL exists. For him, because anything moving at the speed of light has an infinite mass, the speed of light relative to the ether cannot be surpassed. But the combined velocity between two objects moving in opposite directions can exceed the speed of light, because they can both move at close to the speed of light in the ether. Lorentz's velocity composition is Galilei's. This is consistent with the view of T. Flandern, the scientist who in 1998 obtained the result that gravity travels faster than the speed of light, which he thought could be explained by Lorentz's theory of Relativity, but not by SR.

There used to be a common view that, because of Lorentz's formula for mass-velocity, it was obvious that faster-than-light motion was impossible, and that faster-than-light spacecraft were absurd. But we does not agree with this view, because Lorentz mass-velocity formula is derived for the electromagnetic mass of electrons, whether it is applicable to neutral particles and neutral matter has not been directly proved by experiments.

In short, Lorentz's relativity, unlike Einstein's, is superior. But we don't think Lorentz's theory is the best view of time and space. This gives rise to the Modified theory of Lorentz (MOL), that is, the modified Lorentz theory. MOL mainly has two kinds, one is to connect the ether with the gravitational field, for example, the earth's gravitational field around the earth corresponds to the ether; the other does not do much change, length contraction and time dilation and the speed of the earth prevail. The other is generalized GT, which combines length contraction and time dilation into GT, such as Mansouri and Sexl (MS) transformation and Modified Lorentz ether Theory of Ronald Hatch. Then there are also other people's GGT theories.

SEVERAL NON-RELATIVISTIC VIEWS OF TIME AND SPACE

From 1892 to 1904, Lorentz postulated a shortening in length and a delay in time of motion in order to explain the Michelson-Morley experiment. Einstein gave the derivation of length reduction in 1905 and 1952, but these relativistic length reductions were logically contradictory. Lorentz's theory is that there is a relationship between the length of an object stationary in the ether and the length of an object moving relative to the ether. But there are many paradoxes in SR in which the mutual view of physical phenomena causes length reduction. This is because the logical basis of SR is relative motion, which causes a paradox in principle. There is actually no experimental proof of the length reduction theory.

In Lorentz theory, the time delay is caused by the absolute motion of a moving body. A clock with a high absolute speed slows down relative to a stationary clock; This is Lorentz's etheric delay of time. However, when the relative velocity of the moving body is used to replace the absolute velocity in SR, the situation is completely different. Einstein explains length reduction and time delay by replacing the relationship between the observer and the ether with the relative motion of the reference frames of different observers. As a result, many paradoxes arise questioning the self-consistency of SR.

The relativistic principle, one of the laws of physics, was first introduced by Poincarè, and the Lorentz transform (LT) embodies the principle of consistency in terms of arbitrary inertial frames. But Lorentz's idea of relativity, published in 1904, was based on the existence of the ether. Einstein's 1905 paper included a postulate — the principle of the invariance of the speed of light — that there was no need for an ether, that is, a preferred frame of reference. Subsequent discussions have always included the question: which better describes nature, Einstein's special theory of relativity (SR) or the modified Lorentz theory (MOL)? The main difference between the two is that SR considers all inertial frames to be equal and equivalent, while MOL considers the existence of preferential reference frames. Over the years, numerous studies and discussions have shown that SR is logically inconsistent and lacks of truly confirmed experimental confirmation. It is important to note that SR cannot explain the recent advances in gravity propagation and quantum entanglement propagation, while MOL does.

SR no absolute space and absolute motion and absolute reference frame, which denies the absoluteness of material movement, into a mire of relativism—the same events observed different results in different reference frame, there is no judgment standard of the test results, relative motion of two observers say that the other side of the clock is slow, short feet. In addition, SR insists on "simultaneous relativity", "superluminal impossibility", and locality opposite to (QM). All of this contradicts the experimental facts. The internal logic of SR is chaotic, and it cannot get true recognition and support. On the other hand, although Lorentz theory also has obvious defects, it is superior to SR in insisting on the absoluteness of material motion as well as the absoluteness at the same time.

In 1959, F. Tangherlini proposed a space-time transformation called Generalized Galileian Transformation (GGT). Chinese scientist Professor Cao Zhang has been trying to introduce this non-relativistic spatiotemporal transformation since 1979. The space-time coordinate transformation formula is

$$x = \gamma(X - vT),\ y = Y,\ z = Z,\ t = \gamma^{-1} T \tag{21}$$

This means that one particular inertial frame is $\sum_0 (X, Y, Z, T)$, and the other inertial frame $\sum$ is moving in a direction X with a constant velocity v relative to $\sum_0$.

Coefficient $\gamma = (1 - v^2/c^2)^{-1/2}$. Similar to GT, GGT adopts the external synchronization method, that is, when $\Delta T = 0$, there is $\Delta t = 0$. But GGT does not require $t = T$ and allows the moving clock to slow down, unlike GT. Note that GGT asserts absolute simultaneity, which occurs between inertial frames $\sum$ and $\sum_0$. In the space transformation between the two systems, there is a scaling factor; The change of time has a clock slowness factor. In $\sum$, the one-way speed of light is non-isotropic, while the round-trip loop has the same average speed of light.

GGT is a nonstandard form of LT, which is also the inheritance and development of Lorentz's physical thought. As mentioned above, Lorentz believed that there was an absolute frame of reference; And that there is a real time. ...In addition, the MM experiment can be easily explained by GGT. Superluminal motion is also allowed in GGT. From GGT's point of view, there is no problem of causality breaking, time traveling backwards and the like when FTL occurs—the opposite of SR in this respect. When we say that GGT is the inheritance of Lorentz's physical thought, we should not simply understand it as the inheritance of LT; The fundamental point is to acknowledge the existence of a superior reference frame (absolute reference frame). At the same time, GGT is the inheritance of GT — Cao Zhang met Professor Tangherlini to discuss this point in the United States. He agreed to change the original name of the "absolute Lorentz transform" to generalized GT, which he had elaborate mathematically and physically. Therefore, now that Professor Cao Zhang has passed away for several years, the author suggests to call GGT also Tangherlini-Chang Transformation, or TCT for short, as a memorial to the two professors.

In 2007, Shu-Sheng Tan, a professor at National University of Defense Science and Technology, published his theoretical achievement *Standard Space-Time Theory* (SSTT) in the form of a monograph. The book points out that Lorentz theory has three basic assumptions (the existence of an absolute reference frame of the ether; the shortening of its length; and the delay of its time), which view of time and space is logically self-consistent with it? Not LT, but GGT. SSTT is unwilling to adopt Lorentz's length contraction and time delay hypothesis, but takes the theory as the basis of two principles, namely, the absolute reference frame principle and the loop average speed of light constant principle, because the former is the essence of Lorentz's theory, and the latter has been proved by a large number of experiments. Now, SSTT is different from SR, but consistent with QM; It argues for the absoluteness of simultaneity, allowing faster-than-light movements without violating the law of causality. SR denies absolute space, absolute motion and absolute reference frame, thus denying the absoluteness of matter motion and falling into complete relativism. The results (such as the relativity of "simultaneously", the void with nothing, the theory of the speed of light limit, and the theory of locality) are all inconsistent with experiments. Tan said that SSTT derived GGT strictly from two hypotheses and established a complete theoretical system, which is his own independent contribution.

SUMMARY OF THE NEW ETHER THEORY

Fans of faster-than-light research like to point to 1947, when the United States achieved the first supersonic flight of an airplane, as a motivator. But it has been argued that sound waves need compressible media to travel, while light waves do not. They said, this is an essential difference, that the similarity of equations does not translate into the similarity of physical mechanisms, and so on... But we have to ask: does light really need no medium to travel?

Until the middle of the 19th century, it was thought that there was no such thing as a wave that could be transmitted without media. Therefore, since light fluctuates and travels in a vacuum, as evidenced by the fact that sunlight strikes the Earth, there must be a medium of light. It can be invisible but pervasive in the universe, known to physicists as ether. So from the beginning of the 19th century, through the middle of the century and into the later part of the century, the scientific community made a big deal out of studying the ether. To this end, Fresnel, Fizeau, Lorentz, Maxwell, Michelson, et al. The ether was thought to be absolutely stationary, and the speed of the earth relative to the aether was the speed of the earth's revolution around the sun. Measuring this relative velocity would be difficult, but not impossible.

After taking into account the speed of the Earth's orbit around the Sun, it was concluded that the speed of the smooth and inverse ether should not be the same (2.15×10^{-9} difference to be exact). But the Michelson-Morley experiment did not discovered. In July 1887, the two men jointly conducted extremely accurate experiments that denied the existence of the ether.

Historians of science have shown that Michelson had a certain preference for the ether; This contradicts the popular belief that he experimented to deny the ether. In fact, he won the Nobel Prize in 1907 mainly for inventing the very sophisticated interferometer. In 1926-1928, when Michelson was in his 70s, he tried again to find the drift of the ether. However, he never announced that he had given up the ether. He also had reservations about the special theory of relativity (SR) and actually disagreed with it.

Lorentz's physical ideas are getting renewed attention for a reason. In 1977 Smoot reported that it had measured the Earth's velocity relative to the microwave background (CMB) at 390km/s; So the great physicist P. Dirac said, in a sense Lorentz was right and Einstein was wrong. American physicist T. Flandern published in 1997-1998 that the speed of gravity was $v \geq (10^9 \sim 2\times10^{10})c$, and he claimed that Lorentzian relativity could explain these results. On the other hand, SR is can't explained superluminal gravitational velocities.

In 2007, *New Scientist* reported on "ether's high-profile comeback as a replacement for dark matter", saying that G. Starkman and T.Zlosnik et al. were pushing the ether to explain "dark matter" in a new way. The latter was proposed because the Milky Way seems to contain much more mass than visible matter. They argued that the ether was a field that would form an absolute coordinate system, thus contradicting SR.

In recent years, many scientists have proposed the existence of prefered frame, that is, the formation of an absolute coordinate system. Therefore, the Lorentz-Poincarè time-space view has received renewed attention, and further theories have emerged. The "high-profile comeback of ether theory", reported in the science publication *New Scientist* a few years ago, is a reminder that we should not completely dismiss the scientific work done before SR. If there is a shift back towards Galilei, Newton and Lorentz, it is at the highest level in modern terms, not simply backwards.

There are currently three main options for the "new ether": physical vacuum, gravitational field, and microwave background radiation. We think the new ether is better defined as a quantum physical vacuum. Historically, the master J.Maxwell first connected the speed of light c in vacuum with the characteristic parameters (ε_0, μ_0) of vacuum as a medium. In recent years, I has made a profound discussion on the vacuum of quantum physics. In 1865, Maxwell used two physical parameters of vacuum (ε_0, μ_0) to derive the wave equation of electromagnetic wave, and proved that the speed of light in vacuum is

$$c = \frac{1}{\sqrt{\varepsilon_0 \mu_0}} \tag{22}$$

This is a statement that light needs a medium to travel, and that medium is a vacuum. Maxwell was very clever. He put the known values (ε_0, μ_0) into the equation and got a speed of about 3×10^5km/s. It was so close to four existing measurements of the speed of light that he concluded that the waves described by his electromagnetic wave equations were light waves. Here we give Table 2: The four measurements of the speed of light known to Maxwell in 1865 (and the only ones at that time), in which the so-called systematic error is calculated by comparing the measured values with the standard values (c =299792458m/s) stipulated by the International Bureau of Metrology.

Table 2 Measurements of the speed of light from 1676 to 1862

Surveyor and publication time/year	Method of measurement	Measured values c (km/s)	System error	For note
O. Roemer, 1676	Benon observations	214000	- 30%	Visible light waves
J. Bradley, 1728	The star is out of alignment	301000	+ 0.4%	Visible light waves
A. Fizeau, 1849	Method of screw tooth	313000	+ 4.4%	The round-trip distance is 17.2km
J. Foucault, 1862	Rotating mirror method	298000	- 0.6%	One-way distance 20m

Therefore, the author judges that Maxwell's academic thought is similar to Lorentz's later, but different from Einstein's later. Maxwell believed there was an ether, a vacuum. Of course, he couldn't have thought in terms of quantum theory, because QM wouldn't appear for another 60 years. Quantum field theory (QFT) holds that each quantum field in the vacuum state is still in motion, that is, each mode is still oscillating in the ground state, which is called vacuum zero oscillation. In vacuum, virtual particles are constantly produced, disappeared and transformed into each other, because of the interaction between the quantum fields. On March 25, 2013, the website of *Science Daily* reported that French and German scientists respectively proposed research results published in *European Journal of Physics*, which said that the speed of light is a real characteristic constant, while quantum theory holds that vacuum is not empty. This results in the speed of light c not being fixed, but having fluctuating values.

In 2021, I published a long English paper abroad, the title was: "Two Kinds of Vacuum in Casimir Effect". According to the Casimir effect, since its discovery in 1948 and the present situation, it is necessary to make a new discussion on the definition and characteristics of "physical vacuum". The fact that there is an attraction between two parallel metal plates in a Casimir structure, which was experimentally demonstrated in 1997, is not Newton's gravitational force, nor is it a Coulomb force because it has no charge. This peculiar phenomenon becomes apparent when the distance between the plates is small.

Therefore, it can not be ignored in nanoscale scientific research. Since the interplate may be a negative-energy state, even if the outside of the plate is the usual physical vacuum state (called free vacuum), the interplate situation must be a further vacuum structure, which I call it negative-energy vacuum. The calculation shows that the refractive index n is less than 1, so the superluminal phenomenon will occur.

In short, the new ether is a physical vacuum medium with quantum properties. But it is still a matter for discussion. For example, the "new ether" is a free vacuum, what role does the so-called "negative energy vacuum" play? Our view is that the empty space is indeed a medium, and only in this way can it be specially arranged to create local "emptier vacuums" in the medium — the nature is indeed more wonderful than we can imagine!

MAXWELL EQUATIONS (ME) COVARIATION PROBLEM AND EXPANDED ME

When discussing "covariation of ME to LT" in electromagnetics books, they always write the **E**, **B**, ρ, **J** transformation relation between and two reference frames. In fact, a premise has been used that ME covaries with LT, but not with GT. This derivation does not prove with whom ME covaries.

In my paper, I quoted famous theoretical physicist Ling-Jun Wang, who said:

"Maxwell's equations conform to the Lorentz transformation, which presupposes the theory of Relativity. Special relativity proposes a formula for the transformation of electric and magnetic fields in different coordinate systems, which is based on the formula for the transformation of forces in different coordinate systems in Relativity, and the strength of the electric field is simply the force per unit charge. So does the magnetic field strength. If Relativity does not hold, the relativistic field-strength transformation does not hold, and therefore ME does not conform to LT. People hope to use Lorentz covariation of ME to prove the correctness of Relativity, and borrow Maxwell's great achievements in electromagnetic field theory to "endorse" Relativity. They play the trick of logic cycle: according to the formula of relativistic field strength transformation, ME conforms to LT, and conversely, ME conforms to LT to prove the correctness of Relativity. Relativity's many logical contradictions and basic premises (including the principle that the speed of light does not change) have proved that Relativity is an impossible theory, and its logic cycle breaks down.

Even if we take a step back and admit that the relativistic formulation of field strength transformation can make ME obey LT, Lorentz covariation cannot be regarded as a universal physical law. The fact that an equation conforms to a certain covariation is only a mathematical characteristic of the equation, and therefore it cannot be regarded as an iron law requiring all physical theories to conform to the Lorentz covariation."

Another famous physicist Xiao-Chun Mei once pointed out that SR original paper introduced a transformation, called relativistic transformation of electromagnetic field in order to prove that the motion equation of electromagnetic field satisfies the relativistic principle. However, this apocryphal transformation contradicts the LT of the electromagnetic field itself and has no physical basis. In addition, the constitutive equation of dielectric electromagnetic field has no invariance originally, and the classical electromagnetic field motion equation does not satisfy LT invariance originally.

Professor Qing-Ping Ma pointed out that ME obeys Lorentz covariation but not Galilei covariation because Lorentz et al. insist on the absolute rest of the ether in space, and on the other hand have to accept the optical and electromagnetic experiment results that appear when the ether is completely dragged by the earth. The LT is the result of a compromise between these two sides. By abandoning the view that the ether/light medium is absolutely stationary in space and accepting that the light medium is completely dragged by a massive object such as the Earth, ME obeys GT invariance completely and LT invariance not at all.

The analysis angles of the above three scholars are not completely the same, but the conclusions are consistent. This issue has also attracted the attention of other scholars. For example, Guo-Fu Ji published an article on the Internet in 2021 entitled "On the Covariation of Maxwell's Equations Under Galilei Transformation". In this paper, it is very interesting to transform the space-time coordinates with wave frequency and wave vector, and obtain the wave equation satisfying GT invariability. However, it is not proved that the classical electromagnetic field equation satisfies GT invariance. In 2022, Guo-Fu Ji published "Rediscussing the Covariation Problem of Maxwell's Equations". Galilei coordinate transformation and Galilei velocity transformation in absolute space-time view are universal to both classical mechanics and electromagnetics. Secondly, it is proved that under the condition of giving up the principle of constant speed of light, it can be concluded that ME covaries under different inertial frames and obeys GT, and the transformation formula is given. It is considered that the electromagnetic wave velocity (including the speed of light) in different inertial frames obey GT. The principle of constant speed of light of SR and LT are denied. Thus, the relativistic view of time and space was abandoned.

Now let's look at Expanded Maxwell's Equations; Prof. Zhong-Lin Wang said:

"By 1905 it had been realized that Maxwell's equations could not keep their form under Galilei coordinate transformation. Galilei absolute time and space view, however, gives a very good approximation in many cases... Based on Galilei space-time view, Extended Maxwell Equations do not necessarily maintain Lorentz covariation."

In addition, one should not ignore the time-space view based on Galilei transformation, which separates time and space from one another. This is when he uses a term: Relativistic Electrodynamics, in which ME remains covariant to LT.Galilei electromagnetics, "works only at low speeds and does not preserve LT covariation". As for his own theory, LT covariability is not maintained.

What is commendable is that Prof. Wang affirms the value of Galilei space-time view and GT in both papers, and also flatly admits that his theory has no LT covariant. However, he believes that the ME ontology is covariant on LT, and he emphasizes the "approximation" of electromagnetic theory with GT covariant, arguing that as long as the non-relativistic equation is "only used at low speed". The Schrödinger equation (SE) is used as an example to illustrate why this view is wrong... To sum up, Prof. Wang actually believes that there are three types of electromagnetics:

— Galilei Electromagnetism, based on GT;

— Electromagnetism based on LT, namely Relativistic Electromagnetism;

— Wang's Electromagnetism (WE), based on GT.

However, we believe that it is impossible to discuss the theory of Relativity without clarifying its attitude.

CONCLUSION

The covariation between ME and LT is closely related to the space-time view in physics. In 1904, Lorentz first proposed the time transformation relation from the electromagnetic theory, which met the covariation requirement of ME for LT. However, the Michelson-Morley experiment was not explained, so length contraction was proposed. However, Lorentz insisted on the existence of ether, that is, on the existence of absolute coordinate system, which was fundamentally different from SR. Lorentz's theory of length reduction states that there is a relationship between the length of an object stationary in the ether and the length of an object moving relative to the ether. Similarly, Lorentz believed that the time delay was caused by the absolute motion of the moving body; A clock with a high absolute speed slows down relative to a stationary clock. But in SR, Einstein explains length reduction and time delay by replacing the relationship between the observer and the ether with the relative motion of the reference frames of different observers. This creates a series of paradox.

But Lorentz's ideas about relativity today, some of them are wrong. For example, "length shortens in the direction of motion" has never been verified by experiments; "Mass-velocity formula" for neutral particles, neutral matter has no experimental proof. But his "motion clock slows down" formula is here to stay; The formula of mass-speed should be changed to the formula of force-speed. Especially, in the framework of Lorentz theory, superluminal phenomena can be explained, unlike SR.

Poinocarè summarized Lorentz's work in 1904 and pointed out that it could be a transformation group, and named it after Lorentz Transformation (LT). Einstein felt that SR needed LT and independently derived the same formula as LT without mentioning

Lorentz's name. The results were unsuccessful. From today's point of view, LT has its place in the history of science, but it is not an absolute necessity. Therefore, it is a mistake to require LT covariability in any new physical theory. It should be emphasized here that although Einstein became a worldwide celebrity, both H. Poincarè (who died in 1912) and H. Lorentz (who died in 1928) expressed objections to Relativity in their later years — "tacit naysayers," according to Ohanian, a historian of science. That means "refuse to consent".

From Galilei to Newton, this view of time and space is both correct and important. Therefore, Galilei transform (GT) cannot be denied. The generalized GT (or GGT) theory has special value today. We stress that Galilei's view of space-time remains correct and needs to be refined to explain contemporary experiments.

Therefore, Maxwell's equations (ME) can be covariant with GT under certain theoretical conditions. It makes no sense for relativists to self-promote SR and deny the value of some new theories on the grounds of "covariation of ME to LT". In addition, this paper rejects the terms "relativistic electromagnetic field" and "relativistic electromagnetism".

Einstein's attempt to correct Newton's mechanics led in the wrong direction. QM, with a new way of thinking, inherits Newton mechanics in the view of time and space, and corrects Newton mechanics in the aspect of certainty and probability thinking — this is the beneficial guidance. Importantly, both Newton mechanics and QM don't put an upper limit on the speed of the immovable body, allowing for superluminal motion. This draws the line off SR and makes them a useful theory to explain many new experiments. At the same time, this paper shows that it is necessary and meaningful for Chinese scientists to put forward the concept of "Superluminal Light Physics". This paper emphasizes that quantum non-locality is a non-relativistic view of time and space in nature, Bell inequality theory is an important part of the history of physical thought, and the theoretical and experimental confirmation and development of quantum entanglement state is an excellent supplement to the study of FTL.

COMPETING INTERESTS

Author has declared that no competing interests exist.

REFERENCES

[1] Chen B Q, et al. Special Research of Electromagnetism[M]. Beijing: Higher Education Press, 2001.

[2] Tan S S. From Special Relativity to Standard Space–Time Theory[M]. Changsha: Hunan Science and Technology Press, 2007.

[3] Ma Q P. Inquiry into the Self-constancy of Relativity Logic[M]. Shanghai: Shanghai Scientific Literature Press, 2004.

[4] Zhang Y L. Introduction to Relativity[M]. Kunming: Yunnan People's Publishing House, 1980.

[5] 黄志洵. 对 Maxwell 方程组的研究和讨论 [J]. 中国传媒大学学报, 2022, 29(4): 65-82.

[6] Song W M. Dyadic Green Functions and Operator Theory of Electromagnetic Fields[M]. Beijing: University of Science and Technology of China Press, 1991.

[7] Fan D N. Lecture by Schrödinger[M]. Beijing: Peking University Press, 2007.

[8] Dirac P. Lectures on Quantum Mechanics[M]. New York: Yeshiva University Press, 1964.

[9] Cao S L. Relativity and cosmology in Finsler space-time[M]. Beijing: Beijing Normal University Press, 2001.

[10] Cao S L. Exceeding the Speed of Light[M]. Shijiazhuang: Science and Technology Press, 2019.

[11] Liu X G. Electrodynamic study of moving body[M]. Beijing: Beijing Normal University Press, 2010.

[12] Huang Z X. Wave Science and Superluminal Light Physics[M]. Beijing: National Defense Industry Press, 2014.

[13] Huang Z X. Faster-than-light research — The intersection of relativity, quantum mechanics, electronics and information theory[M]. Beijing: Science Press, 1999.

[14] Huang Z X. Introduction to cutoff waveguide Theory[M]. Beijing: China Metrology Press, 1991.

[15] Huang Z X. New advances in FTL research[M]. Beijing: National Defense Industry Press, 2002.

[16] Huang Z X. Theory and experiment of FTL research[M]. Beijing: Science Press, 2005.

[17] Huang Z X. Faster-than-light research and electronics exploration[M]. Beijing: National Defense Industry Press, 2008.

[18] Huang Z X. Research on the physics of superluminal light[M]. Beijing: National Defense Industry Press, 2017.

[19] Huang Z X. Advances in physics of microwave and light[M]. Beijing: National Defense Industry Press, 2020.

[20] Brillouin L. Wave Propagation and Group Velocity[M]. Pittsburgh: Academic Press, 1960.

[21] Tangherlini F. Introduction to General Relativity[M]. Shanghai: Shanghai Science and Technology Press, 1963.

[22] Zhang T. Discussion on Physical Spatiotemporal Theory[M]. Shanghai: Shanghai Scientific Literature Press, 2011.

[23] Huang Z X. Light of Physics—Open physical thought[M]. Beijing: Beijing University of Aeronautics and Astronautics Press, 2022.

Negative Velocity Characteristics in Electromagnetism[*]

INTRODUCTION

Waves are common phenomena in nature, such as wheat waves in the field, waves in the ocean. Of course, electromagnetic wave is a very important kind of wave. Matter must have mass, so a wave is not matter itself, but an external manifestation of its motion. Newton's classical mechanics, which studied the motion of particles, extended to objects on the ground and celestial bodies in the universe. The uses of Newton's mechanics were ubiquitous; But it does not generalize the motion of waves. In fact, Newton didn't study waves. A wave is an external form of motion of matter with no fixed shape and definite mass, which cannot be accurately described by Newton mechanics. For example, its trajectory cannot be found in Euclid space, nor can it be accelerated by force. In modern theories of electromagnetic field and wave, operator theory and wave function space are used to describe its motion state, which is very different from the treatment of macroscopic matter.

With regard to the negative wave velocity, it must be pointed out by M.Born and E.Wolf in their famous book *Principles of Optics* that the velocity of waves (phase velocity, group velocity) is a scalar rather than a vector. Unfortunately, some physicists are not clear about this and insist that "negative velocity is motion in the opposite direction".

"Negative velocity" means the opposite direction of motion in Newton mechanics, but not in wave mechanics. Although Einstein (in 1907), Sommerfeld (in 1914), and Brillouin (in 1960) all discussed negative velocity problems, their theories were either flawed or incomplete; They need to be restated today. In addition, although J.Wheeler and R.Feynman pointed out as early as 1945 that the advanced solutions of Maxwell-Helmholtz wave equation should not be arbitrarily discarded, they did not dare to say that there would be a single advanced wave. Today, we know that there are waves with negative velocity. For example, in 2009, N. Budko experimentally discovered the negative velocity in the near field of an antenna. The phenomenon occurs in free space and does not depend on materials such as anomalously dispersive media.

Since wave motion is a unique manifestation of matter, a new understanding of wave velocity is needed now. The study of wave velocity is a key point and breakthrough in wave

---

[*] The paper was originally published in *Physical Science International Journal*, 2023, 27 (1), and has since been revised with new information.

science. This paper believes that the advanced wave exists and the negative wave velocity is a special form of superluminal velocity. For the advanced wave, this paper defines it as "wave with negative velocity".

PLANE SPECTRUM METHOD IN ELECTROMAGNETIC FIELD ANALYSIS

We will now outline the relationship between the antenna problem solving method and the advanced wave. Several decades ago, the calculation of antenna once attracted the attention of the Chinese scientific community. In the fall of 1969, relations between China and the Soviet Union were strained and war seemed imminent. In October of that year, the Chinese government held a conference on the security of broadcasting and communication systems to examine war readiness. One of the decisions made at the meeting was to ask scientists to solve the problem of reducing the size of antennas, since huge antennas are difficult to conceal. The Chinese Academy of Sciences has set up the Antenna Computing Group to conduct numerical research on dipole antennas. The first thought was the solution of Maxwell wave equation. The calculation includes: (1) the near region—the region near the antenna that is far smaller than the wavelength; (2) the far region—several wavelengths and further away from the dipole. The serious problem is that the wave equation has two sets of convergent and divergent solutions. How to prompt the computer to keep only the outwards divergent solutions? In analytical processing this is known as the Sommerfeld boundary condition. What to do in numerical computation? Later, it was suggested to solve the initial value problem of Maxwell equations of order 1 directly until the steady-state solution was obtained. This simply circumvents the other boundary condition by adding the zero field condition at infinity. ... The traditional antenna calculation method is to first give the current distribution on the antenna, and then use the retarded potential integral expression to calculate the radiation field and antenna parameters. Strict numerical calculation (using the high speed and large capacity computer at that time) has been insufficient, especially to consider the far field from the origin of the antenna (theoretically infinite), it is difficult. A fundamental question is, is it better to start from the wave equation or from the basic electromagnetic field equation?

In free space, if there is no charge source (body charge density $\rho=0$) and no current source (conductivity $\sigma=0$ or current density vector $\mathbf{J}=0$), the electromagnetic wave equation can be written as:

$$\nabla^2 \mathbf{E} - \varepsilon\mu \frac{\partial^2 \mathbf{E}}{\partial t^2} = 0 \qquad (1)$$

$$\nabla^2 \mathbf{H} - \varepsilon\mu \frac{\partial^2 \mathbf{H}}{\partial t^2} = 0 \qquad (2)$$

where **E**, **H** are electric field intensity vector and magnetic field intensity vector; In the monochromatic simple harmonic condition, take the expression of wave $e^{j\omega t}$, then $\partial^2/\partial t^2 \to -\omega^2$, and get the vector Helmholtz equation:

$$\nabla^2 + k^2 \mathbf{E} = 0 \tag{3}$$

$$\nabla^2 + k^2 \mathbf{H} = 0 \tag{4}$$

and we have

$$k^2 = \omega^2 \varepsilon \mu \tag{5}$$

For the sake of a concrete discussion, we may take an electric dipole as an antenna (fed centrally) and consider the related problems of far and near field. Figure 1 shows the electric dipole antenna and the field strength measurement plane.

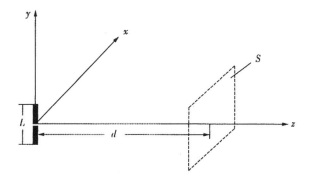

Fig. 1 The dipole antenna and the near-field measurement plane

We now solve the electric field intensity; There is a solution at $r \geq 0$:

$$\mathbf{E}(x,y,z) = \frac{1}{2\pi}\iint_{-\infty}^{\infty} \mathbf{F}(k_x,k_y) e^{-j\mathbf{k}\cdot\mathbf{r}} dk_x dk_y \tag{6}$$

where **k** is the wave vector and **r** is the position vector. Similarly, magnetic field intensity can be derived:

$$\mathbf{H}(x,y,z) = \frac{1}{2\pi}\iint_{-\infty}^{\infty} \mathbf{k} \times \mathbf{F}(k_x,k_y) e^{-j\mathbf{k}\cdot\mathbf{r}} dk_x dk_y \tag{7}$$

where, k_x and k_y, is the component of the wave vector, satisfying the following relation:

$$\mathbf{k} = k_x \mathbf{i}_x + k_y \mathbf{i}_y + k_z \mathbf{i}_z, \quad k^2 = k_x^2 + k_y^2 + k_z^2$$

Therefore may be sought k_z, If $k_x^2 + k_y^2 \leq k^2$, there is

$$k_z = \sqrt{k^2 - k_x^2 - k_y^2} \tag{8}$$

If $k_x^2 + k_y^2 \geq k^2$, there is

$$k_z = -j\sqrt{k_x^2 + k_y^2 - k^2} \tag{9}$$

The imaginary number will correspond to an evanescent plane wave spectrum (PWS), k_z which is characterized by sharp decay from the source location with the increase of r. And this PWS can be determined by the boundary conditions of the $z=0$ plane.

In the above case $\boldsymbol{F}$ is a vector function, which we will not discuss. But here we must consider the near field measurement, near field and far field relationship, evanescent field effect and so on. The outstanding point is how to calculate the evanescent state effect theoretically. The near field measurement technique was developed in the 1990s. This planar scanning near field measurement technique provides an economical and accurate way to determine the antenna pattern and its parameters. Its theoretical basis is the plane spectrum description of the field. There is a simple relationship between the antenna far field pattern and PWS. In most applications, the far field pattern can be obtained from the near field measurement data after Fourier transformation.

Planar scanning is now available in the plane shown in Fig. 1, where d is in the near area. In that plane there is

$$\boldsymbol{E}_y(x,y,d) = \frac{1}{2\pi} \iint_{-\infty}^{\infty} \boldsymbol{F}_y'(k_x, k_y) e^{-j(k_x x + k_y y)} dk_x dk_y \tag{10}$$

In formula

$$F'(k_x, k_y) = \boldsymbol{F}_y(k_x, k_y) e^{-jk_z d} \tag{11}$$

It can be seen that $\boldsymbol{E}_y(x,y,d)$ and $F_y'(k_x,k_y)$ is a Fourier transform pair; When $z=0$, we have

$$\boldsymbol{E}_y(x,y,0) = \frac{1}{2\pi} \iint_{-\infty}^{\infty} \boldsymbol{F}_y(k_x, k_y) e^{-j(k_x x + k_y y)} dk_x dk_y \tag{12}$$

Thus $\boldsymbol{E}_y(x,y,0)$ and $F_y(k_x,k_y)$ is another Fourier transform pair.

The above provides the theoretical basis; Due to the use of computer simulation, the aperture distribution of a dipole antenna can be calculated using the PWS method. The measurement plane is chosen to be close to the antenna, so the evanescent wave is included in the near-field measurement. Although the near-field behavior of the evanescent state has no effect on the far-field pattern, it is important for accurate modeling in the near field. The above principle has been demonstrated experimentally.

Based on the electromagnetic wave equation [Equations (1) and (2) in this paper], the boundary value problem of the second order elliptic partial differential equation is obtained by adding the boundary conditions. Due to some mathematical and physical reasons, it is difficult to calculate in this way. In 1977, a paper jointly published by the Institute of Physics and the Institute of Computing Technology of the Chinese Academy of Sciences analyzed the causes of the difficulties and pointed out that divergent waves tending to infinity had been selected in the analytical processing, excluding convergent waves. Finally, the calculation method chosen is directly discretized from Maxwell equations (first order hyperbolic equations), so that the initial value problem becomes an effective algorithm.

This paper considers that the above-mentioned early work of Chinese scientific community is beneficial. However, at that time, scientists did not realize that convergent wave solutions have special significance in the nearby region, and they could not be arbitrarily "excluded". This is what the author called the advanced wave, it is a solution to the Maxwell wave equation, is the existence of negative velocity physical performance! In the author's opinion, having two sets of solutions to the wave equation is not a "serious problem", but one of the features of nature. Nature may seem stranger than one can imagine, but we must take her for what she is.

POTENTIAL FUNCTION ANALYSIS OF ELECTROMAGNETIC FIELDS AND THE SOLUTION OF THE D'ALEMBERT EQUATIONS

The static field analysis can be simplified by introducing scalar electric potential function, scalar or vector magnetic potential function in static field analysis. This method is even more important for alternating electromagnetic fields. In vector algebra, For vector $\mathbf{A}$, $\nabla \nabla \times \mathbf{A} = 0$, and Maxwell's equations have an equation $\nabla \cdot \mathbf{B} = 0$. Therefore, vector potential can be defined according to

$$\mathbf{B} = \nabla \times \mathbf{A} \tag{13}$$

Given that Maxwell's equations contain a formula: $\nabla \times \mathbf{E} = -\partial \mathbf{B}/\partial t$, so it can be obtained

$$\nabla \times \left(\mathbf{E} + \frac{\partial \mathbf{A}}{\partial t} \right) = 0$$

But in vector algebra, there is $\nabla \times \nabla \Phi = 0$ for any scalar Φ, so it is desirable, so we obtain:

$$\mathbf{E} + \frac{\partial \mathbf{A}}{\partial t} = -\nabla \Phi$$

So

$$\mathbf{E} = -\nabla \Phi - \frac{\partial \mathbf{A}}{\partial t} \tag{14}$$

Therefore, it can be determined $\mathbf{E}$ by vector potential $\mathbf{A}$ and scalar potential Φ; Be known by $\mathbf{B} = \nabla \times \mathbf{A}$, $\mathbf{B}$ may be determined. This is the basic concept of introducing the potential vector function to determine the electromagnetic field.

But in order to uniquely determine **A**, Φ it is also necessary to know the value of $\nabla \cdot \boldsymbol{A}$. This is arbitrary, as given by Lorentz

$$\nabla \cdot \boldsymbol{A} = -\frac{1}{v^2}\frac{\partial \Phi}{\partial t}$$

where v is the wave speed ($v = 1/\sqrt{\varepsilon\mu}$), so the above formula is also written

$$\nabla \cdot \boldsymbol{A} = -\varepsilon\mu\frac{\partial \Phi}{\partial t} \tag{15}$$

This is called the Lorentz condition. By substituting the formula for and into Maxwell's equations and citing Lorentz's condition, the following partial differential equations can be proved **B E**:

$$\nabla^2 \boldsymbol{A} - \varepsilon\mu\frac{\partial^2 \boldsymbol{A}}{\partial t^2} = -\mu \boldsymbol{J} \tag{16}$$

$$\nabla^2 \Phi - \varepsilon\mu\frac{\partial^2 \Phi}{\partial t^2} = -\frac{\rho}{\varepsilon} \tag{17}$$

The above two formulas indicate that, the source of **A** is the current density vector **J**. The source of Φ is the bulk charge density ρ. The above equations are collectively referred to as d'Alembert equation. To facilitate the analysis of the problem, only the scalar equation of the function Φ can be considered. Assuming that the charge source is a point charge $q(t)$, it can be shown that the solution Φ can be expressed in the following form:

$$\Phi = \frac{1}{4\pi\varepsilon_0}\left[q\left(t-\frac{r}{v}\right) + q\left(t+\frac{r}{v}\right)\right] \tag{18}$$

Here v can be the speed of light c or it can be a different value. The first term on the right of the equals sign means that the situation at the time t and space point $P(x,y,z)$ depends on the situation before t, i.e. $(t - r/v)$, so the space points lag behind the source. This is the phenomenon of retarded potential. The second term shows that the situation of the point P depends on the situation after t, i.e. $(t + r/v)$ so the space point is ahead of the source, which is an advanced potential phenomenon. The latter term should be removed from textbooks in the past because "it is impossible for waves to converge from the outside to the source", which does not explain the source of the wave. Moreover, it implies a negative velocity, i.e

$$t - \frac{r}{(-v)} = t + \frac{r}{v}$$

In the past, negative velocity used to mean "in the opposite direction," not that it was actually negative. If the velocity value itself is negative (independent of the direction of the vector), then it does not meet the requirements of causality, that is, cause precedes effect, not effect precedes cause. Today, these views are all wrong!

NEGATIVE VELOCITY AND THE MEANING OF THE CONCEPT OF ADVANCED WAVES

In recent years, the advanced waves has been proved by experiments, or we say the development of experiments has exceeded the theoretical expectation. In 2013, I published a paper "Negative Characteristic Motion of Electromagnetic Waves and Negative Electromagnetic Parameter of Medium", which mentioned that the concept of advance wave originated from the early papers of J.Wheeler and R.Feynman. In 1940, Feynman pointed out to Wheeler that a single electron in space does not emit radiation, but only when both the source and the receiver are present. Feynman analyzed the case of only two particles and asked Wheeler, "Does this force of one affecting the other and acting back on it explain radiation resistance?" Wheeler suggested introducing the concept of a advanced wave to the two-electron model—a previously neglected solution to the Maxwell equation. Wheeler and Feynman develop this concept into the relationship between the electrons and the multiple "absorbers" around them, that is, the radiation damping is viewed as the reaction of the charge of the absorbers on the source in the form of a advanced wave; Their theory now had symmetry, but only in terms of waves moving inwards and backward in time. There was just a new wrinkle — it went back to its source before it could be launched. But they avoided unpleasant contradictions by taking the customary hysteresis waves and cancelling them out with the advance waves in a suitable way; That is, all radiation is guaranteed to be absorbed at some point, somewhere in the universe. This proved that they had not yet dared to use the concept of advanced waves alone.

The inward-moving wave described by Wheeler-Feynman (a wave moving backwards in time) is in fact the negative-velocity wave we are talking about now. There are two manifestations in wave science—negative phase velocity (NPV) and negative group velocity (NGV). In the past, when studying the cutoff waveguide theory, I found that the phase constant is negative ($\beta<0$), which is actually a kind of advanced wave. Later, the British scholars put forward the experimental roof. $v_p<0$ certainly means that the phase refractive index is negative ($n<0$), but it does not mean that the advanced wave must be included in the framework of metamaterials (i.e., left-handed materials, LHM) to be understood. In the condition of ordinary materials, there is also the phenomenon of advanced wave, which can also be seen in the near-field physical state of general antennas. For example, N.Budko published a paper "Local Negative Velocity Observation of Electromagnetic Field in Free Space" in 2009. The theory and experiment show that the near field and mid-field dynamics of vector electromagnetic field are much more complicated than the simple "outward propagation". There exists a region close to the source, where the wave front travels outward at the speed of light, while the core body of the waveform moves inward, that is, the wave travels backward in time, that is, the wave may travel back in time. Experimental observations of negative velocity were provided in this paper. It was believed that negative velocity in the near field area was found, and the first 5 near-field waveform at (3.5~8)mm showed that the internal peak was retrograde to time. Therefore, electromagnetic waves may travel faster than the speed of light in free space under near-field conditions even if there is no medium.

In 2013, I put forward the concept of "negative characteristic motion of electromagnetic waves" and distinguished it from simple "reverse motion". The paper argues that it should be regarded as a normal physical phenomenon inherent in nature.

Wheeler wanted to determine what would happen if retarded and advanced electromagnetic waves always occurred equally. In particular, it means that radio transmitters emit half the wave's power into the future and the other half into the past. All advanced electromagnetic waves can be considered to disappear from observation for the following reason: As retarded electromagnetic waves from a particular source of waves on Earth spread through space and encounter matter, they are absorbed. This absorption process involves the interference of electrical charges caused by the electromagnetic waves, with the result that distant charges produce secondary radiation. According to the hypothesis of this theory, this radiation is also half delayed and half advanced. The advanced component of this secondary radiation travels in the opposite direction of time, and some of it travels to the source of the emission on Earth. This secondary radiation wave is just a weak reflection of the source, but there are countless such weak reflections from space that can have a huge additive effect. It can be shown that under certain conditions,these advanced secondary radiation can be used to strengthen the primary delayed wave to its maximum intensity. At the same time, the leading component of the wave source is eliminated due to interference cancelling. At the end of time, when all these electromagnetic and reflected waves are added together, the net effect appears to be pure retarded radiation.

P. Davies argues that Wheeler-Feynman's theory assumes that the universe is rich enough to absorb all the radiation going into space, so the universe is opaque to all the electromagnetic waves. That's a strict condition. Judging from the surface, the universe seems completely transparent to many different wavelengths of wave, otherwise we can't see distant galaxies. On the other hand, there is no time limit to the absorption process, because reflected waves ahead (in the opposite direction of time) can travel backward through spacetime, and it is just as easy for them to travel back from the distant future as it is from the near future. So the success of the theory depends on whether an outwardly propagating electromagnetic wave can eventually be absorbed somewhere in the universe.

Davies says, we don't know if this is really the case because we can't predict the future. However, we are able to extrapolate current trends in the universe, and the result seems to be no — that the universe is not completely opaque. This would seem to negate Wheeler-Feynman's idea, but there is a curious possibility. Suppose there is enough matter in the universe to absorb most radiation, but not all of it. According to Wheeler and Feynman, this would lead to incomplete cancellation of advanced electromagnetic waves. Could it be that there are some advanced electromagnetic waves that have "walked into the past" — or come from the future — but whose wave intensity is so low that we haven't detected them yet? ...

Now, I must say that there are points Wheeler-Feynman (and Davies) makes that we cannot agree with. For example, the advanced wave is always cancelled out by the retarded wave, so that there is no single advanced wave. Experiments in recent years (mostly since 1998) have made us more confident that the existence of advanced waves has been confirmed by numerous NGV experiments and antenna near-field experiments, and that they are not offset. And logically, why is it that the advanced wave is always cancelled out and the retarded wave is not? That doesn't make any sense. The three scientists said that because of the limitations of their time, we now have a unified understanding of advance wave and negative velocity.

In order to improve life and explore the universe, human beings have made unremitting efforts to improve the motion speed of macroscopic matter; At the same time in the micro-field of exploration, research near the speed of light, the speed of light beyond the particle dynamics. Scientific concepts are also being expanded, such as the in-depth exploration of negative velocity and advanced waves.

THE SUCCESS OF MANY NEGATIVE GROUP VELOCITY EXPERIMENTS IS PROOF OF THE EXISTENCE OF ADVANCED WAVES

The founders of the classical wave velocity theory were A.Semmerfeld and L.Brillouin. Brillouin presented the curve of the relation between c/v_g and f calculated on the basis of this theory, which clearly showed the process of changing v_g from positive to negative. Brillouin said: "This curve presents a curious anomaly in the absorption band, c/v_g can become less than 1, and even less than zero. This means that the group velocity can be greater than the velocity of light, v_g can be infinite and even negative!"

The curious NGV physical phenomena of light pulse propagation was first proposed by C.Garrett. This is theoretical rather than experimental work. He proved that the concept of group velocity can be used even under strong anomalous dispersion (which v_g can be greater than the speed of light c or even negative) and explained the time advance phenomenon. In 1982, S. Chu's paper first proved the existence of NGV by experiment, and the experimental results perfectly gave three states (v_g>0, v_g=∞, v_g<0), which were very similar to the calculation curve provided by Brillouin in 1960. He also pointed out that when the peak of the pulse emerges from the sample at an instant before the peak of the pulse enters the sample.

After many years of exploration and research, scientists from many countries have used various methods to do successful group velocity superluminal experiment and negative group velocity experiments, which can be summarized into the following categories:

① In the short wave and microwave bands, using the combination of transmission lines to create the abnormal dispersion state, so as to obtain the superluminal group velocity and even negative group velocity. For example, in 2002, Hache was cascaded with multiple coaxial segments and obtained v_g=(2~3.5)c. In a similar way, Munday obtained v_g=4c and v_g=-1.2c. In 2003, Huang obtained v_g=2.4c; Zhou won v_g=2.2c and v_g=-1.45c; In addition, in 2012, Yao cascaded 3 rectangular waveguides, and mode effect and interference effect caused group velocities exceeding the speed of light, v_g=10c.

It is worth mentioning that in 2014, I instructed doctoral student Jiang to carry out experiments in microwave, using left-handed transmission line (LHTL) and negative refraction (n<0) on the basis of anomalous dispersion, to obtain negative group velocity, v_g=-1.85c. We use the digital oscilloscope to compare the input waveform and the output waveform, intuitively see the output is ahead of the input.

The above experiments are carried out in the framework of classical physics, without the intervention of quantum theory and technology.

② The method and technique of quantum optics are adopted, but the state of anomalous dispersion is still used. Specifically, the use of electromagnetic induction absorption (EIA) media anomalous dispersion states; In 2000, Wang passed cesium atomic gas with laser pulse at optical frequency, and obtained negative group velocity v_g=-c/310; Stenner used laser pulse to pass through potassium atomic gas in 2003, and obtained negative group velocity v_g=-c/19.6. In 2006, Gehring used erbium-doped fiber amplifier at optical frequency, due to the anomalous dispersion of the gain system, laser pulses through the fiber to obtain negative group velocity v_g=-c/4000.

③ Still using quantum optics, but technically more complicated. For example, anomalous dispersion states in electromagnetic induction absorption (EIA) media are used; In 2011, Zhang used the nonlinear process of excited optical Brillouin scattering in optical frequency to construct an optical fiber ring cavity, through which laser pulses can obtain superluminal group velocities and negative group velocities. It is also observed that the output signal is ahead of the input signal, and the negative group delay $\tau_g = -221.2$ns; In 2012, Glaser uses laser pulse through rubidium gas chamber at optical frequency, and uses 4 wave mixing technology to obtain negative group speed $v_g = -c/880$.

④ Still use the method and technology of classical physics, but introduce some special electromagnetic state. For example, the use of electromagnetic devices in the evanescent state; Enders in 1992 discovered group velocities exceeding the speed of light, $v_g = 4.7c$, when microwaves sent pulses through a cutoff waveguide in the evanescent state; Nimtz obtained $v_g = 4.34c$ by the same method in 1997; Wynne obtained group delay $\tau_g = -110$fs by a similar method.

The above experiments cover the three frequency bands of short wave, microwave and visible light, including classical techniques and quantum techniques, simple methods and very complex ones. The achievements and significance of these works cannot be denied.

A NEW THEORY OF NEAR-FIELD ELETROMAGNETIC PHENOMENA

As mentioned at the beginning of this article, the outside of a source of electromagnetic radiation, such as an antenna, is divided into near and far regions depending on distance. In fact, there is a transitional middle region between the two. Recent studies have shown that the dynamics of the near-zone field and the middle-zone field are much more complex than that of the far-zone field, and several special properties have been discovered successively, mainly in four aspects:
—static property, that is, similar to static field (such as electrostatic field);
—evanescent state, that is, properties similar to the evanescent field, such as the characteristic of almost pure reactance, and the rapid decay in field strength with distance;
—superluminal property, where it may travel faster than the speed of light;
—negative velocity property, that is, the possibility of a negative wave velocity, as well as a reverse motion in time, is actually the appearance of a forward wave.
Why is the near field so different from the far field? Why does it have these strange properties? We can't fully explain it yet, but we'll have to learn more in the future.
In Newton mechanics, the law of gravitation is also called Inverse Square Law (ISL),

which is written in the following form:

$$F = G\frac{m_1 m_2}{r^2} \tag{19}$$

where G is the Newton gravitational constant:

$G = 6.673 \times 10^{-11}$ m³/kg·s²

In the more than 300 years since ISL was proposed, no theory has come close to matching it in accuracy of prediction.

If we notice that the area of a sphere of radius r is calculated by the formula:

$$S = 4\pi r^2 \tag{20}$$

It is easy to understand why the inverse square law appears in different physical phenomena. ISL was proposed by Newton in 1687; In 1785, 98 years after ISL appeared, the French physicist C.Coulomb announced that he had experimentally found that the repulsive force between two spheres with the same number of electrostatic charges is inversely proportional to the square of the distance between the centers of the spheres and proportional to the product of their respective charges, i.e

$$F = K\frac{q_1 q_2}{r^2} \tag{21}$$

This is Coulomb's law, and it's surprisingly similar to the law of gravitation, which has inspired further comparative studies. In fact, Coulomb's law is also ISL. Assuming that gravity travels faster than light, does the Coulomb field (electrostatic field) also travel faster than light? This is possible, and the international research has also been carried out in this way, and the relevant results will in turn promote the research on the gravitational propagation velocity.

R.Tzontchev, a Mexican physicist, carried out research using a van de Graaf electrostatic generator. The radius of the two metal spheres is 10cm, the distance between the centers is 3m, and the height is 1.7m above the ground. Sharp electrical pulses were used. It was measured that Coulomb's action propagated at $v = (3.03 \pm 0.07) \times 10^8$ m/s, which was $v = 1.0107\,c$. It is faster than the speed of light.

R. Sminov-Rueda is a Spanish physicist who supervised papers in 2007; One paper was by A. Kholmetskii "Experiments on binding magnetic field delay conditions", in which two faster-than-light data ($v = 2\,c$, $v = 10\,c$) are obtained by conducting experiments using a ring antenna. The explanation is "non-locality properties of bound fields in near zone". As we know, non-locality is one of the important properties of quantum mechanics (QM), and its meaning is almost equivalent to superluminality. Therefore, the point of view of this paper is significant: it is necessary to use quantum theory to study near-region field.

The other paper of A.Kholmetskii is "Measurement propagation velocity of bound electromagnetic field in the near region", which is more complete in theoretical analysis, calculation and experiment. Both the transmitting and receiving antennas are ring antennas, mounted on a wooden table larger than 3m in size. The experiment gives the relationship between v/c and r. In the far region ($r \geq 80$cm), $v=c$; In the near region, when r =(50~60)cm, $v/c=4.3$; When $r=40$cm, $v=8.2c$. The conclusion is that, when $r<\lambda/2\pi$, the bound field travels at a faster-than-light speed, exhibiting apparent non-locality.

In 2011, the paper of Missevitch seems to be the third study on the near-area bound field of antenna under the guidance of Smirnov-Rueda, and the experimental techniques and methods have been improved. One measurement result given in this paper is v =$(1.6\pm0.05)c$; The authors consider the work in question to be "physics of EM wave propagation at a speed exceeding light."

In his 2014 paper, "Measurement of propagation velocities in Coulomb fields", R. Sangro began by discussing the problem of gravitational propagation velocities. This justifies our judgment that static or quasi-static fields in the universe have similar laws and can be usefully studied in comparison. The source is not merely an isolated charge, but a beam of electrons in uniform motion, that is, an electric charge moving at a constant velocity, whose electric field is still the Coulomb field. The experimental technique is complex and delicate. The results do not provide clear velocity data, but confirm the idea that the electron beam carries the Coulomb field.

The above literature covers the period from 2004 to 2014, and the obtained Coulomb field propagation velocity is in the range of $(1.01\sim10)c$. These advances not only enrich our understanding of the near-field, but also make us firmly believe that "gravity propagates faster than the speed of light".

Now we discuss an important theoretical problem — the "evanescent state" property of near-field. The time phase factor of an electromagnetic wave is $e^{j\omega t-\gamma z}$, where z is the coordinate of the propagation direction (distance) and γ is the propagation constant ($\gamma=\alpha +j\beta$) α is attenuation constant, β is phase constant. For a metal-walled uniform column waveguide, the internal electromagnetic state is with a cutoff field, the cutoff frequency $\omega_c=2\pi f_c$ (subscript c represents cutoff). It can be proved that the corresponding cutoff wavelength is

$$\lambda_c = \frac{2\pi}{h} \tag{22}$$

where h is the eigen value, not the Planck constant. The above equation reflects the characteristics of the transmission system of non-zero eigen value.

Now define a parameter called propagation factor:

$$k_z = -j\gamma = \beta - j\alpha \tag{23}$$

Therefore, the transmission system can be divided into two areas, namely transmission area and cut-off area; Since the cutoff region is almost purely imaginary, the corresponding wave vector k_z is called imaginary wave vector and the corresponding wave is called imaginary waves. Now we can compare the two electromagnetic states in the waveguide:

① Transmission region (transmission state): $f>f_c$, α is small, β is large, So propagation constant $\gamma \cong j\beta$ (approximation); propagation factor $k_z \cong \beta$ (approximation);

② Cutoff region (evanescent state): $f<f_c$, α is large, β is small, propagation constant $\gamma \cong \alpha$ (approximation), propagation factor $k_z \cong -j\alpha$ (approximation).

In the evanescent state, the time phase difference between the electric field and the magnetic field vector is $\pi/2$ (TM mode magnetic field intensity **H** lead, TE mode electric field intensity **E** lead); Poynting vector instantaneous value $\mathbf{E} \times \mathbf{H}^*/2$ is a pure imaginary number, Poynting vector mean value $\text{Re}(\mathbf{E} \times \mathbf{H}^*)/2 \neq 0$, wave impedance is reactance, reflecting the storage of electric energy and magnetic energy. For the near field of electric small antenna, the phase difference of **E** and **H** is $(-\pi/2)$, **H** is leading; The Poynting vector instantaneous value is pure imaginary, and the mean value is zero. It is also the property of energy storage field, which is manifested as electric reactance field. ...The above comparison shows that the property of the two is almost exactly the same!

Moreover, both of them decay rapidly with the increase of distance, but the decreasing law is different — the evanescent field follows the law $e^{-\alpha r}$, and the near field follows the law of inversely proportional relationship with r^3 or r^2. We believe that under certain conditions the two can be very close; The evanescent state of field strength

$$E_e = E_0 e^{-\alpha r} \tag{24}$$

The field strength of the electric small antenna is

$$|E_S| = \frac{K}{r^3} \tag{25}$$

Now let $E_e = |E_S|$, i.e.

$$E_0 e^{-\alpha r} = \frac{K}{r^3}$$

Take the natural log of both sides of the equation and get

$$\ln E_0 - \alpha r = \ln K - \ln r^3$$

So we can get

$$a = \frac{1}{r} \ln\left(\frac{E_0 r^3}{K}\right) \tag{26}$$

As long as the above equation is satisfied, both fields fall exactly the same. This is unlikely in practice, but it is an interesting comparison between two fields.

In addition, both two fields have the character of quasi-static fields. In evanescent field theory, although it is a time-varying field, it can be regarded as a static field with a single electric field for the analysis of some structures (for example, TM mode in cut-off waveguide is analyzed and treated by equivalent capacitor, TE mode by equivalent inductor). In electrically-small antenna theory, a similar situation exists — the field near the antenna follows Poisson's equation and can be treated as an electrostatic field.

For these reasons, faster-than-light propagation has been found in both the evanescent field structure and the near-field structure of the antenna. In 2009, N. Budko conducted experiments in microwave and found the phenomenon of negative velocity propagation in near-field. Using vertical dipole antenna for transmitting antenna, microwave pulse center frequency 4GHz; Waveforms traveling in reverse time were observed.

In 2011, O.Missevitch conducted experiments on meter waves and found that the fluctuation propagated at a faster-than-light speed in the near-field, $v = (1.6 \pm 0.05)c$. In 2013, Fan carried out experiments on short wave. He sent 20MHz sinusoidal wave signals to the ring antenna with a diameter of 25cm. He used two 3cm diameter coils to receive the signal at different distances. Prove that magnetic field line speeds in the near area can reach more than 10 times the speed of light ($v \geqslant 10c$).

Budko believed that there exists a region in which the body of the waveform is in reverse motion at any time — the extreme value of the waveform received by the recipient is progressively advanced as the distance from the source increases. Assuming that the source is a finite sinusoidal beam, the center frequency $f_0 = 4$GHz, then $r = 10$~100mm for the near field and the midfield region, the simulation of the near field waveform shows that there is a negative velocity in the near field region (e.g. 10~13.6mm). In other words, the outer edges shift rightwards (normal phenomenon) and the inner part shift leftwards (negative velocity phenomenon). Note that this is independent of the environment, even in a vacuum.

The near-field electromagnetic phenomena revealed by Budko are amazing, for example, he shows the details of several near-field waveforms as a function of time, which are obtained by progressively increasing the distance from the source (r). Although the edge of the wave packet moves to the right, the inside part moves to the left, traveling back in time. Experiments confirm the above simulation calculation, and the actual negative velocity region is about 8mm. Although the measured retrograde movement against time is small, this can be explained by factors such as the interaction between the source and the receiving antenna. Budko finally tries to explain the observed phenomenon in terms of classical or quantum theory, but it is clear that the paper is weak in this respect. One explanation is that both the near and intermediate field components contain an additional relative delay with respect to the far field. These relative delays gradually disappear with increasing distance from the field source. The overall effect of the radiation field therefore consists of two parts: the usual outward motion at the speed of light, and the relative inward motion. This results in the speed of the selected waveform features in the near and middle fields significantly exceeding the speed of light.

The authors believe that the phenomenon of near-field superluminal can be explained by the evanescent state theory, and the phenomenon of negative near-field velocity can be explained by the advanced solution of Maxwell-d'Alembert equation. With the help of virtual photon theory, all of these are understandable physical facts. Budko, however, seems unfamiliar with these theories and feels a bit at sea.

Now we try to explain near-field faster-than-light phenomena with quantum theory. In 2007, Kholmetskii claimed that they "confirmed non-locality" in antenna near-field experiments, so a deeper understanding of this non-locality should be obtained. The author believes that among the three essential characteristics of quantum mechanics (QM), non-locality is the most important, and the core idea of which is superluminal. The affirmation of non-locality by near-field experiments indicates that there is a deep connection between classical electromagnetic theory and quantum theory, and natural phenomena can be understood and interpreted only when the two are used together.

In 1971, C. Carniglia published his paper "Quantization of Electromagnetic Evanescent Waves", in which he "selected the virtual photon path of evanescent waves to express the field". In 1973, S.Ali published a paper "Evanescent Wave in Quantum Electrodynamics", in which he said that "evanescent wave is actually virtual photons of the interaction between the carrying field and the source", and that evanescent wave will become a virtual particle swarm of quantitative theory. The evanescent field is the same as the virtual photons field, which is not a mode to mode identity. In 2006, A.Stahlhofen published a paper "Evanescent Modes are Virtual Particle Populations", in which he stated that many years of research based on QED agree that evanescent modes are consistent with virtual photons, and that their strange properties (such as non-locality and unobserved) violate relativistic causality. In 2000, Professor G. Nimtz wrote to me that "Evanescent modes can be correctly described and understood only when QM is introduced and considered; The evanescent mode appears as virtual photon, but it cannot be measured. would argue that evanescent modes are Galilei invariant".

Therefore, from the point of view of quantum field theory (QFT) and quantum electrodynamics (QED), Evanescent modes are the result of the overall contribution of virtual photon groups. Since the two components of the near field of an electromagnetic source (bound field and Evanescent field) are Evanescent state, it is useful to use the virtual photon theory as an explanation of the superluminal phenomenon in the ultra-near region. For example, Nimtz has pointed out that in Feynman-type space-time diagrams, the corresponding process of virtual photons is that space distance is changing while time is basically unchanged, which represents potentially extremely high velocities. This makes the author feel that when the classical electromagnetic theory is used to study the cut-off waveguide, the phase constant near zero in the cut-off region is found, which indicates that the evanescent state propagates very fast.

In recent years, faster-than-light propagation and negative wave velocity propagation have been discovered very close to the source, both of which require more profound explanations. However, the comparative study on the connection between the above phenomena and the superlumen propagation of pure Coulomb field and the gravitational field filling the universe brings more enlightenments to people.

In this paper, a number of theoretical duals are given: bound field and evanescent field; Retarded solution and advanced solution; Positive wave velocity superluminal and negative wave velocity; Static electromagnetic and gravitational field propagation; Classical electromagnetic analysis and quantum theory analysis; Real photon and virtual photon. The duality of these dual properties is the manifestation of the nature of things.

EINSTEIN'S NEGATIVE VELOCITY IDEA AND THE PROOF OF BRILLOUIN'S NEGATIVE GROUP VELOCITY THEORY

We will now take the discussion further, first reviewing the early days of the negative velocity concept. At the beginning of the 20th century, A. Einstein thought about negative velocity during his work on special relativity (SR). Einstein believed that it was necessary to determine whether the velocity of physical action and the velocity of signal could exceed the speed of light, he was hesitant to do so. In 1907 Einstein published his article "On the principle of relativity and its conclusions",§5 of which ("The addition theorem of velocities") deals with both signal velocity and negative velocity. The article said, Postulate

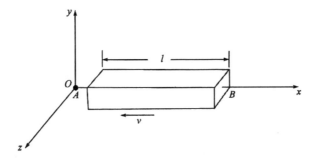

Fig. 2 The graph Einstein used to discuss the speed of signals

Place a strip of objects (Fig.2) along the x axis of the frame of reference S relative to which some action can be transmitted by velocity u (judging from the strip of objects), and have stationary observers not only at the point, $x=0$ on the x axis (point A), but also at the point x (point B) S; The person at the place A sends a signal to the person at the place B via the object, which moves in the direction of ($-x$) with a velocity ($v<c$). Then, according to the SR velocity synthesis formula, the signal velocity is

$$v_s = \frac{u-v}{1-\frac{uv}{c^2}} \tag{27}$$

The transfer time is

$$t_s = l\frac{1-\frac{uv}{c^2}}{u-v} \tag{28}$$

where, l is the object length. If $u>c$, select $v<c$, which can always make $t_s<0$. This results in a negative transfer time, as well as a negative signal speed. According to Einstein, this transmission mechanism causes "the effect to arrive before the cause" and therefore "no such signal can be transmitted faster than the speed of light in a vacuum."

In his opinion, this statement is wrong. Since we have thoroughly analyzed and critically criticized the theory of SR, we will not discuss SR itself nor comment on Einstein's formula of addition, but merely discuss it briefly. It is interesting that Einstein has made the basic judgment that "signal speed cannot exceed the speed of light", but he is not 100% sure. He said, "Although this outcome is logically acceptable, there is no contradiction in it; But it is so alien to the character of all our experience that the impossibility of the hypothesis $u>c$ seems sufficiently well established." Here, Einstein shows that something that violates causality does not violate logic, except that because it violates human experience, the speed of the signal cannot exceed the speed of light.

In 1914, A Sommerfeld and L.Brillouin developed a classical theory of wave speed, which, though not ideal, was far more valuable than Einstein's work. As is well known, there are two different concepts of wave velocity — phase velocity v_p and group velocity v_g, and the meaning of group velocity is generally considered to be greater than that of phase velocity. As for the evaluation of group velocity, there are two different tendencies — overestimation and underestimation. Examples of the former are British physicist Lord Rayleigh. In his book *Theory of Sound*, in 1877, he not only defined group velocity, but also believed that group velocity is consistent with energy velocity and signal velocity. But we now know that this view is true only under certain conditions. There are others who believe that group speed is extremely important. The other view is that the estimation of group velocity is so low that it seems that its study (whether calculated or experimental) is not valuable and meaningful. The author disagrees with both of these tendencies. If you don't study group velocity; what should you use instead? In many cases, energy velocity and signal velocity are complex and practically difficult to master. In Sommerfeld-Brillouin (SB) theory, front velocity is considered as the propagation velocity of a sudden disturbance, and its definition is not clear. It can be said that when discussing the velocity of electromagnetic wave (or electromagnetic signal), the group velocity is still fundamental and important, and can be used as a valuable reference material in scientific research. The situation would be even better if complex (reshaping) measures were taken in the experiment to reduce waveform distortion or to do no distortion at all.

In 1914, A.Sommerfeld discussed the wave velocity problem in detail. He assumed that a sine wave $f(0, 0)$ suddenly appeared on the surface of the medium $z=0$ at $t=0$, and the observer at z would not see the transient phenomenon occur until $t=z/c$. From this point on, no information is transmitted until the steady-state signal is established. Sommerfeld studied in the complex frequency domain ($p = \sigma + j\omega$), taking $f(0, 0)$ as the incident wave, he derived the wave of z described in the following integral equation:

$$f(z,t) = \frac{1}{2\pi j} \int_{\sigma-\infty}^{\sigma+\infty} \frac{\omega_0}{p^2 + \omega_0^2} e^{p(t-\frac{z}{c})} dp \tag{29}$$

In the formula, the real number σ should ensure that the integration path is in a certain region, and ω_0 is the steady-state carrier frequency. The above formula is also written:

$$f(z,t) = I_m \frac{1}{2\pi j} \int_{\sigma-\infty}^{\sigma+\infty} \frac{\omega_0}{p + j\omega_0} e^{p(t-\frac{z}{c})} dp \tag{30}$$

Actually no wave front arrives until $t=z/c$, and the wave front velocity is c; When $t=z/c$, the steady-state and transient components cancel out and the wave is still zero. This indicates that the signal always builds up from zero. When $t > z/c$ there is

$$f(z,t) = e^{-\alpha z} \sin(\omega_0 t - \beta z) \tag{31}$$

α, β are the attenuation constant and phase constant of the medium respectively. The process before steady state is completely established is called precursors, they develop gradually and rapidly to complete a continuous transition.

In order to understand the whole process of signal establishment, the above integral equation must be solved. Also in 1914, L. Brillouint used saddle point integration to solve Sommerfeld's integral equation. After complex calculations and graphing analysis in the complex plane, he obtained a family of curves. Figure 3 is the relationship between sum and frequency given by Brillouin in 1960 according to the integral equation (note that the ordinate is the ratio of the speed of light to the speed of wave, i.e. c/v), showing that throughout the frequency domain, but Brillouin ruled out the possibility of negative phase velocity, which is not true. Brillouin's regular description of group velocity is more meaningful, as can be seen from Figure3: there are cases of $v_g < c$ and cases of $v_g > c$. In addition, near the central frequency, three phenomena will appear, namely group velocity exceeding the speed of light ($v_g > c$), group velocity infinite ($c/v_g = 0$), and negative group velocity ($v_g < 0$). It must be noted that the group velocity increases and increases until after infinity has passed, the negative group velocity is reached. These analytical results have been proved by a number of facts.

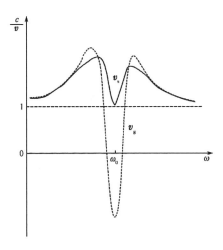

Fig. 3 Phase velocity and group velocity in Brillouin theory

What do we think of the Sommerfeld-Brillouin wave velocity theory today? It would be wrong to say yes or no to it outright. The author points out that it has the following problems: (1) The theory excludes the possibility of negative phase velocity (and thus negative refractive index), which is inconsistent with the known theoretical and experimental results; (2) Although the theory points out the possibility of negative group velocity, it fails to elucidate its physical mechanism and significance; (3) The way the theory studies signal velocity, the ideal step function requires infinite bandwidth, can not be realized in practice, so people doubt whether there is a problem in the definition method and research method, that is, the theory does not construct a reasonable definition of signal velocity; (4) Because SB theory appeared many years before the invention of quantum mechanics, it is impossible for SB theory to estimate the faster-than-light phenomenon when the quantum barrier particles pass through the tunnel; (5) The theory lacks a strict definition of wave front velocity. ...However, SB theory still has some reference value for researchers today.

It must be noted that the ordinate of Figure 3 is c/v, i.e. c/v_p or c/v_g; Thus, the point at which the curve representing the group velocity passes through $c/v_g = 0$ indicates that the group velocity reaches infinity. Thus, the Brillouin diagram is characterized by its description of the group velocity increasing until it crosses infinity to reach negative group velocity. That is to say, NGV represents a negative velocity larger than the infinite velocity. This may seem unfathomable, but it's not particularly surprising. In a 2021 paper, I described a "negative-energy vacuum emptier than the free vacuum", based on the Casimir effect, a similar situation. People can wonder: how can anything be faster than an infinite velocity? It is also possible to wonder "how can a vacuum be emptier than a vacuum?" But natural science research has entered the stage where one has to admit something that is difficult to understand if it is supported by solid theoretical or experimental evidence.

This leads us to a conclusion. Any wave with a negative velocity must be a faster-than-light wave. What about the evidence? Back in 1982, S.Chu did an experiment that proved this idea. His experiment was the first of its kind, and at the same time very sophisticated. In 1979, three years before Chu, R.Ulbrich conducted experiments with semiconductor (GaAs) samples at optical frequencies, and observed that light pulses propagate slowly, and that the group velocity v_g can be reduced from $c/3.6$ to $c/2000$. Although this was not an experiment in quantum physics; the experimental technique was distinctive. The sample thickness is 3.7μm, the area is 200μm×500μm, and the sample is placed at ultra-low temperature (1.3K). In his experimental system, the minimum group velocity can be obtained by adjusting the center carrier frequency ω_0. The maximum delay $\tau_g=35$ps was observed in the experiment. This gives you a way to measure it.

S. Chu, who published the paper "Linear pulse propagation in absorbing media", seems to have been the first person to experimentally prove the existence of NGVS, and it was the negative velocity that made the experimental breakthrough. Following Ulbrich's method, Chu used epitaxial GaP/N samples with thickness of 76μm or 9.5μm; If the thickness is L, then there is

$$v_g = \frac{L}{\tau_g} \tag{32}$$

So measured τ_g and v_g can be calculated, and the sample from the optical path access and take out is the experimental step. Obviously, if the zero delay is measured ($\tau_g=0$), it is measured v_g to the infinite. FIG.4 shows the experimental results when $L=9.5$μm is taken. It can be seen that all three aspects ($v_g>0$, $v_g=\infty$, $v_g<0$) are present, and the transition is smooth. Here we should point out that this experimental curve is very similar to the theoretical calculation curve of Brillouin (group velocity curve in Brillouin diagram) many years ago!

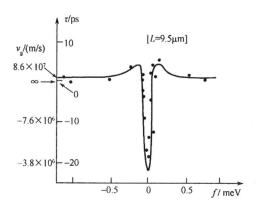

Fig. 4 Earliest experimental results of NGV (Chu, 1982)

The negative pulse velocity obtained by Chu shows that, when the peak of the pulse emerges from the sample at an instant before the peak of the pulse (when the peak of the pulse emerges from the sample at an instant before the peak of the pulse enters the sample). In 2000, Wang also discovered that this phenomenon, which some theoretical physicists thought impossible, was a bizarre theory. But in fact, it was discovered by Chu before Wang 18 years ago!

THE ADVANCED WAVE IS A WAVE WITH A NEGATIVE VELOCITY

Although both Chu and Wang measured NGV, the former did so using classical physics methods and the latter using quantum optics. Interestingly, when Wang's paper was published in the prestigious journal, it raised some eyebrows among relativists because it described itself as a "faster-than-light experiment." For example, [42] was fiercely critical of Wang's paper and its implications. However, although the author of reference [42] is familiar with the theory of relativity, he does not know Born and Wolf that "wave velocity is a scalar rather than a vector", and does not know Sommerfeld-Brillouin theory of wave velocity, and thus makes a wrong judgment. [42] says that the result of Wang's experiment is ($v_g = -c/310$), which after taking the absolute value gives $c/310$, and is therefore sublight speed. This statement is completely incorrect; Wang's result was ($-c/310$) not $c/310$, which is fundamentally different.

Not only do they know this, but the authors, though admirers of Einstein, have no idea what his 1907 paper said. Recalling Einstein's 1907 discussion, it is characteristic that negative transmission time appears simultaneously with negative signal speed; And this discussion shows that negative velocity is one of the characteristics of superluminal speed, so Einstein denied the meaning of negative velocity in order to prove that superluminal speed does not exist. Thus, the key lies in experiments, and if experiments (not just the

Wang's experiment, but many more in recent years) prove the existence of negative velocity, then that indicates the existence of superluminal velocities. That is why Wang insists that his experiment is a faster-than-light experiment. So [42] the main argument doesn't work Moreover, it is wrong to think of the "negative" of a negative velocity simply as "in the opposite direction", since wave velocity should be regarded more as a scalar than a vector.

Literature [42] also has merit; First, it points out that the study of faster-than-light problems is often linked to the concept of negative energy. "The light pulse moves from the exit to the entrance at $(-c/310)$, which is consistent with our calculation of the velocity of the negative energy density; The negative energy density is difficult to explain, and 'negative' may represent energy extraction from the cesium gas."

Secondly, [42] makes a theoretical concession on the relativistic proposition that "SR does not allow faster-than-light speed". The paper states: "SR does not rule out that light does not move faster than other matter. The key is to judge the order of causality in proper time, so that if the law of causality is satisfied in its inertial frame, it will be satisfied in all inertial frames; The modification of Einstein's interpretation of causality allows for the existence of such superluminal velocities."

However, this argument still absolutes the law of causality as an "iron law" that cannot be broken. Since a wave with a negative velocity is a wave, it is also against the classical law of causality for such a wave. We emphasize once again that the existence of a single advance wave is possible, which has been confirmed by many NGV experiments and near-field experiments. The author believes that it is helpful to describe the phenomenon of negative wave velocity with space-time diagram. For example, the experiment of Wang can be described in Figure 5, where L is the thickness of the air chamber. This figure can also be used to illustrate Chu's experiment, then L is the sample thickness.

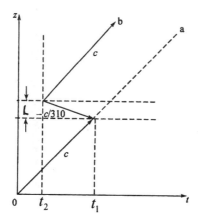

Fig. 5 The space-time diagram of the negative velocity experiment

The abscissa of Figure 5 is time (t), the ordinate is distance (z), and t is the departure time of the motion of the light pulse. At this interval of $0 \sim t_1$, the pulse travels in vacuum at the speed of light (c). When $t = t_1$, the light pulse enters the air chamber, the group velocity becomes negative ($v_g = -c/310$). When $t = t_2$, the light pulse appears at the exit, but $t_2 < t_1$, i.e. $t_2 - t_1 < 0$ (negative time). From the beginning, the light pulse continues to move forward at the speed of light, shown higher up in the diagram. —This figure has some value, but it is more complicated in practical analysis and involves consideration of the width of the optical pulse, which is omitted here.

NEW UNDERSTANDING OF CAUSALITY

The English word "causality" is important, which is different from the English word "causal law". Relativists like to use the term causal law, implying that it is a law that cannot be violated at all. However, there is no such law in physics. Causality, like symmetry, is a conviction. Its meaning is: (1) Everything has a cause; (2) Every cause has an effect; (3) The cause must precede the effect. Taking advantage of the receptivity of everyday experience, some people introduce it into the study of natural science and place it in a sacred place and high place.

Associated with this is certainty, known in English as definity. It is also the belief that nature is intrinsically predictable and that all events are determined by a prior cause and follow certain rules. It is just a matter of finding that law and mastering the initial state so that the future can be deduced precisely from the present. In 1814 P. Laplace said:"The future of the world can be determined by its past; The state of the world at any given moment (expressed mathematically) can be predicted." This is a typical view of deterministic causality. By the 20th century, the poster child for this view as Einstein. On January 20, 1920, Einstein wrote to M.Born:"The question of causality troubles me; Whether the quantum absorption and emission of light may one day be understood in the sense of full causality, or whether a statistical tail is to be left... I would be very unhappy to give up complete causality." On April 29, 1924, Einstein wrote to M.Born:"I shall not give up until there is stronger evidence against strict causality than has so far been presented... I cannot tolerate the idea that an illuminated electron should jump away at a time and in a direction of its own choosing by its free will... It is true that I have repeatedly failed in my attempts to give quantum definite form, but I do not want to give up hope for long." In 1924, in a letter to M.Born, Einstein added: "The theory of quantum mechanics makes a great contribution, but brings us no closer to the mystery of God. In any case, I'm sure he's not rolling the dice." ...His words spread far and wide, but they are not true; "God" (nature) not only rolls the dice, but often in unexpected places.

In March 1927, W.Heisenberg proposed the famous uncertainty principle in quantum mechanics, which tells us that there is always an irresolvable uncertainty in the behavior of microscopic particles, that is, events in the microcosmic world often happen for no reason. In fact, it is quantum theory that poses the greatest challenge to determinism. Beginning in October 1927, Einstein expressed his rejection of the uncertainty relation and devised "thought experiments" to prove that the principle of the relation could be exceeded. This process continued for at least a decade, including the famous EPR paper. In short, uncertainty relationships lead directly to unpredictability. The quantum world breaks free from the tight chains of cause and effect. "According to quantum theory, it is possible to have an effect without a cause," says British physicist P.Daves. Zhang Yongde, a famous quantum mechanic in China, said quantum theory opposes Einstein's objective realism because its view of things is simple and mechanistic, departing from the principle of superposition of states and wave-particle duality. In addition, what QM does not allow is relativistic local causality, the theoretical basis for the invariance of Lorentz transformations. Therefore, quantum theory can also be considered incompatible with relativistic local causality.

The development of science before QM was proposed actually predicted the end of certainty through the study of randomness and chaos. In the late 19th century, H.Poincare found that the solvability and solution value of some differential equations (such as Hamilton equation type) are sensitively dependent on their initial conditions — small changes in the latter can lead to great changes in the value of the solution or no solution. This discovery made predictability a non-rule and was philosophically opposed to Laplace. Therefore, Poincare moved toward the theory of indeterminacy, which holds that any small uncertainty factor in the state of a system may gradually become large and make the future unpredictable. One of Poincare's contributions was the study of the three-body problem, which led to the discovery of a new concept, chaos, in the analysis of celestial orbits. Like other scientists before him, he did not succeed in solving equations and finding quantitative solutions, but broke new ground in qualitative research. He proposed the concept of phase space, a hypothetical dimensional space, in which each point represents a state of the system. The analytical conclusion is that asymptotic solutions have an infinite number of sequences with different periods, as well as an infinite number of aperiodic sequences — the latter being chaos, which is sensitively dependent on initial conditions or states. He was fond of saying that "prediction is impossible".

In 1933, the chairman of the Nobel Physics Committee said in his speech, "In the microcosmic world, the requirement of causality must be abandoned. The laws of physics express the probabilities of events. "I think this could not be more true. In the microcosmic world, if the arrival of a particle is an event, it can be said to have no cause.

Quantum mechanics, though it shows how microscopic particles behave in the quantum world, contrary to deterministic causality. But still some researchers, after publishing a good paper on faster-than-light speeds, nervously ask, "Is causality a line scientists should not cross?" This is not a surprising question, since the term "violation of causality" is a common criticism of faster-than-light research. Some relativity books exaggerate the role of causality, but fall into a logical paradox, when they explain the principle of the speed of light limit. It seems that "superluminal speed is impossible" can be judged from causality. In this way SR does not work.

The timing order (positive order $\Delta t > 0$, reverse order $\Delta t < 0$) has relatively property, only $v < c$, the timing has absolute significance, when $v > c$ timing inversion may have observational significance. On the other hand, it is necessary to distinguish temporal relativistic properly from "backward flow of time", and as long as light is used as the method of observation information transmission to observe faster-than-light motion, the reverse order will appear, which is a new expression of causality under the conditions of faster-than-light motion, rather than the destruction of causality. In short, there is no relative change in timing order under the condition of subluminal motion. Moreover, SR treats ds^2 as an invariant, which in fact only applies to sublight systems, and ds^2 is not an invariant when considering the possibility of superluminal motion. In some literatures, the time-sequence relativistic property that must exist when the superluminal motion occurs is described as the backward flow of time, which causes great confusion in understanding. The practice of sanctifying both causality and SR at the same time has blocked and blocked reasonable discussion.

The author believes that observing tachyon motion with light must produce superficially strange phenomena. Aeroplanes flying at supersonic speeds have been realized for a long time, and no one cries "causality is destroyed" because of the reverse sound, and no one thinks that the sound is "transmitted to the past". The key is to recognize the local characteristics of SR theory — its space-time takes light signal as the observation horizon, and the speed limit is set at the speed of light. Light defines time and takes light as the basis of observation theory. This leads to the conclusion that faster-than-light speeds violate causality.

In fact, the discussion of causality does not have to start from quantum reality. Starting from classical physics, as long as some old viewpoints of laws are corrected, negative velocity and advance wave an be analyzed satisfactorily. Liu Liao's argument is an example of this. He acknowledged the impact of experiments on existing theories and argued that the limitations of theories should be clarified and ways to improve them should be considered. As a noted expert on relativity, he made a notable statement that the Wang's experiment, which contained the possibility of faster-than-light pulses of light, was a challenge to relativity. Specifically, the appearance of negative velocity actually transforms a delayed (conventional) light pulse into an advanced one, resulting in the outgoing pulse being ahead of the incident pulse in time, which seems to violate the conventional temporal causality, that is, effect is ahead of cause in time. Prof. Liu held that the limitation of time sequence should not be absolutized, but the law of causality should be expressed as "effect cannot affect cause in any way". In this way, the objectivity of the law was maintained (people cannot change history) and the new experiment was explained. In addition, Prof. Liu suggested using the concept of "advanced waves" to explain Wang's experiment, which is consistent with the viewpoint of this paper.

In short, it is no longer valid to use the so-called "law of causality" to suppress new ideas in physics. It must be noted that papers published in 2019 have proposed the following: quantum time is in a superposition state, where past, present and future are integrated, cause and effect are reversed, and cause and effect cannot be distinguished. This means the same thing as the speech given by the chairman of the Nobel Committee in 1933.

A DISCUSSION ON "TIME TRAVEL"

"Time travel", also called "time machine", has always been a hot topic among people. There are a lot of strange theories about time travel. Such questions as "can time be turned back?" and "can man go back in time?"

It's important to mention the famous "grandfather paradox" — someone who goes back in time, finds his unmarried grandfather who is in love and kills him; That person would never have been born. If he didn't exist, how could he "go back in time and kill grandfather"? This paradox is often used to prove that it is impossible to travel back in time. —There is also a paradox with time travellers who want to travel to the future — someone knows they will have a car accident in the future and takes measures (such as staying at home) to avoid it, so the accident does not happen. But since nothing has happened, how can that person be sure that "one day he will have an accident"? ... These points actually rule out the feasibility of "time travel".

So what do the relativists think? There are two theories: One is: "it is relativity that makes time travel possible"; The other is "since relativity says that faster-than-light speeds are impossible, time travel is also impossible." A common saying is "going beyond the speed of light leads to going back in time". Again, people say "particles that travel faster than light can move backward in time". In short, some people do not hesitate to equate faster-than-light with "moving backward in time", or "moving backward in time", or "being able to travel backward in time." And they generally say: "This is the result of relativity."

On closer inspection, we find that these statements correspond to Einstein paper in 1907. It first gives a formula for the speed of the signal, then a formula for the transmission time; And it is proved that under certain conditions these formulas may get negative results, that is negative transmission time and negative signal speed. According to Einstein, this transmission mechanism causes "the effect to arrive before the cause" and therefore "no such signal can be transmitted faster than the speed of light in a vacuum." He added: "Although such an outcome is acceptable in terms of its transport alone, there is no contradiction in it; But it is so alien to the character of all our experience that "the impossibility of the hypothesis $u > c$ seems sufficiently well established."

Thus, the so-called "move backward in time", or "move backward in time", is the counterpart of negative transfer time in Einstein's paper. Einstein's conclusion (the impossibility of faster-than-light speeds) should result in the impossibility of both negative travel time and negative signal speed. In this case, one should say that "SR considers neither backward nor backward time to be possible", i.e., the construction of a "time machine" is impossible. Yet people like to say that "Einstein's theory of relativity makes time travel possible".

Much has been written about time travel, and it is generally believed that forward travel (to the future) is hopeful, but that going back in time is impossible. The question is whether negative Velocity (NPV and NGV) experiments have proved its existence. This form of faster-than-light speed exists. Can a time machine be made? It has been suggested in the literature that Wang's NGV experiment can be regarded as a time machine. In 2011, Professor Cao Zhang said that even if the speed of light is proved to be faster, it would be impossible to build a time machine and travel back in time. His argument is that under the generalized time definition of the Generalized Galilei Transform (GGT), simultaneity is absolute and faster-than-light motion does not cause time inversion. By this he meant a kind of true time; Faster-than-light does not destroy causality under this definition, but relative time as defined by SR is no longer valid. ... But what he didn't realize was that the success of the negative-wave velocity experiment seemed to give new impetus to the "time machine".

The quantum post-selection effect, proposed by the American physicist John Wheeler and also called the delayed choice experiment, was published in 1979. It means that the observer's choice can affect the photon's previous behavior, and that the upcoming event interacts with the completed event. In recent years, European experiments have proved that post-selection can affect photon characteristics at the nanosecond level, so it is believed that the post selection could change the entire history. On this basis, some physicists proposed that quantum time machine could be built by using quantum entangle. There are quite a few people who think so.

Wheeler's idea of "delayed selection" can be transformed into an experiment in which a light source is dampened so that it emits only one photon after the previous one hits S.S appears as a random pattern at first, but gradually appears as interference streaks as the number of photons increases. In this regard, if it is considered that the emitted photon interferes with the photon that has reached S, it means that the event that has not yet happened interacts with the completed event, which violates the law of causality. Therefore, it can be said that each photon interferes with itself, and this can only be done if the photon passes through the double slit at the same time. It is impossible for two photons to travel two paths at the same time in classical physics, which shows that the photon has a strange property.

In order to study time travel using quantum principles, it is necessary to understand the nature of the "quantum post selection" phenomenon. The approach taken by Lloyd in 2010 uses the following quantum effects — quantum particles (such as photons or electrons) are not bounded by the arrow of time. For example, the future of quantum particles can affect the past. In heeler's experiment, an unobserved photon passes through two slits at the same time. We know that a photon can pass through both slits at the same time because an interference pattern appears on the terminal screen; If a photon passes through a single slit, it won't.

Europeans have shown experimentally in recent years that post-selection does affect photon properties on a nanosecond scale, hence the saying that "the post selection process could even change the entire history of the universe"; But I think this is a bit of an exaggeration. The method suggested by Lloyd-Steinberg; everyday experience, they argue, tells us that the state of our beginnings determines the future. But quantum particles can't tell the difference between forward and backward time. This also means that determining the future state can determine what happened before it. ... C.Bennett and B.Schumacher had previously suggested that quantum entanglement could be used to build a time machine. Using two particles, such as photons, makes them so closely related that they share existence. Entangled particles are special because a measurement of one affects the other particle, no matter how far apart they are. Now imagine teleporting Athird particle from A to B; The trick is to create A pair of entangled particles and place one in A and the other in B, and then make a series of measurements in both places. If this can be done, it ensures that the second particle will end up in the same state as the space traveler.

Precisely, the traveler's body doesn't move, but the quantum information completely describes the traveler to complete the journey, allowing the second particle B to assume traveler status. What's curious is that this teleportation happens in an instant. In the process, quantum information goes from point A to point B. Therefore, it is natural to think of completing the journey only by taking measurements at point A. But since teleportation is instantaneous, this is merely a basis for considering the trip triggered by point B, even if it happens later. Figure 6 shows Alice, a girl in the "present", asking Bob, a boy in the "future", to send her a message via quantum teleportation. In the figure, Alice and Bob have entangled photon pairs (both called A); After: ① Alice sends a message (which is decoded as photon X by measuring the entangled photon); ② Bob gets a message when he measures his own photon; ③ the post selection action indicates that it is measured by Bob in the future, causing Alice's photon to have this property; ④ according to the post selection effect, it is equivalent to Bob sending a message in reverse time. ... This is post-selection at work, and it has the characteristic that, like quantum computing, quantum physics uses all the time to do things. It is this fuzzy relationship between cause and effect that Steinberg and Lloyd developed their time-travel simulator.

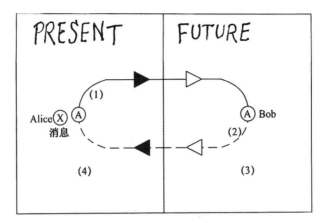

Fig. 6 Diagram of post quantum selection action

It is an exaggeration to call it time travel. Perhaps, in the same way that quantum teleportation conveys the state of the quantum rather than the material itself. However, Lloyd and Steinberg argue that "the logic of post-selective teleportation is consistent with time travel, so our experiment is a time travel simulator. However, it cannot take one back to the age of the dinosaurs, and it has a lot of special things to do." One of the first things the Lloyd and Steinberg team did was simulate the grandfather paradox by sending the photon back to kill itself. Using teleportation to do this is an important twist. Traditional quantum teleportation is guaranteed to give you a copy state of the intended transmission. Lloyd and Steinberg wondered if this would work for photons to kill themselves with quantum guns.

From a practical point of view, their simulation requires two additional features: a quantum gun capable of firing bullets; And a way to spontaneously stop the transmission process. The team also decided that instead of tracking two photons, as is common in quantum teleportation, they would track two properties of a single photon. Specifically, the way a photon is polarized represents its "present" state, while its direction of motion represents its "past" state. Based on this, they would pair the photon with a quantum gun that could either fire a bullet or spontaneously terminate the quantum transport process. This device, also known as a wave plate, changes the way the photon is polarized. This is because the photon's polarization and direction of motion are tracked, and the photon is also outfitted with a quantum gun to influence its "past" state.

Now, how do you make sure that the transmission can stop automatically if necessary? This is actually easier than the previous task, because quantum teleportation itself has a built-in termination mechanism. Unless it is measured in a special way, the quantum is actually working for only 25% of the time during the transmission. So, in this experiment, there are actually four possible outcomes, depending on the transport process and the state of the quantum gun.

And after this experiment, something interesting happens: At each individual point in time, if time travel is set in motion, the quantum gun can't be set in motion. And once time travel doesn't start, the quantum gun works. From the perspective of the grandfather paradox, if there's a reasonable chance that your gun won't fire and the alleged assassination attempt fails, time travel can start. "You can aim the gun, but you can't pull the trigger", Lloyd says.

Lloyd and Steinberg's experiment has piqued interest in time travel, especially since it doesn't rely on the space-time warping, closed time-like curves of general relativity, or on black holes, wormholes, and many worlds. New thinking builds on quantum-post selection and the idea that time travel is theoretically possible.

Finally we must say that "going back in time" is not possible anyway. But we have no objection to continuing the discussion.

"NEGATIVE CHARACTERISTIC ELECTROMAGNETIC WAVE MOTION" IS A UNIQUE PHENOMENON IN NATURE

When we use the classical-mechanics analyze the motion of substance, who has inherent mass and shape, the velocity is vector, then the negative velocity express backward direction. But when we discuss the motion of waves, who has not inherent mass and shape, the velocity is scalar, the negative velocity does not obey that rule, i.e. does not indicate only the direction of movement is flow backward. But what does that phenomenon mean? The negative wave velocity (NWV) means that, for example, a pulse propagates in special medium with a negative group velocity (NGV) of c/n_g ($n_g<0$), then it is not only faster than a pulse travelling in vacuum, but so quickly that if left the medium before it had even finished entering. In 2013, we establish the concept of "negative characteristic motion of electromagnetic waves", and we differentiate it from the meaning of "movement in backward direction". We must receive the advanced solution of D'Alembert equation, and understand the concept of negative wave velocity. The truthful and rich of Nature give a lesson for us, and she will still instruct us continually in the future.

The dualistic nature of matter is basic feature of the world; and the possibility of electromagnetic parameters in modern physics become negative value or negative characteristic motion of electromagnetic waves can exist everywhere. The positiveness and negativeness of physical parameters are one of symmetry in nature. Then, the study on this investigation area is the new way for research of objective laws. As part of the research of wave science, the author not only focuses on the theory and experiment of the study of negative electromagnetic parameters, but also discusses the negative motion of electromagnetic waves.

These rules have not been included in the symmetry and conserved quantities table in physics, but were related only in the concept of chiral symmetry. We believe that the positive or negative physical parameters and the positive or negative motion of electromagnetic wave are one of the symmetry mechanisms in nature. The term "negative characteristic motion" proposed by the author has a profound connotation, which is different from the simplistic "changing direction".

As mentioned above, J.Wheeler and R.Fleynman were the pioneers in the study of electromagnetic field advanced solutions and electromagnetic wave advanced waves, but I think their work is not good enough, and hopefully their view is wrong. Therefore, we try to make a more comprehensive generalization and a deeper elaboration under the heading of "negative characteristic motion of electromagnetic wave". Of course, we do not deny Feynman's contribution.

Feynman noted that electromagnetic radiation had been interpreted before him in terms of "an interaction between a source and an absorber". For example, H.Tetrode believed that if the sun in the solar system were alone and had no other absorber of radiation, it would not radiate. The presence of an absorber is essential to the mechanism of radiation. ... Feynman's analysis is based on the following assumptions: (1) In a charge-free space, an accelerated charge does not radiate energy; (2) The field action on a given particle is caused only by other particles; (3) These fields are represented by the sum of half delayed solutions and half advanced solutions. He called the subject he studied "advanced effects of the theory of action". In the paper's only illustration ("Example of advance effects in an incomplete absorption system"), the general picture is divided into eight smaller graphs with the following instructions:"Incident waves before source acceleration; These incident waves are absorbed; The incident wave will act on the source at the time of acceleration; The source is acted upon by colliding particles or other forces or incident waves; The source radiates waves; Some radiation waves are absorbed; The waves that continue outward leave forever; The outward wave looks like an incident wave that continues through space in addition to the change of sign."... Here, his so-called incident wave does not seem to be an alien wave, because space has only 1 source; It is like the wave that returns to the source, the one that reacts.

In 1965, Feynman said of his view of time (past and future): "We can remember the past, but not the future; What we can do to affect the future, but not what we can do to affect the past. ... The whole world seems to be moving in one direction. But in the laws of physics there is no difference between the past and the future. The laws of gravity, the laws of electromagnetism are reversible in time." It can be seen that Wheeler and Feynman boldly proposed a "time-symmetry theory of half retarded wave and half advanced wave" in 1945, that is to say, when trying to reveal the interaction mechanism of particles, in order to avoid the contradiction between the past and the future, inward moving waves (waves moving backwards in time) must be used to make the theory symmetrical.

Wheeler-Feynman's absorber theory is similar to the cut-off waveguide theory discussed in detail by the author. It can be shown that two reactive storage fields interacting with each other can produce some active power, while a single field cannot. This interaction suggests that nature has some strange properties, just as the presence of a single tree is different from that of placing it in a forest with many, trees! ... Although previous studies have not found the existence of negative Motion (or advanced wave) in simple mutual inductance coupled AC circuits, we have found that the phenomenon of negative phase constant ($\beta<0$) in cut-off waveguide theory is actually a kind of lead wave. In addition, recent studies show that the antenna near-field has evanescence-state like conditions, and the resulting superlumic-light group velocity or even negative group velocity (NGV), which is the "negative characteristic motion" of electromagnetic wave! The above association gives people a feeling of "dawning light". However, we do not agree with the idea of "canceling the retarded wave with the advanced wave".

Someone say inward waves were found on the basis of the rapid development of so-called left-handed material in recent years. However, the phenomenon of inward propagation of waves does not exist only under LHM conditions. For a radiation source, the near-field and midfield dynamics of vector electromagnetic fields are far more complicated than the simple understanding (outward propagation). In the near field region of the source, there may be a phenomenon that the main body of the waveform moves inward. N.Budko had demonstrated this phenomenon with experimental observations, which he believed was negative waveform velocity, and the relevant waveform was travel back in time. These conditions were independent of LIIM. This goes back to R.Feynman and J.Wheeler's 1945 paper, before Vesselago had even been written, let alone LHM. However, the idea of advanced wave had already been proposed, which was worth thinking about.

From 1958 to 2004 some people simply concluded that "backward waves' meant' the vector direction of the phase velocity is opposite to that of the group velocity", but we have pointed out that it is not appropriate to think of negative velocities uniformly as "motion in the opposite direction". Negative group velocities (NGVS) have a vivid physical manifestation that cannot be described by a simple "directional judgment". For example, when an impulse incident into the NGV medium is compared with an impulse coming out of the NGV medium, the outgoing pulse may be timed ahead of the incoming pulse (the outgoing pulse appears before the incoming pulse arrives). This "advanced phenomenon" leads to a new interpretation of causality. If the output pulse is regarded as "effect" and the arrival of the input pulse as "cause", the experiment proves that "effect precedes cause" can occur. ... One cannot understand new things and new phenomena by clinging to the old doctrine of "the law of cause and effect".

THE IMPORTANT CONTRIBUTIONS OF HOHN BELL AND ALAIN ASPECT

A.Einstein, who completed his work on the theory of relativity in 11 years from 1905 to 2016, is regarded as "the greatest genius of the 20th century". However, in my 2022 paper I have pointed out that SR is hardly Einstein's independent creation—H.Poincare's papers in 1900 and 1904, and H.Lorentz's papers in 1904, contain the ideas on which SR is based, but Einstein has not explained them in his papers. Einstein never wrote his papers without a table of references. Whether he had read the two great scientists before 1905 is known only to Einstein himself. We judge, of course, that he did. Although Einstein mentioned Lorentz several times after he became famous, Lorentz's ideas did not coincide with his—hence the recent term "Lorentzian Relativity." Lorentz is known as a modest gentleman who does not compete with Einstein for "priority". As for Poincare Einstein never mentioned it after he became famous, as if he didn't exist; In fact, the formula $E=mc^2$ was derived by Poincare in 1900, and even the Lorenz transform (LT) was named after Poincare. ... We do not mean that Einstein is a plagiarist here, for his own reasoning is often riddled with myths, and only a person who has not worked hard would blindly praise Einstein's greatness.

Quantum mechanics (QM) was established in the three years between 1926 and 1928. It is truly a collective product, with the involvement of E. Schrödinger, W.Heisenberg, P.Dirac, M.Born, N.Bohr, and others. The approach of QM is different from that of relativity — although it is also based on rigorous mathematical analysis, its physical concepts are deep and clear, and not at all intended to show off the esoteric use of mathematics. QM is the epitome of scientific beauty! ... As soon as QM came out, Einstein was immediately alarmed and tried repeatedly to refute it and make it untenable. However, events that followed were completely out of Einstein's control, and QM went on and on, attracting countless physicists and opening up numerous applications. Thus, the creation of QM was a real threat to relativity!

In 1935, Einsteins published his famous EPR paper, which focused on delivering a fatal blow to QM. It was a sign that the conflict between relativity and quantum mechanics had reached fever pitch. The paper was published in the journal *Physical Review* and signed by Einstein, Podolsky and Rosen, so it is called the EPR paper. On May 4, 1935, the *New York Times* published a headline on its front page saying: "Einstein attacks Quantum Theory." This is a sensation. The EPR thesis is against quantum theory (QM), SR and QM these two views of the world, time and space the decisive moment has arrived.

Some of the content in the EPR paper is merely foreshadows (e.g., saying that the physical theory must not only be correct but also "complete"; Or that the description of reality given by the wave function in quantum mechanics is "incomplete"). The fundamental thing lies in the analysis of the interaction of "two-system systems" (systems of two subsystems). Here, subsystems I and II should be understood as microscopic systems, such as particles. The states of the two subsystems before $t=0$ are known. Between $t=0$ and $t=T$, they interact with each other. When $t>T$ they no longer interact with each other (for example, move away—separate in different directions). Let $\Psi(x_1, x_2)$ be the quantum state of the system, which can be expanded in terms of the eigen function system for measuring A physical quantity (such as a mechanical quantity) of I, or in terms of the eigen function system for measuring a physical quantity of I. According to quantum mechanics, the wave packet $\Psi(x_1, x_2)$ will collapse during measurement, and will be condensed after measurement, resulting in a state that will affect II when measuring I. But I and II have been separated, so this strange effect of action at a distance is impossible. Because the special theory of relativity stipulates that the interaction of nature can only be realized at less than the speed of light, a system separated by space should be local, but quantum mechanics gives the non-local case, so quantum mechanics is not self-consistent and incomplete. These are the most important things in the EPR paper.

It follows that there is an invisible thread that connects special relativity and EPR; It can also be said that EPR thinking is based on special relativity. Secondly, we say that special relativity is in sharp contradiction with the world view of quantum mechanics, precisely in the question of "local realism or non-local realism". The EPR paper was Einstein, at the age of 56, giving quantum mechanics as much of a blow as he could possibly hope for with his wisdom.

Einstein believed that the EPR paper would refute Heisenberg's uncertainty principle and prove that quantum mechanics was imperfect. The discussion of "two systems" (I and II) in EPR seems to imply that it is possible to "know both position and velocity", since the velocity of I is the velocity of II.

After the article was published, N.Bohr began to refute it. Bohr meant that the setting in the EPR paper could be rejected — uncertainty affected both I and II and II was immediately affected when I was measured to make the result consistent with Newton's law; This effect occurs immediately, even if I and II are far apart. ... But younger scientists, such as Heisenberg, can't argue with Einstein the way Bohr can.

Russian academician V. Fok said: "It is particularly amazing that Einstein, who did a lot of work for quantum theory in its early days, has taken a negative attitude towards modern quantum mechanics. There is no direct force interaction between the two subsystems of EPR thinking, and one can also affect the other, which Einstein considered incomprehensible, and thus considered quantum mechanics incomplete."

According to Fok, the interaction (influence) of Pauli's principle in QM is an example of a non-force. The interaction (influence) between two particles with a common wave function (EPR system) is another form of non-force interaction (influence) in quantum mechanics. The existence of non-force interactions (influences) is beyond doubt, and it would be wrong to negate such interactions.

So how do you decide whether the EPR paper is right or wrong? This is a difficult thing to do in the early days. John Bell, an Irish physicist [who has since worked at CERN, the European Organisation for Nuclear Research], was not sure whether the EPR paper was right or wrong. He even wanted to come up with a new theory to make it more convincing. Since the EPR centers on the-analysis and judgment of two-particle systems and argues that quantum entanglement cannot be maintained, this is where it begins. In 1965, J. Bell proposed a hidden variable model compatible with quantum mechanics, holding that "no local variable theory can reproduce all the statistical predictions of quantum mechanics", and proposed the inequalities that should be satisfied between some correlation functions when two particles do spin projection along different directions of space-time:

$$|P(\mathbf{a}, \mathbf{b})-P(\mathbf{a}, \mathbf{c})| \leqslant 1+P(\mathbf{b}, \mathbf{c}) \qquad (33)$$

and he also said, was not only experiment proves this type, to prove QM correct and relativity (SR) is wrong. So, it is J.Bell make conflicts and arguments between SR and QM had to judge by experiment. Therefore, Bell's 1965 paper was a landmark.

Bell's theorem says that a hidden variable theory cannot reproduce all of QM's predictions. ... But exactly how this is the case must be determined by experiment. The breakthrough was made possible by precise experiments by French physicist Alain Aspect, he led experiments that demonstrated, with high precision, that the results greatly violated the Bell inequality and were in good agreement with the predictions of quantum mechanics. The fact that the Bell inequality was disproved by precise experiments meant that the EPR paper was wrong and quantum mechanics was right. In a manner of speaking, Bell opened the door to quantum informatics technology research!

Bell had been a staunch supporter of Einstein, believing in physical reality and locality. He believed that some hidden variable was responsible for the mysterious action at a distance in quantum mechanics. In fact, it is possible to construct a theoretical inequality to which observations of particles must follow, thus proving the incompleteness of quantum mechanics described in the EPR paper. Bell's analysis builds on Bohm's spin-dependent scheme and the theory of hidden variables. We now dispense with the mathematical analysis and simply point out that Bell's inequality is inconsistent with quantum mechanics!

Bell was still a proponent of Einstein's theory when he presented his theory in 1965. Twenty years later, in 1985, he was an opponent. He gave a clear answer to the BBC, arguing for the existence of a preferred frame of reference, i.e., the ether. He argued that superluminal speeds of light were possible, that relativity was an obstacle to the development of quantum theory, and that Einstein's worldview was untenable. In short, he argued for going back to that before special relativity, i.e. Poincare and Lorentz situation.

Bell made his comments in connection with special relativity and the EPR, because both theories are relevant to what view of nature and the universe we adopt. The EPR began as an argument against quantum mechanics and ended in failure. The turning point was the Aspect experiment of 1982; Quantum informatic technology developments over the next 40 years further doomed Einstein's theory.

As can be seen from the progress in the study of quantum entangled states, the world view of quantum mechanics has completely defeated the world view of special relativity. The entanglement distance between the two photons in the successful experiment has gradually developed from 15m at the beginning (Aspect) to 25km, and even 144km (10 years ago). The June 15, 2017 issue of *Science* magazine reported that a team of Chinese scientists have made a new achievement with a quantum satellite—quantum entanglement on the scale of the order 1,000 kilometers (1,303km from Delingha Station in Qinghai Province to Gaomeigu Station in Yunnan Province). The result has taken the world by surprise. All in all, a series of experiments perfectly demonstrated that special relativity had a problem with space-time.

Since 1965, the Bell Inequality has been widely verified and has become an important means for identifying entanglement that can be described by discrete measurements. For example, measuring the spin direction of one quantum particle and then determining whether this measurement is related to the spin of another particle. If a system violates this inequality, then entanglement exists! In short, the Bell inequality became an iconic test to see if it was obeyed. Both theory and experiment show that nonlocality is a fundamental characteristic of quantum mechanics — experimental results that violate Bell inequality indicate the existence of nonlocality. Bell's name entered the history of science when his inequality was hailed as "one of the greatest scientific discoveries in human history".

At the opposite end of the spectrum is the classic, macroscopic, and locality nature of relativity (mainly special relativity SR). The main contents of this locality realism are: believe in classical physical reality, believe in localized causal rate, and oppose probabilistic thinking; That the speed of light is the limit of the speed of moving bodies and the speed of information propagation in the universe; And does not admit the possibility of entanglement in physics.

In the first week of October 2022 came the exciting news that the 2022 Nobel Prize in Physics had been awarded to physicist Aspect and two others. They were honored, the awarding body said, "for their experiments on photon entanglement, their determination that the Bell inequality is not true, and their groundbreaking work in quantum information science." "All three scientists have performed quantum entanglement experiments," Reuters news agency reported. In quantum entanglement experiments, two particles stay connected no matter how far apart they are. This bothered Einstein, who called it spooky action at a distance.

Now, Aspect is an old man. The significance of this award is not only to support quantum mechanics (which is the same as in 1933), but also to encourage the research and development of quantum communication, quantum computer, quantum radar, which is a landmark significance. Of course, it is also a blow to those who have denied Aspect experiments for a long time, and they have always been unwilling to accept the failure of relativity. Bell might have been entitled to the prize, but he died in 1990, and the Nobel

Prize can only be awarded to living people. In fact, whether the prize itself is not important, he is key to establish a correct view of time and space.

John Bell put forward the theory of hidden variables in 1965 and gave inequalities, which was to support the theory of relativity. The results were contradicted by Aspect's precise experiments (and later supported by multiple experiments). In addition, considering the implications of the discovery of the cosmic microwave background radiation, he finally announced his departure from relativity in 1985, showing the integrity and courage of a scientist. J.Bell was one of the great scientists of the 20th century. Fig.7 shows him giving a lecture.

In short, quantum entanglement, a physical phenomenon known as action at a distance (infinite velocity), was finally recognized by the great developments in quantum information technology that led the Nobel Committee to recognize the groundbreaking significance and value of the 1982 Aspect experiment, so to award the 2022 Nobel Prize in Physics. Therefore, some people say that this award is a rejection of Einstein's "light speed limit theory", and I think this view is correct.

Fig. 7 John Bell, an Irish physicist with outstanding scientific ideas, is giving an academic lecture (His famous inequality is written on the blackboard).

However, quantum entanglement is not action at a distance. In other words, the entangled state does not travel at an infinite speed, but at a finite speed. And this speed is superluminal, which Salart proved experimentally in 2007, $v=(10^4 \sim 10^7)c$. That's a lot, of course, but it's by no means infinite. As for the negative velocity, it has not yet been observed in quantum entanglement. ... All in all, Aspect's winning the Nobel Prize after 40 years is a criticism or even a denial of Einstein.

CONCLUSION

In classical electromagnetic theory, Maxwell's equations are a set of several first-order hyperbolic partial differential equations. From this, the electromagnetic wave equations can be derived, but this is a second order elliptic partial differential equation of the system. To solve these two kinds of problems, the former can be discretized, so that the initial value problem becomes an effective algorithm. The latter is solved by the boundary value problem. Either way, two wave solutions are found — divergent waves moving toward infinity and convergent waves moving toward an electromagnetic source. Today, we call the former a retarded wave and the latter a advanced wave. Although as early as in the 1970s, the Chinese Scientific Community considered how to treat these two types of waves when designing the antenna, and decided to abandon the converging wave (advanced wave) solutions and only use the retarded wave solution, today the understanding and processing were wrong, although the original intention was to avoid the confusion of the data.

The matter of advanced wave is related to negative velocity, here the negative velocity mainly refers to the negative wave velocity (NWV). It is difficult to discuss the negative velocity of macroscopic matter, so this article will not focus on it. When the motion of a material object is analyzed by classical mechanics, the velocity is a vector, and the negative velocity generally indicates the opposite motion. But for the motion of waves without mass and invisible, the velocity is scalar. It cannot be said that the negative velocity only represents the reverse direction of the flow. For a negative wave velocity, such as a pulse passing through a particular medium with a negative group velocity (NGV), it is not only faster than the pulse passing through a vacuum, but it is faster than leaving the medium before it enters it. Therefore, we put forward the concept of "negolive characteristic electromagnetic wave motion", which is rich in connotation. It is necessary to accept the advanced solution of d'Alembert equation in order to understand the negative motion. Nature, so to speak, teaches us a lesson in her truth and richness.

When the pulse entering the NGV medium is compared with the pulse coming out of the NGV medium, the outgoing pulse is preceded in time by the incoming pulse (the outgoing pulse appears before the incoming pulse arrives). This "advanced phenomenon" leads to a new interpretation of causality. Its emphasis is no longer "cause precedes effect", but rather "effect cannot affect cause or react on cause". This is a new understanding within the framework of classical physics. In the realm of quantum physics, causality is inherently disproven. It is pointed out in this paper that there are two kinds of basic electromagnetic environments in the near field — bound field and evanescent field; The former includes static field (which decays according to r^{-3} regularly) and inductive field (which decays according to r^{-2} regularly); The latter contains the evanescent plane spectrum, with the field decreasing exponentially as the distance from the source increases. The bound field is referred to as the evanescent field in this paper. In recent years, the phenomenon of electromagnetic wave traveling faster than the speed of light in free space has been found in both of them, and the negative wave velocity has been observed experimentally. The results of recent experiments do not support the commonly held view that propagation at the speed of light ($v=c$) is retarded for bound fields. Based on the observation of no hysteresis in the near area of the antenna, the non-locality experimental evidence of the bound electromagnetic field is provided. The non-locality property is a quantum mechanical concept, so the non-locality property of bound field can be closely related to classical electromagnetism and quantum mechanics.

This paper emphasizes the new phenomena found in the near-field measurement and gives several theoretical duals. As for the quantum explanation of the near-field faster-than-light phenomenon, it holds that the idea of "evanescent state is virtual photon" should be applied theoretically.

Finally, it must be noted that the concept of "negative characteristic motion of electromagnetic wave" proposed by the author has three main contents in a broad sense: (1) negative refraction index phenomenon (NRI); (2) negative wave velocity phenomenon (NWV); (3) negative Goos- Hänchen shift (NGHS). Limited in length, this paper mainly discusses (2); For (1) and (3), please refer to literature [61]. In addition, the author's mew book of 2022 can also be used as a reference when reading this article.

COMPETING INTERESTS

Author has declared that no competing interests exist.

REFERENCES

[1] Born M. Wolf E. Principle of Optics[M]. Cambridge: Cambridge University Press, 1999.
[2] Brillouin L. Wave Propagation and Group Velocity[M]. Pittsburgh: Academic Press, 1960.
[3] Huang Z X. Introduction to the Theory of Waveguide Below Cutoff[M]. Beijing: China Metrology Press, 1991.
[4] 黄志洵, 姜荣. 量子隧穿时间与脉冲传播的负时延 [J]. 前沿科学, 2014, 8(01): 63–79.

[5] 黄志洵. 电磁源近场测量理论与技术研究进展 [J]. 中国传媒大学学报, 2015, 22(5): 1-18.

[6] 黄志洵. 爱因斯坦的狭义相对论是正确的吗? [J]. 中国传媒大学学报, 2021, 28(5): 71-82.

[7] Ma Q P. Inquiry into the Self-constancy of Relativity Logic[M]. Shanghai: Shanghai Scientific Literature Press, 2004.

[8] Huang Z X. Wave Science and Superluminal Light Physics[M]. Beijing: National Defense Industry Press, 2014.

[9] Huang Z X. Light of Physics—Open Physical Thought[M]. Beijing: Beijing University of Aeronautics and Astronautics Press, 2022.

Research and Discussion of Quantum Theory[*]

INTRODUCTION

Quantum mechanics (QM) was established between 1926 and 1928. In 1935, A.Einstein, B. Pridolsky, and N. Rusen published an article entitled "Is Quantum Mechanics a Complete Description of Physical Reality?" The principle of locality echoes Einstein's theory of special relativity (SR), but is inconsistent with quantum mechanics (QM). In 1951, D. Bohm made a modern statement of EPR thinking, which actually started the study of quantum entangled states. On this basis, in 1965, J.Bell proposed the hidden variable theory, which was later called Bell inequality, and in 1981-1982, A. Aspect did a number of accurate experiments, and the results were inconsistent with Bell inequality, but consistent with QM. Therefore, there is a singular correlation of QM expectations in the two-particle system, but the hyperspace (overdistance) action is contradictory to EPR thinking. In the following decades, the Bell experiment flourished, and the interval of entangled photons gradually increased from 15m at Aspect time to 144km, and in 2017, the Chinese quantum satellite expanded to 1200km, which is very surprising. The errors of EPR papers provide profound lessons for scientific research. The development of quantum communication technology in recent years is based on quantum non-locality and quantum entanglement.

Quantum informatics (QIT) has three main research directions — quantum computing, quantum communication, and quantum radar. The key point of quantum communication is that there must be absolute confidentiality. But this is very difficult in practice, so we cannot say that the problem has been solved so far. The research and development of quantum computers has made great progress in the United States, Japan and China, which are already in a fierce competition with each other. As for quantum radar, the technology to design it entirely around photon entanglement does not yet exist.

In this context, quantum theory has attracted a lot of attention from the scientific community in recent years, and many people who are not physics majors want to understand the meaning of some proper terms — such as wave functions, entangled states, Bell inequalities, hidden parameters, and so on. And it has revived interest in questions about the history of science. It is no accident that the theory of quantum science and

---

[*] The paper was originally published in *Global Journal of Science Frontier Research: A Physics and Space Science*, 2023, 23 (6), and has since been revised with new information.

related application technologies are developing, and China has not only launched quantum satellites, but also invested huge resources in the research of quantum communication technology on the ground. As for quantum computers, together with artificial intelligence, they have become two hot spots in the world, and their development is related to the future of all mankind. If electricity, nuclear energy, computers, and the Internet are the landmark milestones that humanity has achieved in the past, then we must now pay attention to the development of quantum information technology and artificial intelligence, because they are bound to dramatically change the face of society and human life.

The basic theory of quantum mechanics was formed in the early 20th century (1926-1928), and its theoretical system has not changed much. But there has always been a lot of theoretical debate, which has been stimulated by developments in recent years. In particular, in 2022, the Nobel Prize in Physics was awarded to three physicists who studied quantum oddities, a development that further boosted interest in studying quantum theory. This paper summarizes the author's views and opinions.

THERE IS A FUNDAMENTAL CONTRADICTION BETWEEN RELATIVITY AND QUANTUM MECHANICS

Relativity (SR, GR) and quantum mechanics (QM) are two of the most important scientific theories of the 20th century. Yet relations between the two have been strained. In 1998, UNESCO published the World Report on the Development of Science, with an introductory section entitled "What is the Future of Science?", in which it was stated: "The theory of relativity and quantum mechanics are two of the great academic achievements of the 20th century, but unfortunately the two theories have so far proved to be contradictory. This is a serious problem." It is rare for a disagreement between two scientific ideas to find its way into a UN document.

As we all know, E.Schrödinger created the wave mechanics of QM in the first half of 1926, the core of which is the basic motion equation of QM—Schrödinger equation (SE), which describes the motion change law of the microscopic particle system. According to M. Planck, this equation laid the foundation for quantum mechanics, just as the equations created by Newton, Lagrange, and Hamilton did for classical mechanics. It must be pointed out that SE is derived from Newton mechanics; This fact makes some relativists uncomfortable and therefore insist that SE "only applies to low speed cases (particle velocity $v \ll c$)".But they were wrong — the development of optical fiber technology is theoretically supported by SE, and the photons in the optical fiber travel at the speed of light (c), which is not a slow condition at all. Relativists are afraid that SR and GR will be negated one day, so they insist on "splitting the world equally": macroscopic and high-speed phenomena are governed by relativistic tubes, and microscopic and low-speed phenomena are governed by quantum theory. But what about the fact that quantum theory is also valid in the macro sense? !

Some physicists say that the fusion of SR and QM has long been solved in quantum field theory (QFT), with Dirac's successful derivation and application of the quantum wave equation (DE) in 1928. Our view is that these statements are not only false, but have been misleading for years. Although DE's derivation is not directly based on Newton mechanics like SE, it does not really use SR's space-time view and world view. The derivation of DE is derived from two equations about mass, the mass-energy relation and the mass-velocity relation, but both of them can be derived from classical physics before the advent of relativity. The mass-energy relation was put forward by H. Poincare in 1900, and the mass-velocity relation was put forward by H.Lorentz in 1904. Therefore, DE is not actually derived from relativity. Since DE is not necessarily related to SR, it is unacceptable to say that it "represents the combination of SR and QM".

In this case, what reason is there to say that "the Dirac equation represents the establishment of relativistic quantum mechanics"? In fact, in-depth analysis has shown that SR and QM are opposing theoretical systems, and Einstein himself was indeed "lifelong" in his opposition to quantum mechanics. As a result, Weinberg's claim that "the only theory that can make quantum mechanics compatible with relativity is quantum field theory (QFT)" is empty.

Dirac's Nobel lecture, at the age of 31, exuded relief and triumph that he had solved what Schrödinger had failed to do and Klein and Gordon had failed to do, namely, "to derive the wave equations of microscopic particles under the guidance of relativity". But later, although in 1964 (Dirac was 62 years old) he still had the sense that SR was dominant and QM was subordinate, he clearly stated that "there are insurmountable difficulties in establishing relativistic quantum mechanics". In 1978 (at the age of 76) Dirac showed a strong sense of confusion and dissatisfaction: he was fundamentally disenchanted with "the coherence and harmony of relativity and quantum mechanics"; No longer think quantum electrodynamics (QED) is a good theory; He called for a "really big change" in physics.

In short, in his later years Dirac lost his fascination with relativity and gradually distanced himself. This is highlighted by the disparagement of QFT and QED. He said the success of QFT, which includes quantum electrodynamics, has been "extremely limited" and simply does not suffice to describe nature.

Quantum field theory (QFT) was proposed and shaped in the decades after 1927, when the physics community generally accepted relativity as a guiding theory. It was believed that both QM and QFT were subject to the requirements of relativity until 1982, when the famous physicist J. B. ell (among others) publicly criticized Einstein's views in 1985 and strongly supported QM. He also suggested that physics thinking should "go back to before Einstein." By this time, however, elementary particle physics had taken shape, and there was no further study of fundamental questions such as whether the interactions of microscopic particles really had Lorentz transformation (LT) invariance. However, serious analysis and calculation can prove that LT transformation invariance may not exist in the process of particle physics, and there is a fundamental problem with QFT. The principle of relativity in SR does not hold.

A great debate about QM broke out in the 1920s and 1930s, and it was Einstein who started it. Einstein anticipated the crisis of relativity early on from the rise of QM, and began to deal with it. It is well known that W. Heisenberg won the 1932 Nobel Prize in Physics for his work on matrix mechanics and the uncertainty principle, which were important for the establishment of QM. Einstein, however, was against QM; This began to emerge in 1926 and culminated in 1935, when he published the EPR paper with B. Podolsky, N. Rosen. The localization principle in this paper corresponds to SR; For a separate system (I and II), there can be no out-of-range effect between them. N.Bohr refuted the EPR paper by pointing out that the effect of the uncertainty principle on I and II—II will react when I is measured, regardless of the distance between them. Of course, this discussion is all about microscopic particles.

The author has sorted out the situation of the great debate on quantum mechanics, in fact, only to give a partial contradiction and disagreement (in fact, more than these). Now let's list the two schools of thought on science; Q stands for quantum mechanics (Copenhagen School) view, R stands for relativity view.

a) Wave function
Q. It is believed that the wave function reflects the probability distribution and evolution of microscopic particles in space and time, and actually accurately describes the state of a single system (such as particles).

R. objects to the notion that "wave functions accurately describe the state of individual systems," and to arbitrary, statistical explanations (" God does not play dice ").

b) Uncertainty relation (Uncertainty principle)
Q. It is believed that the operation of microscopic particles has uncertainty that cannot be eliminated, and the law of uncertainty relationship is not only important but also causes unpredictability contrary to causality.

R. rejects the uncertainty relation, arguing that quantum emission and absorption of light can one day be theorized on the basis of complete causality.

c) Quantum mechanical completeness
Q. That quantum mechanics is complete and correct; QM is a statistical theory, so it can only determine the probability of possible outcomes. There are no hidden variables. It is considered useless to use hidden variables, because these so-called hidden variables do not

appear when describing the real process. In fact, no local hidden variable theory can derive all the statistical predictions of QM.

R. believes that quantum mechanics is incomplete and that there may be deeper physical laws—for example, there may be undiscovered hidden variables that can determine the laws of individual systems. If hidden variables are found, causality still exists. In short, there must be deterministic descriptions of nature, and efforts should continue to pursue better (but now unknown) theories.

d) Wave-particle duality and complementary principle

Q. It is believed that all microscopic particles (whether they have mass or not) have wave-particle duality, sometimes manifested as particles (with a definite orbit), sometimes manifested as waves (can produce interference fringes); It depends on the experimental method of the observer. But it is impossible to observe both at the same time, in fact, the fundamental point is a mutually exclusive and complementary quantum relationship, and any experiment will lead to uncertainty about its conjugate variables; Therefore, the complementarity principle is consistent with the uncertainty relation.

R. As the originator of the photon theory, Einstein has long recognized that it is a contradictory phenomenon that light is both a wave and a particle. However, he did not agree with the uncertainty principle, and thus could not accept Bohr's complementarity theory, which saw uncertainty relations as an illustration and result of the complementarity principle.

e) Quantum entangled states

Q. Bohr immediately refuted the EPR paper after it came out; The author holds that QM has the same mathematical expression form at the beginning and the end, and accuses QM of being incomplete and unconvincing. The so-called "reality criterion" is not strict. It is suggested that the existence of the interaction of separate systems (I and II) is possible.

R. 1935 EPR paper, the first part of which argues that the QM hypothesis wave function determination contains a complete description of the physical reality of the system. The second part is intended to show that this assumption, together with the criterion of reality, will lead to a contradiction. In general, it denies the completeness of QM, and denies that the system will interact when divided into two parts.

From the above, it can be seen that the local description in relativity is incompatible with particle volatility in QM, and it is also incompatible with allowing particle transformation in QM. In particle physics, the non-relativistic QM is a logically self-consistent single-particle theory, but the premises of relativistic QM are logically inconsistent and difficult to use as an equation of single-particle motion like SE. So what does relativistic local reality mean? It contains two aspects: physical realism and relativistic local causality. But quantum theory is essentially a non-local theory of space.

FROM WAVE FUNCTIONS, QUANTUM STATISTICS TO THE UNCERTAINTY PRINCIPLE

Max Born (1882-1970) was a German who taught at universities in Germany and the United Kingdom. In 1954, he was awarded the Nobel Prize in Physics for his work in quantum mechanics, particularly the statistical explanation of the wave function. Born's theory, which appeared in June and October 1926, states that the states of microscopic particles are mainly described by the wave function $\Psi(r, t)$, and the probability of finding a particle in the volume element $d\tau$ at space r at time t is given as $\Psi(r, t)|^2 d\tau$ given as the probability density of the particle given as (r, t) given as the probability of the particle occurring $\Psi(r, t)|^2$. Therefore, waves describing microscopic particles are probability waves. In short, when calculating the scattering process, Born realized that the probability of finding a microscopic particle is proportional to the square of the modulus of the wave function, so the wave of a microscopic particle is described as a probability wave. Born's statistical interpretation of the wave function can be applied both to the single behavior of a large number of particles and to the repeated behavior of a single particle many times. Born's theory has been supported by numerous experiments, and also well embodies the wave-particle duality of microscopic particles.

In the first half of 1926, E. Schrödinger proposed non-relativistic quantum wave mechanics. In 1953, Born recalled: "When Schrödinger wave mechanics appeared, I immediately felt that it required a non-deterministic explanation. I guess it Ψ^2 was the probability density, but it took a while to figure out the physical basis. Obviously, a return to determinism is no longer possible." "It is impossible to determine the position of the particle according to the Schrödinger equation, because it is a group of waves with blurred boundaries."

Born realized that the new QM did not allow for deterministic interpretation. Uncertainty relations also emphasize this point. This does not mean that there is no causal relationship in some aspect of nature, but that it cannot be calculated quantitatively. I note, incidentally, that P. Dirac made a similar argument-causality only applies to undisturbed systems (such systems are usually expressed in differential equations);However, under microscopic conditions, it is impossible to disturb the object carelessly while observing (measuring), and the expected causal link cannot be expected.

In "Letter 71" of his 2005 book *The Born-Einstein Letter*, Einstein said, "I still do not believe that the statistical approach to quantum theory is the final answer, but I am the only one who holds this view." Born commented: "At the end of the letter Einstein again rejects the quantum theory of statistics, but admits that on this point he is isolated. I was pretty sure I was right about that. All theoretical physicists at that time were in fact working in terms of statistical concepts, especially for N.Bohr and his school, which made an important contribution to the clarification of concepts."

In Letter 88 (April 5, 1948), Einstein wrote:

"I am sending you a short article, which I have sent to Switzerland for publication in accordance with Pauli's suggestion. I implore you to overcome your long-held aversion in this regard and read this short article as if you were a guest who had just arrived here from Mars and had not yet formed any opinions of your own. I ask you this not because I am under any illusion that I can influence your opinion, but because I think this essay will help you to understand my main motivation better than any other article of mine you know. ... In any case, I shall listen to your counter-argument with great interest."

Einstein's essay, titled "Quantum Mechanics and Reality," does not contain any mathematical analysis, but rather, in a speculative manner, implicitly criticizes uncertainty relations and proposes that physical ideas are established by such things as objects and fields, and that they are real beings independent of perceptual subjects. Objects separated from each other in space maintain their independence; For example, for two objects (A and B), the outside world acting on A has no direct influence on B, which is the principle of contiguity. However, the interpretation of QM is incompatible with this principle. For a physical system S (S consists of two local subsystems S_1 and S_2), they may have been interacting earlier. At the end of the action, when describing the system in terms of wave functions Ψ, it can be seen in the analysis that it is impossible to maintain both the QM principle and the independent existence of two separate parts in space. Einstein has stated that he insists on the independent existence of different parts of physical reality in space, and that QM is an incomplete description of physical reality. That is to say, the quantum mechanical approach is fundamentally unsatisfactory.

Einstein's essay is similar to the EPR paper in that it's not very new. Only in 1935 it was with N.Bohr, and now (1948) it is with M. Born. Born's reply of May 9 was lengthy, stating: "It seems to me that your axiom of 'the mutual independence of spatially separated objects A and B' is not as convincing as you understand it." It does not take into account the fact of coherence. Spatially distant objects are not necessarily independent of each other if they have a common origin.

Born added: "At the root of Einstein's and my differences of opinion is the axiom that events at different locations A and B are independent of each other, in the sense that an observation of A state of affairs at B tells us nothing about A state of affairs at A. My argument against this assumption is taken from optics and is based on the concept of coherence. When A beam of light is split by reflection, birefringence, etc., the two beams take different paths, and one can deduce the state of a beam of light at distant point B by observing it at point A. It is strange that Einstein did not recognize this objection to his axioms as valid, even though he had been one of the first theorists to recognize the significance of de Broglie's work on wave mechanics."

Born's scientific work is closely related to Heisenberg's. Born was 19 years older than Heisenberg, who had been his research assistant. The quantum conditions of the old quantum theory were laid down by N.Bohr and A. Sommerfeld, which defined momentum p and position q for the motion of particles. In ordinary mathematics multiplication is subject to exchange rate $-p \cdot q = q \cdot p$; However, now (July 1925) a breakthrough new formulation of quantum conditions was proposed, in which quantum multiplication does not obey the exchange rate $-p \cdot q \neq q \cdot p$, which is called non-commutativity. Heisenberg proposed the bizarre quantum multiplication rule, which comes from the product of the amplitudes of two quantum transitions. Born realized that this could be the key to creating new mechanics (QM), and that this was nothing more than the case of two matrices multiplied together. Born helped create the fundamental relations of QM matrix mechanics, and it is definitely quantized; The following formula is actually the same as (2):

$$[p] \cdot [q] - [q] \cdot [p] = \frac{h}{j2\pi} [I] \qquad (1)$$

Here $[\cdot]$ denotes the matrix, but the $[I]$ identity matrix; In the Planck constant of zero ($h=0$), i.e., non-quantized conditions, $p \cdot q = q \cdot p$, return to the familiar situation. For this contribution, Born was inscribed(1) on his tombstone when he died in 1970.

Werner Heisenberg (1901-1976) was a German physicist who taught at the University of Gottingen in 1923 at the invitation of M.Born, and later went to Denmark to study at the University of Copenhagen. It should be said that he learned a lot from the guidance of Bohr and Born. In 1927, Heisenberg proposed matrix mechanics to explain the spectrum of hydrogen atoms, and discovered and explained the strange double-line phenomenon. In March 1927 he sent out a paper entitled "Kinematic and Mechanical Contents of Quantum Theory", which contained one of the most attractive principles, the indeterminacy principle, also known as the uncertainty relation. Published in *Zeitschrift fur Physik*, Vol.43, 1927, 172-198, the paper shook up causality and remains a matter of debate today.

Heisenberg's uncertainty principle is a fine theory, and let's see what he says. In his 1933 Nobel Prize citation, Heisenberg stated that in the study of atomic phenomena, the unverifiable part of the measurement of disturbances to the system prevented the precise determination of classical properties, but permitted the application of QM. The analysis shows that there is a relationship between the accuracy of determining the position of a particle and the accuracy of simultaneously determining its momentum:

$$\Delta p \cdot \Delta q \geq \frac{h}{4\pi} \qquad (2)$$

where Δp, Δq is the error when the two are measured, and h is the Planck constant. In this case, p, q are the regular conjugate variable. Since the uncertainty relation specifies the range of these accuracies, there is no visual picture of an atom that is completely unambiguous. Heisenberg stresses that the pattern of QM is statistical. The uncertainty relation provides an example of how accurate knowledge of one variable in QM excludes accurate knowledge of another variable. He therefore highly values Bohr's principle of complementarity — the complementary relationship between different aspects of the same physical process that characterizes QM as a whole.

For microscopic particles, any experiment to measure momentum or coordinates inevitably leads to uncertainty about their conjugate variable information. Therefore, it is impossible to know the coordinates and momentum of the particle at the same time. The uncertainty relation shows that the smaller the uncertainty of the coordinates, the greater the uncertainty of the momentum, and vice versa. Therefore, it is impossible to accurately measure the coordinates and velocities of particles at the same time. In other words, a particle with a definite velocity does not have an exact position in space. From this, it can be further proved that the probability of finding a free particle at any place in space is the same, so the position coordinates of the free particle are completely uncertain... Moreover, the inverse relationship between the inaccuracies of this measurement holds true for other conjugated variables such as energy and time, Heisenberg said, because nature has such a precision limit that causality is no longer true. The Nobel Committee praised Heisenberg's work at the time; They pointed out that the new theory (QM) has greatly changed people's understanding of the microcosmic world composed of atoms and molecules; In particular, here QM must abandon the requirement of causation and accept that the laws of physics express the probability of an event.

For the EPR paper, Heisenberg argues that quantum mechanics itself is complete, that it describes the most fundamental laws of nature, that reality and local nature are non-existent physical properties, and that studying them is as worthless as studying the age-old question of how many angels can stand on the tip of a needle... Heisenberg, however, avoids positive criticism of Einstein.

HIDDEN VARIABLE THEORY AND BELL INEQUALITY IN QUANTUM MECHANICS

The term "hidden variables" was first proposed by de Broglie in 1928 to describe situations in QM that are difficult to explicitly describe analytically. After the publication of the EPR paper in 1935, the physics community was full of opinions, and did not know whether to support the article's criticism and severe attack on QM. The physicist D.Bohm came forward and did two things: First, he proposed the thought experiment model of EPR thinking as a singlet particle, which was done in 1952, and he did not know that he could actually do the experiment successfully. The other is to use hidden variable theory to explain

QM causally under the encouragement of Einstein. Although Bohm does not explicitly say that he opposes QM, his bias is on Einstein's side. In addition, Bohm introduced the concept of quantum potential and participated in discussions in the physics community.

A. Einstein pursued a definitive theory of complete representation of physical realities. He still gave a classical statistical interpretation of the quantum mechanical concept of probability. In this case, it seems to imply unknown variables, that is, hidden variables exist; The current probability is the result of some average of these hidden variables.

The EPR paper of 1935 is a challenge to the Copenhagen school, and its core contents include physical reality, completeness, and localization. Locality refers to the fact that if the two systems are no longer interacting at the time of measurement, no intervention in one of them will affect the other system, so the separable system has the paradox of distance correlation. From this, EPR determined that quantum mechanics is incomplete. In 1951, D. Botham changed the momentum-position correlation in the EPR experiment to the correlation between two spin 1/2 particles. In 1952, D.Bohm proposed hidden variables in quantum mechanics.

The term quantum potential also first comes from de Brogle (1927), but Bohm gave the analytical expression; He takes the wave function

$$\Psi = R\, e^{j2\pi S/h} \tag{3}$$

Then write the Schrödinger equation (SE):

$$\nabla^2 \Psi += \frac{8\pi^2 m}{h^2}\left(\frac{jh}{2\pi}\frac{1}{\Psi}\frac{\partial \Psi}{\partial t} - U\right)\Psi = 0 \tag{4}$$

where h is the Planck constant, and $\Psi(x,y,z,t) = \psi(x,y,z)f(t)$. By substituting the wave function defined by Bohm into SE, we get two equations, one of which is

$$m\frac{dV}{dt} = -\frac{\partial}{\partial t}(U+Q)$$

where $V = m^{-1}(\partial S/\partial t)$, and Q is the quantum potential function:

$$Q = -\frac{h^2}{8\pi^2 m}\frac{\nabla^2 R}{R} \tag{5}$$

Bohm's quantum potential is supported by some physicists.

Although quantum mechanics continued to be confirmed by experimental observations, the "EPR paradox" continued to plague quantum mechanics until the next generation of physicists came along and put an end to one of the most enduring and famous debates in the history of science. The man who solved this problem was the Irish physicist John Bell, whose "Bell inequality" has been described as "one of the greatest scientific discoveries in human history."

Bell has been a staunch supporter of Einstein's belief in the reality and locality of physics. Bell was unimpressed by N.Bohr's statement that "any fundamental quantum phenomenon is only a phenomenon after it has been recorded", saying, "Have cosmic functions been waiting for eons of time for a monomeric organism to appear before collapsing?" "Or will it have to wait a little longer until a qualified observer with a doctorate becomes available?" He believes that the mysterious action at a distance in quantum mechanics is determined by "hidden variables" that are not yet understood.

Bell argues that for at least one QM state (singlet), the statistical prediction of QM is incompatible with the divisibility hypothesis; In other words, no local hidden variable theory can reproduce all the statistical predictions of quantum mechanics. This is called Bell's theorem. It will be recalled that in a previous article Bell suggested that EPR thinking could be refuted only by finding impossible proof of local conditions or divideability of distant systems. In the latter article, it is actually possible to deal with a two-particle system such as two reverse photons emitted from a common source, and the possible correlations between the results of the simultaneous measurement of the two particles. For example, when the polarization of two photons is measured separately, the Bell theory states that there is a limit to the correlation.

In summary, Bell proposed an inequality that observations of particles must avoid, thereby proving the incompleteness of quantum mechanics. Most importantly, this inequality is not a thought experiment; it can be proven experimentally. Let us now elaborate.

In his 1964 paper "On the EPR Paradox", J.Bell stated that "no single theory of hidden variables can reproduce all the predictions of quantum mechanics," and Bell was enamelled with the idea of introducing hidden variables to make up for what was then thought to be a "major deficit" in QM.

Let two identical particles with a spin of $\hbar/2$ form a singlet with a total spin of zero and a wave function is

$$\psi = \frac{1}{\sqrt{2}}[\alpha(\mathrm{I})\beta(\mathrm{II}) - \beta(\mathrm{I})\alpha(\mathrm{II})] \tag{6}$$

where α(i) and β(i) are the eigenfunctions of the spin S of the i th particle taking the value $\hbar/2$ and $(-\hbar/2)$ in a certain direction. Since $S=\frac{\hbar}{2}\sigma$, they also represent the eigenfunctions for which the projection operator $\sigma_n(i)$ of σ in the direction n takes values of 1 and (-1). When the two particles move away from each other, each maintains its own spin, so that the product of the two is (-1) forever. Therefore, if the spin measured for particle i is 1 in the direction n, the particle must have a value of (-1) in the same direction. If (-1) is measured for particle I, the value of particle II must be 1;This means that the value of particle II depends on the measurement of I.But they have been separated to no interaction, EPR paper believes that should be unrelated to each other, the measurement of I should not affect the state of II; This is a contradiction!

In order to study the correlation of pairs of singlet particles, the average value of the product of spin projections of particle I in the a direction and particle II in the b direction can be calculated

$$P(a, b) = \langle \psi | (I)\sigma_a \cdot \sigma_b(II) | \psi \rangle = -a \cdot b = -\cos\theta \qquad (7)$$

σ is the projection operator in the *a* and *b* directions, and θ is the angle between the unit vectors *a* and b. This is an indication that the two particles are related. If *a=b*, P(*a, a*)=−1, this is the case discussed above. If *a=b*, θ=0, P(*a, b*)=−1.

In order to solve the problem, Bell introduced a set of hidden variables based on local realism, i.e., $|\lambda|$, as a description of the state. The measurement result can be determined by a single value. It is assumed that there is a probability distribution $p(\lambda)$ for different hidden variable states. At this point, Bell's tools for inference are in place.

Bell's purpose is to prove with the local hidden variable theory that the local requirement is inconsistent with the statistical prediction of QM. He starts with the following three premises:

① In a system consisting of two spin binaries, the measurement of the spin components $\sigma_1 \cdot a$ and $\sigma_2 \cdot b$ of each particle in a pair of correlated particles has only two possible values:

$$A(a, \lambda) = \pm 1, B(b, \lambda) = \pm 1$$

where *a* and *b* are unit vectors and λ is hidden variables; Latter satisfaction

$$\int p(\lambda) d\lambda = 1$$

② The ideal correlation conditions of the total spin singlet exist in any direction is:

$$A(a, \lambda) = -B(a, \lambda)$$

③ The locality hypothesis is that when two particles are separated without interaction, the measurement result A(*a*, λ) of particle I does not depend on the measurement orientation *b* of particle II; Similarly, the measurement B(*b*, λ) for particle II does not depend on *a*. It must be noted that the derivation of Bell inequality is based on Bohm's spin dependent scheme (spin two-valued particle system). The premise ① assumes that there are only two possible values for the spin components of the related particles. The premise ② is that, under ideal correlation conditions, A(*a*, λ)=−B(*a*, λ) in any direction. Premise 3 assumes the independent properties of the two particles when measured after they are separated. So, the three premises are actually three assumptions—spin two state system, perfect correlation, and locality condition. The following correlation functions are also defined:

$$P(\boldsymbol{a}, \boldsymbol{b}) = \int \rho(\lambda) A(\boldsymbol{a}, \lambda) B(\boldsymbol{b}, \lambda) d\lambda$$

where $\rho(\lambda)$ is the probability distribution function for λ; From the above, Bell derives the following inequality:

$$|P(\boldsymbol{a}, \boldsymbol{b}) - P(\boldsymbol{a}, \boldsymbol{c})| \leq |1 + P(\boldsymbol{b}, \boldsymbol{c})| \qquad (8)$$

$\boldsymbol{c}$ is the unit vector.

This is one of the great inventions in the history of science—John Bell's inequality can be used to test whether the QM or EPR paper is correct. In other words, in a contest between quantum physicists and Einstein, who would win? Although there was still a long way to go, J.Bell had blazed a trail and made his name enter the history of physics. Figure 1 shows the Irish-born physicist giving a talk at an academic conference, with his famous inequality written on the blackboard.

Now let's do a simple test using equation (7). There are 3 unit vectors $\boldsymbol{a}$, $\boldsymbol{b}$ and $\boldsymbol{c}$ coplanar, and the angle between $\boldsymbol{a}$ and $\boldsymbol{b}$ is 60°, and the angle between $\boldsymbol{b}$ and $\boldsymbol{c}$ is 60°, according to formula (7), there is

Fig. 1 Dr. J. Bell giving an academic presentation
(Handwritten inequalities on the board)

$$P(a, b) = P(a, c) = -\frac{1}{2}$$

$$P(a, c) = \frac{1}{2}$$

If you plug in the Bell inequality, you get

$$1 \leq \frac{1}{2}$$

This is clearly not true; It can be seen that the inequality is inconsistent with QM.

Inspired by Bell's work, other physicists have derived different inequalities. However, later experimental progress has proved that the Bell inequality is the best result, and no further derivation is needed.

BELL INEQULITY TRANSITION FROM THEORY TO EXPERIMENT

Obviously, inequality means that local realism limits the degree of correlation so that the correlation lies in a certain interval; QM's prediction of the degree of correlation, on the other hand, is a strict formula, and it falls on a cosine curve. So it would seem to be expected that the Bell inequality would be easier to satisfy.

The transition from theory to experiment is not a simple process. The initial experimental attempt was 7 years after Bell's paper was published (that is, in 1972), Dr. John Clauser of the United States did a real test following Bell inequality at UC-Berkeley. John Bell was a theorist who didn't know how to design experiments to test his theories. This transition was spearheaded by J.Clauser, another scientist is A.Aspect, who is better known than Clauser for his later elaboration of the experiment. Born in 1942, Clauser will be 80 years old when he wins the Nobel Prize in 2022, which is not easy! … In college, he was a student of renowned physicist Richard Feynman, but Feynman was not enthusiastic about the subject of EPR and Bell's theorem. In 1967 Clauser came across J.Bell's paper, which immediately caught his attention. In order to develop the experimental plan, Clauser read the paper of D Bohm ten years ago and visited the Chinese-American physicist Jianxiong Wu (J.X.Wu), both of whom had experience with two-photon experiments, but these activities did not bring about the experimental plan he needed. But these readings and visits are beneficial, because Bohm has long believed that entanglement occurs between two twin photons, which is the opposite of EPR!

To verify the Bell inequality, it is necessary to measure the pair by pair polarization correlation. In 1969, Clauser made a breakthrough in his approach, and the showdown between locality, hidden variables (on behalf of Einstein) and quantum mechanics (on behalf of many people) was approaching. The experiment must be carried out under many different polarization angles. Figure 2 is a theoretical comparison, where the ordinate is the correlation and the abscissa is the polarization angle; HV stands for hidden variables and QM stands for quantum mechanics.

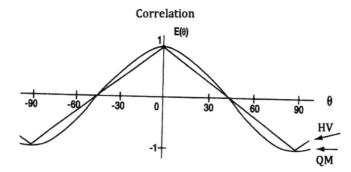

Fig. 2 Relation between correlation and polarization angle

As you can see from Figure 2, the difference between quantum theory and hidden variable theory is very slight. Only by accurately measuring the correlation of pairs of photons at different angles of polarization can researchers tell which theory is correct.

In summary, the polarization angle of each photon in the entangled photon pair must be measured. A 1969 paper with Clauser as the first author opened the door to experimental research. At this time, though, Clauser actually believed in Einstein, not quantum mechanics!

In a series of experiments, Dr Clauser emitted thousands of photons to measure polarization properties, which can only have two values — up or down. The detector results are a series of seemingly random ups and downs. But when the results of the two detectors are compared, the fluctuations have a compatibility that neither classical physics nor Einstein's laws can explain. Something strange is at work in the universe, and entanglement seems to be a real thing.

In a 2002 interview with the American Physical Society, Dr. Clauser admitted that he himself thought quantum mechanics was wrong and Einstein was right. He said: "Obviously we got the 'wrong' result. I had no choice but to report what we saw, and you know, 'This is the result.' But it went against my instincts, and I thought my instincts must be right." He added: "I hope we can overturn quantum mechanics."

One of the quirks of Dr Clauser's findings, and of the quantum-mechanical description of this strange effect, is that the correlation only emerges when individual particles are measured—that is, when physicists compare their measurements after the fact.

Dr Clauser spent much of that decade trying to figure out what holes he might have overlooked. One possibility is called a "positional vulnerability."

And now, Dr. Alain Aspect. began to research; He was born in 1947 and was 76 years old when he won the Nobel Prize in Physics in 2022. In 1982, Dr. Aspect and his team at the University of Paris tried to plug Dr. Clauser's hole by changing the direction in which the photon's polarization was measured every 10ns. He also thought Einstein was right.

Dr. Aspect's results have made entanglement famous, making it a real phenomenon that physicists and engineers can exploit. The quantum prediction holds true, but Dr Clauser has found other possible holes in the Bell experiment that would need to be plugged if quantum physicists were to declare victory over Einstein.

Dr Aspect's experiments, for example, change the direction of polarization in a regular, and therefore theoretically predictable, way that a photon or detector can sense.

It was then that Anton Zelinger, a professor at the University of Vienna, picked up the baton. In 1998, he added more randomness to the Bell experiment by using a random number generator to change the direction of polarization measurements while entangled particles were in flight.

Quantum mechanics has once again decisively defeated Einstein, closing the "positional loophole." Still, there are other possible sources of criticism or prejudice. In recent years Dr Zelinger and his collaborators have been experimenting with "cosmic clocks", which use fluctuations in light from distant objects called quasars, billions of light years away, as random number generators to set the detector's direction.

Fig. 3 The Three Musketeers who won the 2022 Nobel Prize in Physics
(From left to right: American J. Clauser, French A.Aspect, Austrian A.Zelinger)

The "Three Musketeers" of quantum entanglement who won the 2022 Nobel Prize in physics are given in Figure 3, and they are well deserved. Now, let's go back to the 1970s. Early experiments could be done with two-photons, as well as with other subatomic particles. The tests that have been done fall into three categories. One is the singlet proton-on-spin correlation experiment, which is very similar to the original thought experiment. A low-energy proton is hit at a target composed of hydrogen atoms, and the incident proton and the hydrogen nucleus, the proton, enter a single state after a brief interaction. Then both protons leave the target, they're still in a singlet state, and the protons are measured. The second is the experiment of polarization correlation between two gamma photons produced by annihilation radiation. Because annihilation radiation photons are not only emitted in opposite directions, their polarization (corresponding to the spin component) is also opposite, respectively expressed as ±1. The third is the experiment of photon polarization generated by atomic cascade radiation. When an atom of an element rises to an excited state by absorbing laser light, and then returns to the initial energy level in two steps, each step radiates a photon, which leaves in opposite directions and has opposite polarization, denoted by ±1.

The earliest experiments, published in 1972 by S.Freedman and J.Clauser (Freedman was Clauser's graduate student), used calcium atoms to radiate cascades of photon pairs. Since then, most experiments have used photon pairs, with only a few experiments using the singlet proton pair method. ... The results of this first experiment violate the Bell inequality and are consistent with QM.

Between 1973 and 1976, there were eight published experiments. Of these, two are consistent with Bell inequality and support EPR, and six violate Bell inequality and agree with QM.

A.Aspect et al. published three experimental results between 1981 and 1982, all using the calcium atomic cascade radiation photon pair method. These experiments prove with high precision that the results violate Bell inequality and agree with QM.

Aspect's experiment is the most famous. The design of the experiment (FIG. 4), tested by checking whether the photons emitted simultaneously in a single atomic transition follow the Bell inequality, uses a pair of lasers to excite calcium atoms (two-photon excitation) to the ground state to become a light source, with an acousooptic switch at 7.5m on each side of the source. The polarizer passes through or blocks photons with a certain probability. The fate of photons is monitored by electrons and the level of association is assessed. The results show that there is a strong correlation between the measurements of photons, even though there is a distance of 15m between the two measuring instruments. Aspect is very serious and makes the experimental components himself.

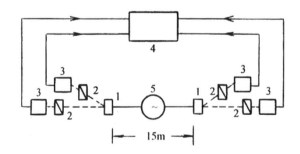

1. Switch　2. polarizer　3. photomultiplier
4. Electronic coincidence monitor　5. two-photon source
Fig. 4 Aspect experiment layout

To sum up the above situation, it is no doubt that the experimental results negate the inequality and support quantum mechanics. Because there are only two experiments that meet the inequality, and it is still early, the precision is not high enough. Later experiments, especially the last three, were more accurate and reliable. Moreover, it is no coincidence that 10 independent experiments not only violate the inequality, but also violate it in exactly the same way that quantum mechanics predicts. Such an outcome would be expected for quantum mechanics, and would seem to shake little, but it would be an unexpected event for physics as a whole and for philosophy as a whole,

It is said that the relevant experiments of Aspect from 1982 to 1986, a total of 15 cases. At the time, it was believed that his experiments provided evidence against EPR.

Aspect's experimental results vindicated the correctness of quantum mechanics and prompted a change in Bell's thinking. In the past, he has called quantum mechanics "expedient and ambiguous"; By the 1980s, he said quantum mechanics was "so accomplished that it's hard to believe it's wrong." As for faster-than-light, Bell said in response to a question from the BBC that the EPR experiment does "contain something faster than light"; To his dismay, this would violate the law of causality: at this point, he said he disagreed with Einstein's worldview. As for the Aspect, he said in response to a BBC question that a simple picture of Einstein's concept of separability could no longer be maintained and could also contain some kind of faster-than-light entity. Apparently, both Bell and Aspect were cautious in 1985, given the mainstream status of relativity. I think it's important for Bell to say that physics should go back to before 1905, to Poincare and Lorenz's theory. In fact, whether Bell decided and abandoned relativity (SR)! He also said that he "wanted to go back to the etheric concept"; The attitude of this distinguished scientist could not have been clearer. J. Bell died in 1990, D.Bohm in 1992;This makes it impossible for these two famous scholars to consider and evaluate the subsequent series of new faster-than-light experimental advances, which is regrettable!

So what do other physicists think? Nobel Prize-winning physicist B.Josephson says that perhaps one part of the universe "knows" another, a distant correlation. P.Coveny and R.Highfield say that Aspect experiments show that two particles that are far apart in the universe can form a system, and there does appear to be a faster-than-light connection at work in distant spacetime. In short, after Aspect proved that quantum mechanics was correct and that the limitations of the Bell inequality did not hold, most of the international scientific community agreed that Einstein's local realism was wrong. For example, B.d'Espagnat says "Einstein's separability assumption must be abandoned, which Bohr has long criticized." K.Copper said: "The possibility of action at a distance should be considered. If action at a distance existed, it would have opposed the special relativistic interpretation of formal systems in favour of the Lorentz interpretation and Newton's 'absolute space'. Therefore, these new experiments based on Bell's theorem can be seen first and foremost as conclusive experiments between Lorentz's theory and Einstein's special theory of relativity."

RATIONALITY OF BOHM EXPERIMENTAL SCHEME

During World War II, theoretical physics was at a standstill. For example, N.Bohr also came to the United States to participate in the development of the atomic bomb so that the Allies could defeat the fascist countries. After the end of World War II, the relevant research was again paid attention to, for example, in 1951 D.Bohm gave a new interpretation of the expression of EPR (D.Bohm, Phys Rev, 1952, 85:166, 180): A microscopic particle with spin zero in a proper position M is separated by decay into two spin 1/2 particles, i. e. I and II. assume that they immediately fly away in the opposite direction and are detected at the same distance but opposite directions (A and B). According to quantum mechanics, when the spin of I(or II) is measured at A(or B), the probability of the measured value being ±1/2 is each 0.5; However, if the spin of I is measured as 1/2, then II must be in the eigenstate of spin (-1/2). Although I and II can be very far apart, the measurement of I can determine the state of II, or the correlation between I and II.

However, some have questioned whether Bohm's improved description is indeed representative of the EPR paper, and here it is repeated to quote Bohm's book *Quantum Theory*: "Let us now describe the hypothetical experiment of Einstein-Rosen Podolsky." We have modified the experiment slightly, but the form is essentially the same as what they proposed, although it is much easier to work with mathematically. Let's say we have a diatomic molecule in a state where the total spin is equal to zero, and let's say the spin of each atom is equal to I /2. Now suppose that a molecule is broken down into atoms in a process in which its total angular momentum remains constant. The two atoms then begin to separate and soon cease to interact significantly.

The EPR thinking described by D.Bohm suggests a strange quantum correlation. When two spinning particles interact far apart, their spins are equal and opposite, so one can be inferred from the other. According to quantum mechanics, the spin of both is uncertain until measured. The measurement determines the spin direction of one particle, and the quantum correlation causes the other particle to immediately accept the determined spin. This is true even when the two are light-years apart. This long-distance interaction suggests that there is a faster-than-light interaction between the particles. This was unacceptable to Einstein — it was the sort of thing that turned him against quantum mechanics. Einstein famously referred to this phenomenon disparagingly as "spooky action at a distance." The scientist, of course, does not recognize celestial spirits, so he thinks such a situation is impossible.

Notably, Bohm's system targets any microscopic particle, not just photons. That is, it could be two electrons, or two atoms that were originally part of the same molecule, and so on. This is important for researchers today. ... Now to the accusation: the EPR paper is saying that I and II no longer have any interaction after separation, whereas Bohm is saying that there is no significant interaction. In the language of modern physics, "there is no longer any interaction" is called local, and "there is no significant interaction" implies the possibility of non-local between particles. After all, the EPR experiment must be thoroughly relativistic, while the Bohm experiment must be non-relativistic. It follows that Bohm's thought experiments are not, as he claims, "essentially the same form as they propose." His subsequent non-local interpretation of quantum potential is in line with this thought experiment. In the interpretation of quantum potential, the expression of the wave function $\psi = R\exp(\frac{j}{\hbar}S)$ does not contain "significant" interactions between quantum particles, but there are indeed non-local interactions between quantum particles due to the existence of quantum potential $Q = -\frac{\hbar^2}{2m}\frac{\nabla^2 R}{R}$.

Our answer to this accusation is as follows: Einstein has long studied gravitational and electromagnetic interactions, and was aware of both the weak interaction proposed by E. Fermi (1932) and the strong interaction proposed by Hideki Yukawa (1934); when he published his EPR paper. Therefore, the "no interaction" in the EPR paper refers to any of the above four kinds of effects; Since EPR is a local realist, he will never acknowledge the existence of any other non-local non-force action. However, D.Bohm is a holist who, in addition to understanding the above four kinds of actions, also acknowledges the existence of non-force interactions (or mutual influences, correlations) in quantum systems, and he proposes that the quantum potential theory is the proof, although the effects of this non-force interaction are not as significant as the above four kinds of effects. Bohm made a distinction, calling those four "significant interactions". In fact, from the EPR paper and Bohm's argument, whether it is "any" or "significant", it says the same thing. In other words, there is no difference between what Einstein and Bohm are claiming. ... It would be futile to deny Bohm's contribution for the purpose of "defeating quantum mechanics." Moreover, many developments since Bohm have proved EPR wrong after all. Theoretically speaking, when a certain part (subsystem) of a composite system with many degrees of freedom is measured, it is incomplete. In this case, the quantum state of the subsystem is described by reduced density matrix; For a two-particle system with spin 1/2, after the spin measurement of I, the above matrix is used to describe I, and the result is a completely unpolarized spin state. The same is true when the system is entangled in other spin states. In summary, the key is that the measurements made on the subsystems of a composite system are incomplete measurements. Therefore, EPR's accusation that "quantum mechanics is not self-consistent" is untenable.

At the EPR thesis stage, the whole thing is very abstract. Bohm made a major contribution to the theoretical visualization of quantum entanglement between microscopic particles; This is undoubtedly a good thing!

One might say, why do we have to choose Bohm's proposal? Why does it have to be ideally related? Let's wait. I don't think it's appropriate to ask questions like this. Bell made some assumptions and derived the results; One can do experiments, and if the experimental results of a two-particle (two-photon or otherwise) system agree with the inequality, then the EPR paper is correct, QM is an incomplete theory, and entangled states (as Schrödinger calls them) do not exist. If the experimental results do not agree with the inequality, then the EPR paper is wrong, QM is complete, and the entangled state exists. So Bell, while subjectively inclined to agree with EPR, is objectively rigorous and impartial. If the experiment can be carried out, then these assumptions are valid. If the experiment doesn't work, then Bell's theoretical work is meaningless. In fact, it has been passed down to posterity, so that we still have to recount it today.

DEVELOPMENT OF BELL TYPE EXPERIMENTS

A. Einstein's opposition to quantum mechanics (QM) began in 1926 and culminated in his 1935 joint paper with B.Podolsky and N. Rosen, and the EPR paper later promoted the development of science from the opposite side. This paper is based on the theory of special relativity (SR), and both SR and EPR deny the possibility of faster-than-light. However, QM allows the existence of superluminal speed, and is consistent with the premise of the study of superluminal speed, that is, QM non-locality. In 1985, John Bell stated, "The Bell inequality is a product of the analysis of the EPR inference that there should be no action at a distance under the conditions of the EPR article; But those conditions lead to the curious correlations that QM predicts. The results of Aspect's experiment were expected, as QM has never been wrong and now knows it can't be wrong even under demanding conditions; Experiments have certainly proved Einstein's ideas untenable." Bell saw the dilemma as a return to Lorentz and Poincare, whose aether was a preferential reference frame in which things could travel faster than light. Bell pointed out that it was the EPR that gave the faster-than-light expectations.

For a long time, scientists have been puzzled by the phenomenon of "quantum entanglement," which seems to defy the classical laws of physics. The phenomenon seems to suggest that pairs of subatomic particles can be secretly linked together in a way that transcends time and space. "Quantum entanglement" describes how the state of one subatomic particle affects the state of another, no matter how far apart they are. This offended Einstein because it was considered impossible to transmit information faster than the speed of light between two points in space. ... Scientists are now acting — out of a sense of duty, but also out of intense curiosity.

The first problem with Bell type experiments is how to create the two-particle system required by Bohm's discussion. Nature seems ready for human experiments, and a common approach is to produce two-photons using atomic cascades of radiation. When an atom of an element descends two specific energy levels (e.g. by absorbing laser light from level $4S^{21}S_0$ straight up to the excited state $4P^{21}S_0$). It then drops to $4S4P'P_1$, then back to $4S^{21}S_0$), radiating one photon at each step, and the two appear on either side of the parent atom and leave in opposite directions, with opposite polarizations (±1). Such photon pairs are connected at birth, like human twins; It's an entangled photon pair. The two are forever entangled in each other, and if one changes, the other changes immediately (or almost immediately), even if they are light-years apart and in different places in the universe.

Another common method is to use positrons to produce double gamma photons, which are not only emitted in opposite directions, but also have opposite polarizations corresponding to the opposite components, expressed as ±1. Another method is to use a singlet proton pair-bombarding a hydrogen nucleus (proton) with low-energy protons, which briefly interact to become a singlet state; The two protons leave and remain singlet, effectively forming an entangled photon pair.

Let's look at what happened after the 1982 Aspect experiment. In 1996, G.Weihs conducted an experiment with two photons with a wavelength $\lambda=702nm$, which proved to violate the Bell inequality at 400m distance and was consistent with QM. Later, the Gisin team in Switzerland added the successful distance of 35m(1997), 10.9km(1998), 25km(2000), also technically using two photons. In 2007, scientists from Austria, the United Kingdom, and Germany joined forces to achieve two-photon entanglement between two distant islands (144km) apart. In 2008, D. Salart achieved two-photon entanglement between two villages in Switzerland, at a distance of 18km. In 2015, a team of researchers in the Netherlands conducted a close-range (200m) dual-electron experiment on a university campus, which is said to have filled the holes in two Bell experiments. In 2017, a team of Chinese scientists achieved thousand-kilometer quantum entanglement in an experiment, setting the highest record.

In the 25 years since 1982, the distance between the two particles in the entangled state experiment, from 15m→400m→25km→144km→1300km, has made amazing progress. Most experiments rely on fiber-optic technology, but China's highest record is the use of quantum satellites. In a 2007 multinational experiment, the research team first created polarimetric entangled photon pairs on the island of La Palma in Spain's Canary Islands, and then left one photon in the pair on La Palma, while the other photon was transmitted via an optical path to Tenerife, 144km away. What is difficult to explain is that this interaction is independent of distance, reaching 144km.

The Dutch experiment is remarkable. The experiment is notable for two things; First, two electrons are entangled, and electrons are particles of matter. Second, although the two (I and II) were not far apart, the experiment closed a loophole that someone could use to attack the Bell experiment. So the experiment broke new ground, electrons have magnetic properties, the so-called "spin". This property causes the electrons to either face up or down. And until they are observed, there is no way to tell which of these two states they are in. In fact, due to the bizarre nature of quantum, they will be in an "overlapping" state facing both up and down at the same time. Facts are only revealed when they are observed. When two electrons get entangled. They all face up or down at the same time. But when observed, one is always facing down and the other is always facing up. There is a complete correlation between them, and when you look at one electron, the other electron is always in the opposite position. The effect is immediate, even if the other electron is on the other side of the galaxy.

QUANTUM ENTANGLED STATES DO NOT ACT AT A DISTANCE BUT PROPAGATE FASTER THAN LIGHT

The space-time representation of quantum theory does not conform to the spirit of SR, and Einstein is sensitive to this fact, which is why he stubbornly opposes QM. However, the wave function of the two-particle system in the EPR paper is an entangled state. This is a special form of (but also universal) quantum state, in addition to the properties of the general quantum state (such as similarity, uncertainty), but also its unique personality — related indivisibility, non-local and so on. N.Bohr had earlier pointed out in his debate with Einstein that separability does not hold in the quantum domain. Einstein would not accept that two subsystems in a system, even if separated, no longer exist independently of each other. The Bell inequality means that the local reality limits the degree of correlation to a certain interval, while the QM is a strict equality for the degree of correlation. The experiment yielded correlation results, which Aspect says "negates Einstein's simplistic picture of the world."

To get some insight, we refer again to a 1985 talk by J.Bell to the BBC. He thinks QM is such an accomplished branch of science that it's hard to believe it could be wrong, so the results of Aspect's experiment were expected. "QM has never been wrong and now knows it can't be wrong even under very demanding conditions; To be sure, the experiment proves Einstein's worldview is untenable." ... At this point, the questioner said that the Bell inequality presupposes objective reality and local (indivisibility), the latter indicating that there is no faster-than-light transmission of signals. After the success of the Aspect experiment, one of the two must be discarded. "It's a dilemma," says Bell, "and the easiest way to do it is to go back to before Einstein, Lorentz and Poincare, who argued that the ether was a preferred frame of reference." It is possible to imagine such a frame of reference in which things move faster than light. There are many problems that can be easily solved by assuming the existence of ether.

Bell repeated, "I want to go back to the ether concept because there is this revelation in the EPR that there is something behind the scene that is faster than light, but this ether does not show up at the observation level."... "In fact, it is Einstein's theory of relativity that makes quantum theory so difficult."

One of Bell's sayings in 1985 was "behind the scene something is going faster than light" (after the scene something is going faster than light). The remark was so striking that it is still quoted by researchers years later. ... J.Bell died in 1990, and there have been many advances in faster-than-light research since then that he failed to see. If he were alive today, he would be the world leader in faster-than-light research.

The fundamental difference between SR and QM is whether to admit the existence of non-local, whether to admit that the superluminal can exist. In recent years, a team of Swiss scientists has done an excellent job of answering with facts. We know that the Swiss physicist Nicholas Gisin (1952 -), who worked at CERN, was a great admirer of his predecessor J.Bell and believed that the Bell Principle was a major breakthrough in theoretical physics. His team first confirmed the violation of the Bell inequality by two-photon entanglement at a distance of 35m in the laboratory at the University of Geneva, thus proving the existence of quantum non-locality. They then extended the experiment to 10.9km in 1997, and were the first to use fiber optic technology in this Bell experiment. Aspect in France congratulated themselves on the news—10km is much better than the original 15m. Gisin firmly believes that quantum entanglement completely violates the spirit of relativity; Next, his team solved another prominent problem—in the theory of quantum entanglement, one particle can change the properties of another particle instantaneously, no matter how far apart they are; So how fast is "instantaneous"?

In 2000, Gisin's team used optical cables under Lake Geneva to send photons 25km away and found the opposite of Bell's inequality. Gisin's group has a very interesting result—the experimental measurement of quantum entangled states (QES) acting at a speed of $(10^4$-$10^7)c$. This is an important case, indicating that the speed of action is not infinite, but faster than light. In short, Gisin believes that some kind of influence appears to be traveling faster than light; Gisin thinks this means that "relativity's description of spacetime is flawed." The 2008 paper says they performed a Bell type experiment with two entangled single photons spaced 18km apart (roughly east-west, with the source precisely in the middle). The rotation of the Earth allowed them to test all possible hypothetical superior reference frames in the 24h period. At all times of the day, two-photon interference fringes above the threshold determined by the Bell inequality are observed. From these observations, it is concluded that the observed non-local correlation is truly non-local, as shown by previous experiments. In fact, it should be assumed that this magic effect will spread even faster than the experimental results $(10^4$-$10^7)c$. That is, Salart et al. have consistently observed two-photon interference that is significantly higher than the Bell inequality threshold. Taking the advantages of the Earth's rotation allows a low limit of the acting speed to be determined for any assumed superior frame of reference. If such a superior frame of reference exists and the earth's velocity is less than $10^{-3}c$, then the action velocity must be $\geq 10^4 c$.

Until 2000, there were two hypothetical superior reference frames in the Swiss Bell experiment, one was the 2.7K microwave background radiation and the other was the Swiss Alps reference frame. The latter is not a cosmic reference frame, defined by the context of the experiment. In these analyses, the hypothetical superluminal influence is defined as the speed of quantum information (SQI), which differs from classical signaling; However, one should know how to obtain the limit (boundary) of SQI in any reference frame.

In an inertial reference frame on Earth, events A and B (in the case of the experiment, two single photons are detected) occur at time, in time t_A, and in time t_B, on the r_A and r_B, Consider another reference frame F (the assumed superior reference frame, moving at speed v relative to the Earth reference frame); When a correlation that violates Bell's inequality is observed, the SQI of the F system (denoted by symbols) creates a correlation with a bound of v_{qi}

$$v_{qi} \geq \frac{\| r'_B - r'_A \|}{| t'_B - t'_A |} \tag{9}$$

where (r'_A, t'_A) and (r'_B, t'_B) are obtained by Lorentz transformation; by simplified it, we obtain:

$$\left(\frac{v_{qi}}{c}\right)^2 \geq 1 + \frac{(1-\beta^2)(1-\rho^2)}{(\rho+\beta_0)^2} \tag{10}$$

where $\beta = v/c$, is the ratio of the speed of the Earth reference system in the reference system F to it, so from the above formula we know $v_{qi} > c$.

In Salart's experiment, the source of the signal, located in the laboratory in Geneva, was the generation of entangled photon pairs in a nonlinear crystal, using fiber Bragg gratings and light loops, each photon pair was separated with certain separation and sent to two villages via a Swiss fiber network system, with a linear distance of 18km. energy-time entanglement is used, which is a suitable state for quantum communication in standard telecommunications cables.

In short, the experiment was complex and sophisticated. Swiss scientists have shown that the speed at which quantum entangled states interact is not light speed, nor is it infinite, but superluminal — at least 10,000 times of c. Therefore, quantum entangled states are not acting at a distance (i.e., $v \neq \infty$), but rather superluminal broadcasting. Could quantum entanglement itself be considered a special form of faster-than-light communication?

Some physicists think yes, some physicists think no. In 2010, physicist Prof. Zhiyuan Shen pointed out: "There is a faster-than-light interaction between two entangled photons, and when the spin of one photon is measured, the spin of the other photon at a distance immediately changes accordingly." Einstein called this "weird action at a distance." Recently, a team at the University of Geneva in Swiss has measured the speed of photons in an entanglement experiment at at least 10,000 times the speed of light. Strangely enough, many authors of physics textbooks and papers say that this does not violate special relativity (SR) because people cannot be used to transmit information. But photons do transmit information, otherwise how would an entangled photon "know" that another photon far away has changed its spin?

Physics is not anthropology, so why does it have to be people who transmit information to count? This view is actually another version of the humanistic principle, which takes the subjective role of man as the criterion of objective law. However, science, especially physics, is objective, and entangled photons have faster-than-light effects between them, which is proved by many experiments to exist objectively, which cannot be denied. We must abandon our subjective biases and accept faster-than-light transmission of information in entangled states as an objective fact.

These are very good words from Professor Shen; In my opinion, many objective laws (including faster-than-light information transfer between entangled particles) existed before there were humans on Earth. The problem is that we have not yet been able to use this phenomenon to enable human communication in space and space exploration. But not today doesn't mean never.

Quantum entanglement is the greatest mystery in physics, and for good reason. Quantum interaction may be called "the fifth fundamental physical interaction in addition to the four fundamental physical interactions (electromagnetism, gravity, weak force, and strong force)." The mistakes made in the EPR paper have profound lessons for people, reminding us that the world is stranger than we can imagine. The fact that entangled particles interact at faster-than-light speeds regardless of spatial distance is fundamentally lacking in theoretical explanation. Scientists know this is so, but they don't understand why it is so; In general, the strength of the effect varies with distance, but the expectation of quantum entanglement is that it will have the same strength no matter how far away; Why is this? No one can answer that at the moment. And this entanglement does not dissolve automatically after a period of time. ... Curiosity motivates us and is an inexhaustible source of thought and exploration.

Now, we say that superluminal signaling based on quantum nonlocality. We believe that this phenomenon has always existed, and the question is only how to implement it in human communication. Although no one can be sure when they will succeed, it is certain that someone will keep trying. It must be pointed out that faster-than-light information transmission and faster-than-light travel are two major pursuits of human beings. If we take a broader view, we will not doubt the significance of studying the "remote transmission of faster-than-light information".

ABOUT THE COPENHAGEN INTERPRETATION OF QUANTUM MECHANICS

The history of physics books tell us that the so-called Copenhagen interpretation (CI) of quantum mechanics consists of three main aspects: the Max Born probability interpretation of the wave function; Werner Heisenberg's uncertainty relation; Niels Bohr's principle of complementarity. The famous Bohr-Einstein debate took place at the 5th Solvy Conference in October 1927 and culminated in the 6th Solvy Conference... Why are we bringing this up now? Because of the controversy surrounding the development of modern quantum communication technologies, some physicists have revived the argument that QM's Copenhagen interpretation is "problematic even today, and Einstein is not wrong." Some scholars logically conclude that "quantum communication is something that has no physical basis at all." If the foundation is not good, there must be something wrong with the house. In this way, the discussion and reflection take people back to 1927.

One theory is that Einstein is not against QM, but rather against the Copenhagen interpretation of QM; I don't agree with that. Because this interpretation of QM mainly comes from Bohr, Born and Heisenberg, and their theory is the main content of QM. In my opinion, the anti-QM and anti-QM Copenhagen interpretation are essentially consistent. While most physicists recognized the work of M. Born and W.Heisenberg, Einstein found the work of both men objectable — he considered both Born's and Heisenberg's work "deviant from the normal path." He believed in the certainty of the objective world; For example, if the track is clearly visible through the cloud chamber, its orbit should not be ignored. In short, Einstein explicitly stated at the 5th Solvy Conference in October 1927 that "the certainty principle is not accepted." He also opposed the idea of quantum mechanics as a complete theory of a single process because it could act at a distance. Einstein said that he did not think of de Broglie-Schödinger waves as individual particles, but rather as ensemps of particles distributed in space. In effect, Einstein viewed waves as the average behavior of a large number of particles. On March 22, 1934, Einstein again objected to the probability interpretation in a letter to Born.

Einstein's 1948 article "Quantum Mechanics and Reality" published in the journal Dialectics can be seen as a statement of his later years. Although he acknowledged that quantum mechanics was "a significant, even decisive, advance in the knowledge of physics", he insisted that "the methods of quantum mechanics are simply not satisfactory". On the one hand, this contradictory statement is due to the fact that the depth and wide application of QM have made him unable to deny its significance, but he is unwilling to admit that he is wrong in academic opinion. Therefore, I do not believe that Einstein changed his attitude against QM in his later years. But some people still say that relativity (SR, GR) can be combined with QM, isn't that ridiculous?... More than 20 years after Einstein's publication, two leading physicists made sobering comments: P. Dirac, in his late years, said that "there is a real difficulty in reconcicating relativity with quantum mechanics"; According to S.Weinberg, "Theoretical physics has big problems, such as the requirement for Lorentz invariance that QM simply cannot meet." It should be said that these two statements are very clear and correct.

The years 1926-1927, when QM appeared, were 21-22 years after special relativity (SR) was proposed, and 11-12 years after general relativity (GR) was proposed. It can be said that relativity on the one hand achieved Einstein's great prestige, but at the same time made him tend to be conservative; This is regrettable.

The formation of QM's Copenhagen School has a process; In the spring of 1912, N.Bohr went to work for the British physicist D.Rutherford, and returned to Copenhagen in the same year to think about the experimental law of the spectrum of the hydrogen atom. In 1913, Bohr proposed the theory of the quantized orbital motion of electrons in atoms orbiting the nucleus, and proposed two new concepts — light radiation or absorption is the result of quantum transitions in atoms and the angular momentum quantization of electrons in orbit, proving Bohr to be a very outstanding innovative scientist. In 1916 Bohr became Professor of theoretical physics at the University of Copenhagen. In 1920 he founded the Institute of Theoretical Physics, where many European scholars came to work. The entrance of W.Paul in 1922 and W.Heisenberg in 1924, both students of the famous A. Semmerfeld, was a landmark event. In addition, people who came to Bohr to do research were P. Dirac, P.Ehrenfest, L. Braillouin, L.Landau, G.Gamov, etc., as is well known, they all made important contributions later on. Of course, the fundamental point is that people under Bohr's leadership (especially W.Heisenberg and M.Born, etc.) proposed a new theoretical system — quantum mechanics (QM), whose unique mathematical expression and physical thinking are completely different from classical physics, and its correctness is gradually proved; This has made the Copenhagen School famous and has many admirers and followers.

The leading figure of the Copenhagen school is N.Bohr, and the leading figure of the opposition is A.Einstein. When QM came out, Einstein was 47 years old and a world-renowned scientist, winner of the Nobel Prize in physics for explaining the photoelectric effect with his theory of photons. Einstein used classical physics to derive the theoretical formulation of photons, but he was able to refer to Planck quantum theory to complete the derivation of photons, which is a revolutionary work. But after the appearance of QM, he insisted on opposing it; His attitude remained unchanged until his death in 1955.

To deepen the understanding, the author proposes a formula:
$$QM \cong CI+SE+DE \tag{11}$$

QM at the left end of the above equation represents all that constitutes quantum mechanics; Right: CI represents the main content of the Copenhagen interpretation (Bohr, Heisenberg, Born), SE represents Schrödinger's quantum wave equation, and DE represents Dirac's quantum wave equation.

SE is one of the core theories of QM and is as important as Newton's equations of motion in classical physics. It has the ability to predict natural phenomena and is widely used. But Schrödinger is all about volatility; According to de Broglie and Schrödinger, the velocity of a moving particle is the same as the group velocity of a wave packet, so their theory implies that a wave packet and a particle are one and the same. It is wrong to view the relationship between microscopic particles and corresponding waves as exaggerating the status of waves. We start with non-relativistic free particles and make a simple derivation; It can be shown that the dispersion equation of de Broglie waves is:

$$\omega = \frac{\hbar}{2m}k^2$$

where $\hbar = \frac{h}{2\pi}$, $k = 2\pi/\lambda$. So we can find the group velocity

$$v_g = \frac{d\omega}{dk} = \frac{\hbar k}{m} = \frac{p}{m} = v$$

So the group velocity is equal to the particle velocity. The derivative of group velocity v_g with respect to wave number k is calculated from the above equation:

$$\frac{dv_g}{dk} = \frac{\hbar}{m} \neq 0 \tag{12}$$

Therefore, it is relevant to indicate that the wave packet will spread (gain weight) during transmission. But particles are stable in transit, so the scientific community rejected their idea; He joked that "Schrodinger's equation is smarter than Schrodinger".

It was Bohr who pointed out that the wave packet "gets fat" during the wave transmission process, while the particle has undoubted stability, so simply thinking of the particle as a wave packet does not make sense. Nevertheless, Schrödinger did not accept the "wave-particle duality" and "wave function collapse" of CI. It is said that Einstein encouraged him to design a thought experiment to disprove CI. In a 1935 article (Naturwissenchaften, 1935, Vol. 23, 807, 823, 844), Schrödinger proposed the so-called "Schrödinger cat state" paradox — a hypothetical device that triggers a small hammer with the decay of atoms. The vial containing the poison gas is broken, and the vial releases the poison gas to kill the cat, in which the decay of atoms is a random quantum event. The problem is that the decay of an atom is a superposition of multiple states, called superpositions, which means that the cat is both dead and alive at the same time. Once the measurement is made, the quantum superposition state is destroyed. In other words, once we open the box to see the results, the cat is only in one state, that is, alive or dead. But this does not mean that the cat was already in this state before opening the box — before the observation, the cat was in a "life and death superposition" state, which is ridiculous. A quantum system in two states at the same time determines whether a cat lives or dies. This experiment shows that quantum theory goes against our intuition. The Schrödinger cat paradox is a blow to the Copenhagen school, because a cat cannot be "both dead and alive."

But Schrödinger's thought experiment was based on the premise that wave functions could describe macroscopic objects (including living organisms), and this was not proven. However, this "cat paradox" discussion is not without merit, and it is intrinsically linked to the EPR paper published in the same year (1935).The inseparable state of a composite system (two-particle system) discussed in EPR is actually an entangled state, and this term happens to appear in Schrödinger's paper, so the entangled state problem is also called Schrödinger's cat paradox. Schrödinger used the term entanglement to describe superposition states of a composite system that could not be represented as direct products, and to illustrate with thought experiments that the wave-function probability interpretation would lead to absurd conclusions when applied to the macroscopic world.

Although Bohr's complementarity principle is widely used and not limited to the wave-particle duality of light, people are used to view the complementarity principle from this duality problem. "Interpretation" holds that both massless and massless particles have wave-particle duality; They sometimes appear as particles (with definite paths, but without interference fringes) and sometimes as waves (with no definite paths, but with interference fringes). It depends on how the experimenter observes, but it is impossible to observe both properties at the same time, i.e. not knowing the path of the particle and having interference fringes at the same time. The complementarity principle of N.Bohr is roughly the same. However, in 2014, the situation changed — recent advances in wave-particle duality research have demonstrated that it is possible to observe both particularity and volatility at the same time by installing two good measuring devices (path information and interference fringe detectors) in the same interferometer device, each of which performs different functions, does not interfere with each other, and works together in the right way. This means that the traditional belief that "two properties are never observed at the same time" may be broken. Prof. Zhiyuan Li, a researcher at the Institute of Physics of the Chinese Academy of Sciences, has been doing research on the "wave-particle duality of microscopic particles and the possibility of violation of the complementary principle".... However, the author believes that even if the complementarity principle is not complete, it will not damage the QM as the physical basis of quantum communication (QC).

QUANTUM COMMUNICATION AND WOOTTERS THEOREM

Now we first give the definition of quantum entangled states in mathematical form; A composite system (I and II) is provided, where the common eigenstates of a complete set of mechanical quantities of I are, and the corresponding eigenstates of II are, and respectively represent quantum numbers. $|n\rangle_I$ and $|m\rangle_{II}$. If the quantum state of the composite system =, it is separable. $|\Psi\rangle_{III} = |n\rangle_I \otimes |m\rangle_{II}$. If not, it is an inseparable state (or entangled state), written

$$|\Psi\rangle_{I,II} = \sum_{xm} c_{xm} |n\rangle_I \otimes |m\rangle_{II} \tag{13}$$

Here I and II are entangled quantum states, indicating that the measurement of I is related to the measurement of II, regardless of the distance between I and II. This is caused by the superposition of quantum states of the composite system. This quantum entangled state is one of the physical foundations of quantum informatics.

The age of quantum information seems to be suddenly upon us. Can we really use quantum communication methods in the same way as we use smartphones? Many people are asking that question. Since there are a large number of cases in which "physicists do not understand communication and communication experts do not understand quantum physics", people engaged in quantum communication experimental research should make a realistic explanation of their work results and international trends, and must not use the ignorance of the public to exaggerate propaganda and even mislead. In particular, one should not promote a "quantum theology" that would plunge oneself and others into the mire of idealism. For example, what is "quantum teleportation"? Caution should be exercised in presentation and promotion. In short, quantum communication (QC) must explain its existence and significance with the results of practice, the fundamental point of course is its security, confidentiality of the actual effect, and come up with the industry that is most concerned about communication security (such as military, banking) has accepted QC and achieved good results to prove themselves. Unfortunately, there doesn't seem to be any information on that at the moment.

Why is Quantum Communication Secure? The most popular explanation is this: the Heisenberg uncertainty principle (uncertainty relation) causes the following situation, when the eavesdrover does not know the sender coding basis, it is impossible to accurately measure the information of the quantum state; In addition, the principle that quantum states cannot be cloned (Wootters' theorem) prevents eavesdroppers from making a copy of a quantum state to measure after knowing the coding base, so eavesdropping causes errors. At this time, the two parties knew that they were being bugged and stopped communicating.

Entanglement is not mentioned in the above statement; Actual quantum communication systems are diverse, and it seems that entangled photons were not used in QC technology until

2004. Thus entanglement appears to be a necessary condition for unsecured communication... In conclusion, quantum communication researchers believe that it is Heisenberg's uncertainty principle and Wootters' quantum non-cloning theorem that guarantee the "unconditional security" of the BB84 protocol. It is assumed that the secretor intercepts the photon from the quantum channel and measures it, and this eavesdropping behavior will interfere with the quantum state, so that the operator at the sending and receiving end will feel that someone is eavesdropping and stop the communication. But instead of measuring, the secret keeper copies the same thing (with the cryptographic information). However, in 1982 W. Wolotters proposed the "theorem that quantum states cannot be cloned", which denied the possibility of this method. This maintains the authority of quantum encryption and is considered unbreakable. To quote a document from the Chinese Academy of Sciences, "Quantum key distribution uses single photons in a superposition state to ensure unconditional security between two parties that are far away from each other."

Wootters' theorem states: "In quantum mechanics, there is no physical process that achieves an exact copy of an unknown quantum state such that each copy is identical to the initial quantum state." By using the linear property of state space, we can simply prove the theorem that single quantum states cannot be cloned, which is very famous in quantum information. Two methods of proof are proposed:

① The input quantum state $|\psi\rangle$ and $|\phi\rangle$ are not exist, and the initial state is the standard pure state $|s\rangle$.

from $U(|\psi\rangle|s\rangle)=|\psi\rangle|\psi\rangle$, $U(|\phi\rangle|s\rangle)=|\phi\rangle|\phi\rangle$, obtain

$$U\big[\alpha(|\psi\rangle+\beta|\phi\rangle)|s\rangle\big]=(\alpha|\psi\rangle+\beta|\phi\rangle)(\alpha|\psi\rangle+\beta|\phi\rangle)$$
$$=\alpha^2|\psi\rangle|\psi\rangle+\beta\alpha|\phi\rangle|\psi\rangle+\alpha\beta|\psi\rangle|\phi\rangle+\beta^2|\phi\rangle|\phi\rangle \quad (14)$$

In addition, there are

$$U\big[\alpha(|\psi\rangle+\beta|\phi\rangle)|s\rangle\big]=\alpha U(|\psi\rangle|s\rangle)+\beta U(|\phi\rangle|s\rangle)$$
$$=\alpha|\psi\rangle|\psi\rangle+\beta|\phi\rangle|\phi\rangle \quad (15)$$

The two are contradictory. So quantum states cannot be cloned.

② There are two quantum systems: A is the quantum state to be cloned, and the initial state is; $|\psi\rangle$. B means we started out in the standard pure state $|s\rangle$. Cloning is described by A unitary operator on a and B complex system, i.e. $U(|\psi\rangle\otimes|s\rangle)=U(|\psi\rangle\otimes|\psi\rangle)$ for $\forall|\psi\rangle$ is true, And for $|\phi\rangle\neq|\psi\rangle$, we obtain

$$U(|\phi\rangle\otimes|s\rangle)=U(|\phi\rangle\otimes|\phi\rangle)$$

Take the inner product and $U^+U=I$; for the pure state $|s\rangle$, from $\langle s|s\rangle=I$, so

$$(|\phi\rangle\otimes\langle s|)\,U^+U\,(|\psi\rangle\otimes|s\rangle)=(|\phi\rangle\otimes|\phi\rangle)(|\psi\rangle\otimes|\psi\rangle)$$

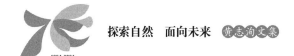

$$<=> \langle\phi|\psi\rangle \langle s|s\rangle = \langle\phi|\psi\rangle \langle\phi|\psi\rangle$$
$$<=> \langle\phi|\psi\rangle = (\langle\phi|\psi\rangle)^2 \tag{16}$$

now we see, $\langle\phi|\psi\rangle = 0$ or $\langle\phi|\psi\rangle = I$, that is, the two states are orthogonal or equal.

The above derivation shows that a quantum cloning machine with a success rate of 1 can only clone a pair of mutually orthogonal quantum states. That is, if the cloning process can be represented as a unitary evolution, then unitary requires that two states can be cloned by the same physical process if and only if they are orthogonal to each other, that is, non-orthogonal states cannot be cloned.

However, in 2018, Xiaochun Mei gave a proof that "the theorem that quantum states cannot be cloned is not true." In the original paper that proved the "theorem that quantum states cannot be cloned," Wootters first assumed that any quantum state could be cloned. Then a quantum state cloning operator is defined, and two conditions under which another quantum state can be cloned are derived. One is orthogonal and the other is non-orthogonal, that is, the integral of the product of these two quantum states is equal to zero or equal to 1. The quantum states that meet these two conditions can be cloned, but cannot be cloned if they do not meet them. Therefore, there is no question of a quantum state that cannot be cloned, but of what quantum state can be cloned. The study also found that for a general quantum system, there can be an infinite number of quantum states satisfying these two conditions, the so-called quantum state can not be cloned is wrong.

In addition, the quantum state cloning operator defined by Wootters has serious problems. Apply this operator to a cloned wave function, and the result remains the same. Applying it to a standard pure state wave function can turn it into a cloned wave function. Such a result is obviously paradoxical, since the pure wave function is also a wave function, and therefore the quantum clone operator is mathematically untenable.

If Mei's derivation analysis is correct, the statement that "absolute secrecy can be obtained unconditionally by quantum communication" is not valid. However, some people think that Mei said in the article that "lasers can clone a large number of photons" is wrong, because although the laser uses stimulated radiation to work, it will inevitably emit spontaneously, so it cannot be said that it can be cloned. They believe that quantum states cannot be cloned for a long time... The author has a different opinion on this matter — even if Wootters' theorem is impeccable, QC cannot be "absolutely confidential"; Otherwise, we would not have used decoy to build QC systems since 2004, because in that year science community use other method to build QC system!

DISCUSSION

Quantum mechanics (QM) has been proposed for nearly a hundred years. Now, it has become the foundation and core of modern physics, and its great influence is still expanding. A series of related experiments, such as discriminating experiments on Bell inequality, new experiments on wave-particle duality, experiments on faster-than-light properties presented by quantum tunneling, and recent experiments on the propagation speed of quantum entangled states, and various experiments on quantum communication, etc.; They have gone beyond the discussion of philosophical speculation, and revealed a series of new non-classical physical phenomena, which have aroused great attention. In recent years, not only are many scientists engaged in the research of QM basic theory and quantum information theory and experiment, but also new books on QM are published constantly. This is very welcome.

At the same time, there are some arguments, even fierce arguments; This is normal. However, some articles attempt to negate QM theory system without factual basis, so far do not recognize the greatness of QM theory, causing some confusion in the physical concept. In 1965, R.Feynman famously said, "I can safely say that nobody understands Quantum Mechanics," perhaps illustrating the difficulty of learning and understanding QM. However, if we do not hold the opinion of the family, we can have a definite grasp and correct understanding of the basic theory of QM. The progress and achievements of quantum information are also obvious and undeniable. This is the view of the vast majority of physicists. QM is a successful theory, Einstein's attitude is wrong, these are obvious facts. Even if it is not quite complete, it is enough to be the physical basis for QIT (including quantum communication QC). As for the publicity that QC is absolutely safe and confidential, we cannot agree!

The author emphasizes that the theory of QM is broad and profound, and its application is both extensive and fruitful. Only by acknowledging these two points can a calm and objective discussion take place. The author believes that the mathematical form of quantum mechanics has been established since 1926 to 1928, although it has been refined and generalized from time to time, it can withstand the test of theory and experiment, and has been finalized in theory. But the physical explanation, the physical reality behind the mathematical laws, has long been debated. De Broglie said, "Physicists today almost unanimously agree with Bohr and Heisenberg's explanation, because it seems to be the only one that fits all the known facts." These calm and objective comments should wake people up now. This article is positive about the Copenhagen interpretation.

However, we must also see that some people are still making criticisms at the basic theoretical level. Some claims are specious; For example, regarding the source of the non-locality of QM, some articles on the one hand say that this source is "due to the fact that the QM equation does not completely satisfy relativity", but then say that even the Klein-Gordon equation and the Dirac equation are also non-local, and these two equations are generally recognized as relativistic equations. This is paradoxical and contradicts Einstein's condemnation of "QM as non-local." Einstein never mentioned that non-locality refers to equations, and from his speech at the 5th Solvay Conference (1927) to the EPR paper (1935), he explained the non-locality caused by the way QM is described. It seems that some authors wanted to follow Einstein, but failed to understand the original meaning of Einstein.

The equation of QM is local, and the description is non-local, which is an inevitable result of the basic principle of QM. There are several principles that constitute the QM framework, which cannot be replaced by one equation. For example, the existence of entangled states is due to the following principles: (1) the wave function Ψ completely describes the particle state and its statistical interpretation; (2) Ψ satisfy the principle of state superposition (which is the embodiment and requirement of volatility) and the measurement hypothesis; (3) The principle of homogeneity (identical particles are indistinguishable, requiring that the wave function of their system must be symmetric or antisymmetric). None of the above requirements is necessary, but it does not matter which QM equation it satisfies.

As for the article that "the other source of non-locality is from Fourier expansion", it is also incorrect. QM is only used with Fourier expansion. Apparently, he mistook the mathematical theorem for QM's superposition principle or measurement hypothesis. The expanded terms of mathematical theorems do not necessarily represent quantum states, whereas the terms of physical principles must be quantum states. It would be a mistake to confuse the two.

In addition, some people use the "Dirac story" to create the atmosphere that "QM is going to die." However, all mechanical quantities in QM are defined by operators, as is angular momentum ($\hat{L} = \hat{r} \times \hat{p}$). Dirac does the same in his book. The so-called "Dirac story" does not mean that "the QM problem is serious" or "Dirac is incompetent."... In addition, we emphasize that "wavelength λ is a spatial range, not a local area", which is common sense.

IMPROVEMENTS AND DEVELOPMENTS IN QUANTUM THEORY

Quantum mechanics is the crystallization of human wisdom and a great scientific creation. But logically, it also needs to be improved and developed. In the author's opinion, the serious problem is that there are always people who want to lead quantum theory with relativity and control quantum theory; However, QM is mainly devoted to the analysis and understanding of the micro world, and the theory of relativity can not deal with the problem of the micro world. Einstein himself developed SR and GR between 1905 and 1915, a period in which he was completely ignorant of the structure of the atom; So how is it possible to use relativity to rule quantum theory? Historically, it is the theory of relativity that has hindered the progress and development of quantum theory, and this is the view of many heavyweight physicists — one example is John Bell, another example is P. Dirac in his later years.

Some say that the combination of SR and QM leads to quantum field theory (QFT), which in turn gives rise to the so-called Standard Model. We do not share this view. This paper has pointed out that SR and QM have conflicting theoretical viewpoints on fundamental issues, which is not only impossible to "integrate with each other", but also incompatible with fire and water. "Relativistic quantum mechanics" simply does not exist (see [14])! It is very ridiculous for someone to insist on such an impossible "marriage". As for the Standard Model, because it is built on the basis of "point particles", it is full of loopholes and unconvincing! The so-called "renormalization method" is to fix these loopholes, but it is also futile.

The problem of infinity used to exist in classical physics. For example, Coulomb's law in electromagnetism:

$$F = k\frac{q_1 q_2}{r^2} \tag{17}$$

F is the electrostatic force between the charges q_1 and q_2, the distance r is distance between the charges; If you try to reduce it, F will keep increasing to an unreasonable degree. If $r=0$, then F becomes infinite. In reality, of course, there is no such force. QFT is said to be an "improvement" on QM, but it is fraught with infinite divergence problems. As Professor Lingjun Wang said, this is a wrong theory, which deals with the infinite very casually and simply inexplicably. If one comes across infinite divergence in classical physics and asks QFT for advice, he will be disappointed! QFT takes relativistic covariance and gauge covariance as its basic principles, which makes it into confusion and cannot solve infinite divergence.

Professor Wang also said: "Another aspect of relativity's influence on QFT is to treat symmetry and covariance as the cornerstones of theoretical physics." Theorists "boldly assume" at the first sign of a problem that things are so absurd in QFT that they would rather have microscopic particles with no mass than stick to their canonical covariance. Yukawa Hideki later realized that there was a problem with QFT, I'm afraid it was too late!

For a long time, large-scale theoretical physics should abide by relativity and Big Bang cosmology, and small-scale theoretical physics should abide by QFT and Standard Model (SM). To do otherwise is heresy. This situation has seriously hindered the development of international theoretical physics — we might say it has not developed at all for many years. SM is also a hypothesis that is increasingly being questioned... In short, here we advocate the original QM, oppose the use of relativity to interfere with everything, even Einstein himself does not understand high-energy particle physics, but also use relativity, QFT and other very suspicious theories to "guide"; In this way, quantum theory will not only not improve and develop, it will only get worse!

Now let's talk about the so-called quantum theory of gravity. As we know, there is no separate time or space in relativity. There is only "space-time"; Although this is a concept that lacks physical meaning, one must accept it. Moreover, this "spacetime" is bendable, although no one knows what that "curved spacetime" looks like. For gravity, relativity no longer recognizes it as a force, but as a manifestation of curved spacetime. In this way, GR turned physics problems into mathematics, and GR was even called geometrodynamics. This treatment gives relativity a cloak of mystery, but it does nothing to explain what gravity is. Einstein gravitational field equation (EGFE) is a problem, the author has a special article to discuss, here omitted.

The term quantum gravity implies that quantum theory is combined with GR. But this is impossible because there is no such thing as "bendable spacetime" in QM. Although there are treatises on quantum gravity, they do not solve practical problems. The so-called "gravitons" were nowhere to be found despite vigorous searches. The current talk of quantum gravity is formal and superficial.

After getting rid of the interference of relativity, I think there are two problems in particular that need to be studied. First of all, some people admit on the surface that "the essence of quantum non-locality is faster-than-light", but they insist on SR's "light-speed limit theory", in fact, they still adhere to Einstein's stuff. Second, what is the nature of quantum entangled states? It is not clear yet, and this is a big problem related to how we understand the universe! John Bell had intended to explore these two questions in depth, but died young (in 1990), leaving us to wonder.

CONCLUSION

Since the birth of quantum mechanics, it has been continuously doubted, criticized and suppressed. This is particularly true of Einstein, who has used his theory and his immense personal prestige to try to nip QM in the bud. If not, then it cannot be allowed to grow naturally, because the development of quantum theory is a threat to relativity. This self-interested critique of QM reached its climax in 1935. Einstein seems to have forgotten that he made some contributions in the early days of quantum theory; Einstein's theory of photons, for example, is still recognized today for his work on the field that earned him the 1921 Nobel Prize in Physics. His explanation of the photoelectric effect was beyond Maxwell's electromagnetic theory! ... But Einstein has forgotten this and spent 30 years criticizing and attacking QM, as if hoping to put it to death. However, the development of history shows another situation — QM continues to advance in theoretical depth and breadth, and its application continues to expand, and finally the world has entered a historical period of great development of quantum information (QIT).This reminds us of the saying, "He who laughs last laughs best!"

Bell's theorem is a general local theory with implicit supplementary parameters. The theorem assumes that quantum mechanics is "incomplete" and preserves Einstein's local view for the time being. It may be assumed, then, that there is a way to complete the quantum mechanical description of the world while satisfying Einstein's requirement that the physical reality occurring at A cannot affect the physical reality occurring at B unless B receives a signal from A (which, according to SR, cannot travel faster than light). In this case, completing the theory would mean discovering hidden variables and describing how they determine the behavior of particles or photons. (Einstein had suspected that distant particles were related to each other because their common origin gave them some local hidden variables.) These hidden variables are like instruction sheets; When there is no direct correlation between particles, they can show correlation as long as they act on instructions. If the universe is inherently local (that is, there is no faster-than-light communication or faster-than-light effects, as Einstein believes), then the information needed to make quantum mechanics complete must be conveyed by some predetermined hidden variable.

But by 1985, John Bell had completely abandoned these views. In effect, he abandoned both EPR and SR. Many physicists believe that entanglement violates the spirit of relativity because there is "something" (whatever it is) between two entangled particles that does indeed travel faster than the speed of light (even if its speed may be infinite), a view later held by J. Bell, which is the affirmation of faster-than-light.

In the past, many people in the international community believed that the theory of relativity was the highest achievement of Western science, which was wrong. The logic of relativity is so confusing and flawed that it is hard to trust. We believe that if we are to choose the highest achievement of Western science, it should be Newton's classical mechanics and quantum mechanics constructed by many people, their success is the triumph of human intelligence!

REFERENCES

[1] Schrödinger E. Quantisation as a problem of proper values[J]. Ann d Phys, 1926, (4): 1–9.
[2] Schrödinger E. Collected papers on wave mechanics[M]. London: Blackie & Son, 1928.
[3] Heisenberg W. Ueber die grundprinzipien der quantenmechanik[J]. Forschungen und Forschritte, 1927, 3(11):83.
[4] Heisenberg W. The principles of quantum theory[M]. Chicago: Univ. of Chicago Press, 1930.
[5] Einstein A, Podolsky B, Rosen N. Can quantum mechanical description of physical reality be considered complete?[J]. Phys. Rev, 1935, 47: 777–780.
[6] Bohm D. Quantum theory[M]. London: Constable and Co., 1954.
[7] Bell J. On the Einstein–Podolsky–Rosen paradox[J]. Physics, 1964, 1: 195–200.
[8] Bell J. On the problem of hidden variables in quantum mechanics[J]. Rev. Mod. Phys, 1965,

38: 447–452.

[9] Aspect A, Grangier P, Roger G. The experimental tests of realistic local theories via Bell's theorem[J]. Phys. Rev. Lett, 1981, 47: 460–465.

[10] Aspect A, Grangier P, Roger G. Experiment realization of Einstein–Podolsky–Rosen–Bohm gedanken experiment, a new violation of Bell's inequalities[J]. Phys. Rev. Lett. 1982,49: 91–96.

[11] Einstein A. Zur elektrodynamik bewegter körper[J]. Ann d Phys. 1905, 17:891–921.

[12] Einstein A. The Field Equations for Gravitation[J]. Sitzungsberichte der Deutschen Akademie der Wissenschaften. Klasse f'ur Mathematik, Physik und Technik, 1915: 844–847.

[13] Einstein A. Die grundlage der allgemeinen relativitätstheorie[J]. Ann. der Phys., 1916, 49: 769–822.

[14] Huang Z X. Does the "Relativity Quantum Mechanics" really exist?[J]. Frontier Science, 2017, 11(4): 12–38.

[15] Dirac P. Lectures on quantum mechanics[M]. New York: Yeshiva University Press, 1964.

[16] Dirac P. Direction in Physics[M]. New York: John Wiley, 1978.

[17] Born M, Einstein A. The Born–Einstein Letters[M]. New York: Palgrave Macmillan, 2005.

[18] Gisin N. L'impensable hasard, non–localitè, tèlèportation et autres merveilles quantiques[M]. Genevai: Odile jacob, 2012.

[19] Salart D, et al. Testing the speed of "spoky action at a distance" [J]. Nature, 2008,454: 861–864.

[20] Shen Z Y. Three questions of physics[J]. Science, 2010, 62(2): 3–4.

[21] Aczel A. Entanglement—The Greatest Mystery in Physics[M]. New Jersey: Avalon Pup., 2001.

[22] Lu H F. Intepretation of Copenhagen group on quantum physics[M]. Shanghai: Fudan Univ. Press, 1984.

[23] Pai C X, et al. Quantum Communication[M]. Xi'an: Electronic Science and Technology Univ. Press., 2013.

[24] Wootters W, Zurek W. A single quantum can not be cloned[J]. Nature, 1982, 299: 802–803.

[25] Mei X C. The proof that the non–cloning theorem of quantum states does not hold[J]. Fund. Jour. of Mod. Phys., 2022,18(1): 27–44.

[26] Kiefer C. Quantum Gravity[M]. Oxford: Oxford Science Publication, 2012.

Study on the Essence of Photons and the Wave-Particle Duality of Light*

INTRODUCTION

Since the birth of natural science in Europe hundreds of years ago, the question of "what is light" (what is, the nature of light) has attracted the attention and thought of scientists. In 1672, I. Newton described his experiment in which he obtained a spectrum of seven colours by dividing sunlight into light at different angles of refraction using a triangular glass prism. Around the same time Newton explained the corpuscular theory of light to explain the reflection of light at the interface. In 1690, C. Hugens proposed the theory that light is a wave, which includes the concepts of "wavelet" and "wave front". In 1802, T. Young did the experiment of two-slit interference of light, which provided the experimental proof of "light fluctuation". In 1818, A. Fresnel calculated the diffraction pattern caused by obstacles such as slit, round hole and round plate, which was consistent with the experiment; Fresnel is credited with developing Huygens' principle. Until now, most physicists believed in the wave theory of light. In 1865, J. Maxwell proposed that light was an "electromagnetic disturbance propagating through a field according to electromagnetic laws", i.e., an electromagnetic wave; In 1887, H. Hertz discovered electromagnetic waves through experiments. The development of the wave theory of light has came to an end, and the whole process took about two hundred years.

Then the salient event was the discovery of photons. The photoelectric effect was discovered by P. Lenard et al. in the late 19th century; However, Maxwell's electromagnetic theory could not explain it. So in 1905 A. Einstein postulated that the energy of light was quantized, that is, made up of "quanta". With the photon quantum hypothesis, A. Einstein explained the photoelectric effect and derived the photoelectric equation. In short, Einstein said that "waves have particle-like properties" according to light, and the energy and momentum of a particles can be the determined by the parameters of the wave (frequency and wavelength), namely $E = hf$, $p = h/\lambda$. From 1905 to 1914, R. Millikan proved the correctness of the photoelectric equation with long-term experiments. Einstein and Millikan were awarded the Nobel Prize in Physics in 1921 and 1923, respectively. In 1924, A. Compton measured the lengthening of the wavelengths of X-rays scattered by graphite.

---

* The paper was originally published in *Global Journal of Science Frontier Research: A Physics and Space Science*, 2023, 23 (10), and has since been revised with new information.

The momentum of the photon was taken into account in the interpretation, and the change in the wavelength of the secondary X-ray generated when the X-ray was scattered could be calculated When the photon collides with the electron, and the calculation was consistent with the actual measurement. At this point, the photon hypothesis was further proved, and Compton was awarded the 1927 Nobel Prize in Physics.

By 1924, the theory of "light fluctuations" had not been refuted by anyone, and at the same time it was established that "light consists of many photons". Thus, a complicated situation arises.

Photons are very strange particles. So far we have only been able to describe photons in terms of abstract physical parameters (frequency, power, dynamic mass, polarization, etc.), but we have not been able to visualize them. What does a photon look like (round, square)? We can't tell. Does a photon have a volume (i.e., geometric size)? We don't know. The erratic nature of photons, which normally exist for a short time, makes them harder to grasp. The current approach to understanding photons indirectly, by looking at pulses of light, is also problematic.

EINSTEIN'S THEORY OF PHOTONS

In 1905, before quantum mechanics (QM) existed, Einstein's light quantum hypothesis was, of course, based on classical physics. Einstein argued that the wave theory of light, which operates in terms of continuous spatial functions, had proved so superior in describing purely optical phenomena that it seems difficult to replace it with any other theory. But optical observations are all about time averages, not instantaneous values. Although the theories of diffraction, reflection, refraction, dispersion, etc., are fully verified by experiments, it is conceivable that the theory of light, operated by continuous spatial functions, will lead to a contradiction with experience when applied to the phenomena of the production and transformation of light. So Einstein hypothesized that the energy of a beam of light emitted from a point source is not continuously distributed over an increasing amount of space in its propagation, but consists of a finite number of quanta confined to each point in space, which can move, but can no longer be divided, but can only be absorbed or produced in their entirety.

Einstein's theory of photons differs from Planck's theory of quantum energy. In 1900 M.Planck's work had only quantized the vibrational energy of the oscillators that make up the wall of the black body, or his quanta had been a computational tool used to derive the radiation formula for radiation. Einstein regards light quanta as a physical reality, and believes that quantization was the basis of electromagnetic radiation and light. Einstein's starting point was the difficulties faced by the theory of blackbody radiation, and the determination of the fundamental quanta by Planck. Starting with Wien's law of Blackbody radiation:

$$\rho = \alpha f^3 e^{-\beta f/T}$$

where f is the frequency and T is absolute temperature; The above formula is fully effective when $hf \gg kT$. From the above formula

$$\frac{1}{T} = -\frac{1}{\beta f} \ln \frac{\rho}{\alpha f^3}$$

Assuming that a radiation of energy E has a frequency between f to $(f+df)$ and occupies a volume v, the entropy of the radiation with volume can be derived if the energy is constant:

$$S - S_0 = -\frac{1}{\beta f} \ln \frac{V}{V_0}$$

S, S_0 are the entropy when the radiation occupancy volume are v, v_0. Therefore, the entropy of monochromatic radiation varies with volume. The above formula can also be written as

$$S - S_0 = -\frac{R}{N} \ln \left[\left(\frac{V}{V_0} \right)^{NE/R\beta f} \right]$$

R is the gas constant, which N is the number of molecules in 1 mole; Citing Boltzman's principle (which states that the entropy of a system is a probability function of its state):

$$S - S_0 = -\frac{R}{N} \ln p$$

We have

$$p = \left(\frac{V}{V_0} \right)^{NE/R\beta f}$$

It is concluded that the monochromatic radiation image with low energy density is composed of a number of unrelated energy quantums of size. $R\beta f / N$. So Einstein said, "Not only the incident light, but also the generated light is composed of the magnitude $R\beta f / N$ of the energy quanta."

With these ideas, Einstein successfully explained the photoelectric effect, photoluminescence, and the effect of ultraviolet light on the ionization of gases. The energy of Einstein's quantum of light is

$$E = \frac{R\beta}{N} f \qquad (1)$$

Compared to the energy quantum proposed by Planck when he formulated the radiation formula in 1900:

$$E = hf \tag{2}$$

If the two theories fit together, there is

$$hf = \frac{R\beta}{N} \tag{3}$$

In formula $\beta=4.866\times10^{-11}$; N is the Avogadro number, which is now customary writing N_A, while the contemporary exact value (standard value) is $N_A=6.02214199\times10^{23}$. The derivation is based on the theory of the motion of gas molecules, where R/N is a universal constant.

It can be seen from Einstein's derivation that light radiation is a large number of photons composed by energy hf. This not only explains the photoelectric effect, but also advances our understanding of how light interacts with matter. This is reflected in the concepts of stimulated and spontaneous radiation that he introduced when he derived the Planck formula for blackbody radiation. Einstein is thus largely responsible for the early development of quantum theory. But after the birth of quantum mechanics (QM) in 1925-1926, Einstein continued to oppose it as we'll see later. Einstein didn't talk about the mass and size of photons and so on. But because of the mass-energy relation $E=mc^2$, it can be obtained in conjunction with equation (2), we obtain

$$m = \frac{hf}{c^2} \tag{4}$$

Therefore, Einstein theory holds that the dynamic mass of a photon depends only on its frequency f; And in the direction of propagation, the momentum of the photon is

$$p = mc = \frac{hf}{c} \tag{5}$$

This mass and momentum derivation makes the photon image particle. Because c is very large, it is small unless f is very large. As for the photon energy, it is calculated using the following formula:

$$E = hf = \frac{hc}{\lambda} \tag{2a}$$

where and f, λ are the corresponding wave frequency and wavelength of the photon respectively. Therefore, we can calculate the wavelength, dynamic mass and energy of the photon. In a wide range of wavelengths, from radio waves to X-rays, the photon has a lower mass than the electron (electron rest mass $m_0=9.109534\times10^{-28}$g). Photon energy can be used as a unit of electron volt (eV), and 1eV=1.60217733×10^{-19}J; Therefore, at the midpoint of the visible light spectrum ($\lambda=5\times10^{-5}$cm), E=2.48eV can be calculated. The energy of the single photon is very small.

In Einstein's theory, the rest mass m_0 of a photon is zero. So does a photon have a volume? In special relativity (SR) there's something called the length shortening formula:

$$l = l_0 \sqrt{1 - \left(\frac{v}{c}\right)^2} \qquad (6)$$

l_0 is the length of the moving body when it is at rest; When the velocity $v = c$, $l = 0$, so the Einstein photon has no volume (the ruler shrinks to zero and becomes a point). The idea that a photon, as a particle with mass, and momentum, and energy, but has no volume. This is controversial.

THE PHOTONS IS NOT A CLASSICAL THING

Einstein won the 1921 Nobel Prize in Physics for his theory of photons, which perfectly explained the photoelectric effect. But in 1951, at the age of 72, he wrote to his old friend Besso: "Fifty years of conscious reflection have brought me no closer to the answer to the question 'What are photons of light?' It is true that every scoundrel now believes that he knows it, but he is deceiving himself." In a 2004 article, historian of science M.Chown said that not many people today know about Einstein's visit to Brazil. The trip left Hamburg on March 5, 1925, for a three-month tour of South America. Brazilian scientists had gathered in Rio de Janeiro to hear Einstein speak about his theory of relativity. But Einstein himself had other ideas; For Einstein, relativity was just an extension of the classical physics of the nineteenth century, but the revolutionary thing in his life was the idea of photons, and that's what he was going to talk about. But waves are spread out across space, and particles are discrete entities. How do you unify the two? Einstein didn't find the answer. Because Einstein used classical physics, it was impossible. —A month after Einstein's lecture in Brazil, W.Heisenberg in Germany invented a new kind of physics, quantum mechanics (QM). The point Einstein couldn't see was that photons were not a classical thing. The night of his presentation to the Brazilian Academy of Sciences on May 7, 1925, marked the end of Einstein's career as a leading scientist. Until his death, Einstein refused to accept quantum mechanics, which replaces certainty with uncertainty. Einstein's speech in Rio de Janeiro said he still desperately hoped that the "monster" (photon) he released in 1905 could still be tamed by old classical physics.

Several decades have passed since Einstein said in 1951 that "we have no idea what a photon"; Do we now have a clear idea of what a photon is? The fact that the "wave-particle duality" is still being debated suggests that the problem remains unresolved. In fact, the particle nature of photons is a difficult subject. It is acknowledged that it is incorrect to regard photons as point particles. On the other hand, according to special relativity (SR), photons have no rest mass ($m_0 = 0$), so classical mechanics is difficult to apply to such "massless particles". Relativistic mechanics is also difficult to apply. For photons,

$$m = \frac{m_0}{\sqrt{1-(v/c)^2}} = \frac{m_0}{0} = \frac{0}{0} \tag{7}$$

m becoming any size is not acceptable; If $m_0 \neq 0$, $m = \infty$, this is also unacceptable. Relativistic mechanics is incapable of dealing with objects such as photons.

A. Compton's 1923 paper on scattering caused by X-rays incident on an element suggested that the incident photons were scattered by collisions with electrons in the element, and the electrons recoxed in the other direction. Let the static mass m_0' of the electron be, the static energy $m_0'c^2$ be, and then form the energy conserved, we obtain

$$m_0'c^2 + hf_i = \frac{m_0'c^2}{\sqrt{1-(v'/c)^2}} + hf \tag{8}$$

The left end of the equal sign above is the incident case and f_i is the incident photon frequency; The right end of the equals sign is the case after collision, v' is the recoil electron velocity, f_i is the photon frequency after scattering by electrons; And by adding conservation of momentum, we can combine these equations

$$\lambda - \lambda_i = 2\lambda_c \cdot \sin^2\frac{\theta}{2} \tag{9}$$

where $\lambda = c/f$, $\lambda_i = c/f_i$, $\lambda_c = h/m_0'c$ (Compton wavelength); The above formula indicates that the wavelength of the photon increases after scattering (the frequency becomes smaller). The above formula was confirmed by Compton experiment, indicating that the photon does participate in the collision process as a particle, and the momentum formula ($p = hf/c$) can indeed be applied to the photon. However, the SR theory can only be applied to massive particles (electrons), and there is no evidence that SR is applicable to photons. This process has echoes of classical mechanics, since momentum is a Newton concept for the motion of macroscopic matter.

As a result, there are only two ways to understand photons—through experiment; Or by quantum mechanics (QM) and quantum electrodynamics (QED). Since there is limited information available from the measurement of photons (only frequency, energy, polarization, and survival time), it is most important to rely on quantum theory to understand photons. Although the Compton experiment proved that photons, like electrons, are physical entities with positive real dynamic mass, it also proved that momentum and energy are conserved in a single collision event of microscopic particles. But the photon is not a billiard ball and cannot be dealt with primarily by classical physics. For example, if we hope to "measure the diameter of the photon", we will probably never get results.

From the point of view of the experimental technology of generating and applying single photon, the condition of generating single photon is that the incident light power is extremely weak (such as less than 10^{-16}W), and the acquisition of single photon depends on the optical pulse signal (or even the electronic pulse signal transformed from the light pulse). We have never seen such reports that scientists have obtained single photon in the form of particles in the experiment. It has been pointed out that the energy of a single photon is (1.6~3) eV, which is calculated according to the frequency of visible light. If considered in terms of 2eV, there is

$$2eV = 3.2 \times 10^{-19} J = 3.2 \times 10^{-12} erg$$

Biophysicists say that the number of photons that enter the pupil and cause vision is 54 to 148. Assuming 100 photons, the energy is equivalent to 3.2×10^{-10}erg. This is the sensitivity of the entire eye; Taking into account some complex reflection and absorption processes, it can be calculated that the number of photons absorbed by the sensor in the absolute threshold time is only 5 to 14, indicating that the human eye is highly constructed and surprisingly sensitive. In fact, only a single photon can activate a cylindrical cell, but in 5~14 cylindrical cells are activated at the same time to cause light perception. But these discussions are all at the energy level; So there is no point in putting too much weight on the number of photons in this discussion.

The fact that a photon has mass (moving mass), but its shape and size cannot be taken into account, is difficult for many people to accept, and probably one of the reasons why Einstein lamented that "I don't know what a photon is." So what can QM tell us? Look at Schrödinger's equation:

$$j\hbar \frac{\partial}{\partial t} \Psi(\mathbf{r}, t) = -\frac{\hbar^2}{2m} \nabla^2 \Psi(\mathbf{r}, t) + U(\mathbf{r}, t)\Psi(\mathbf{r}, t) \tag{10}$$

where mass m is the parameter of particle and wave function $\Psi(\mathbf{r}, t)$ is the parameter of wave; Therefore, the equation combines the description of the wave state with the description of the particle motion state. However, we must note two points: ① Schrödinger equation has non-relativistic properties; ② The equation was proposed for electrons, but later a series of evidences show that it applies to photons. As for ①, according to a 2007 study, Schrödinger equation was found to have the least error when accurately measuring the two-photon transition of a hydrogen atom, followed by Dirac's equation, and the worst Klein-Gordon equation. With respect to ②, although the Schrödinger equation was originally proposed for the motion of electrons in hydrogen atoms, there is evidence that the Schrödinger equation applies to photons—first, the one-dimensional Schrödinger equation analysis of microscopic particles incident towards the barrier (which proves that the barrier is a evanescent state), where "microscopic particles" include photons; And the effect can be seen in the SKC experiment of 1993. Secondly, the WKB method in the analysis of slow-varying index fiber is calculated by using Schrödinger equation, and the motion of photons is the basic process in the fiber. Therefore, the Schrödinger equation has accurately described the behavior of electrons and photons, and the quantum language of wave function has replaced the classical language of particle orbits.

The non-classical nature of photons can also be seen from the principle of indistinguishable identical particles in QM. The homogeneity principle causes the wave function to remain unchanged after the exchange of two particles of the same class. The particles with this symmetry are Bosons, such as photons and mesons, which have integer spin values and do not have to satisfy the Pauli exclusion principle. These are questions that classical physics does not consider. In short, the so-called wave and particle properties actually come from classical physics, but now we can't use classical physics.

THE ESSENCE OF PHOTONS

Now let's try to do something that is not easy—give the essence of the photon. Why do photons exist? Classical physics gives derivation, but cannot explain the essence and strange properties of photons. Quantum theory suggests that photons are non-local particle and differ significantly from ordinary microscopic particles. However, no theory can provide a concrete image of the photon, which causes difficulties in the definition of a single photon. Photons can be obtained from light pulses, the number of which follows the Poisson distribution, and can also be defined from the light power, the photon is actually a kind of weakest light source.

Photon is the energy quantum, with material properties, but physics can not give the size and volume of the photon; Special relativity actually gives photons the image of point particles. In addition, quantum mechanics (e.g. Schrödinger's equation) describes electrons in terms of the wave function $\Psi(\mathbf{r}, t)$, which is spatially located as a probability distribution, which $|\Psi|^2$ is the probability density; However, it is not possible to define a self-consistent wave function for photons, nor to write the corresponding wave equation. And it is well known that the classical Maxwell wave equations do not describe photons satisfactorily.

Rapid advances in quantum informatics technology have forced scientists to consider the behavior of a small number of photons (say, 10,000 or fewer). The success of momentum of light has made it possible to experimentally determine the number of photons contained in a laser beam. Quantum secure communication requires ideal single photon sources (PSPS), in which each light pulse contains only one photon. However, no such sources have been made so far, and the only ones used are approximate PSPS, so the communication is not completely secure. Due to light interference during the day, satellite quantum communication and quantum radar are difficult to work well, such as the microwave frequency band will have a good effect, but the microwave single photon energy is very small, it is very difficult to achieve.

The use of special relativity (SR) to explain photons has failed miserably. The treatment of SR makes the photon a point particle; This is not a good theory by any means. The history of physics offers a lesson in treating electrons as point particles. As we all know, quantum field theory (QFT) and quantum electrodynamics (QED) are considered highly accomplished disciplines; However, the shortcoming of QFT and QED is the famous divergence problem, which is rooted in the fact that it is a point particle field theory. As early as 1940 R.Feynman noted that the "infinite energy of the electron" posed a significant problem for the theory of the electromagnetic field, because the model describing the electron was a point particle. This means that the self-action of a point charge has difficulty diverging. If the electron is regarded as a point without structure, the electromagnetic mass caused by the action of the field on itself is infinite. ... In his lecture on in 1964, P.Dirac, talking about renormalization, he first discussed this problem of electron quality. The mass of the electron is not infinite, of course, but the mass of the electron interacting with the field will change; Dirac points out that it is impossible to assign any meaning to "infinite mass". When people continue to calculate "without the infinitely large term", the results (such as Lamb shift and the anomalous magnetic moment) agree with observations; So say "QED is a good theory" and don't worry about it. Dirac resents this, because a "good theory" is obtained by ignoring some infinity—which is both arbitrary and unreasonable. Sound mathematics, says Dirac, allows you to ignore small quantities, but not the infinite (just because you don't want it). Overall, Dirac believes QFT's success has been "extremely limited."

In fact, the electron is not a point particle, the famous experimental physicist Zhaozhong Ding has long been concerned about the size of the electron measurement problem, not long ago he gave the data is still the electron radius r ⩽ 10^{-17}cm. ... A photon is different from an electron in that the latter has a definite rest mass while the former does not (according to the prevailing view of physics); In addition, electrons have an electric charge and photons have no charge. Still, specifying that photons are point particles can be problematic.

So, what exactly is a photon? Here's our take:

① A photon is a special microscopic particle with no electric charge.

② A Photons is a Boson, which has an unlimited number of particles in each quantum state; The spin is $\hbar$.

③ There is no experimental data about the volume and size of the photon.

④ The upper limit of photon static mass $m_0 \approx 10^{-52}$g; The new theory thinks that may $m_0 \neq 0$.

⑤ It is generally believed that the corresponding wave of photon is electromagnetic wave; However, if the photon is identified as a kind of microscopic particle, it should have the probability wave properties, but there is no photon wave equation at present. In connection with this, it is difficult to define a wave function for a photon.

⑥ There is a view that the wave function of the free photon is the electromagnetic plane wave function; Correspondingly, Maxwell electromagnetic equation is the wave equation of the free-state photon. However, this is only a simplistic view and does not provide a dynamic representation of the physical image of photons. The problem of photon wave equation still needs to be studied.

⑦ It is not clear from SE alone that photon is a microscopic particle similar to electron.

A further discussion is now necessary. We begin by pointing out that photons are much more different from electrons than they are similar. Now, write the classical Maxwell wave equation:

$$\left(\nabla^2 - \varepsilon\mu \frac{\partial^2}{\partial t^2}\right)\Psi(\mathbf{r}, t) = 0 \tag{11}$$

where the wave function Ψ is electric field intensity $\mathbf{E}(\mathbf{r}, t)$ or magnetic field intensity $\mathbf{H}(\mathbf{r}, t)$ is a vector function; Respectively ε, μ are the dielectric constant and magnetic permeability of the medium through which the electromagnetic wave passes; This is the wave equation derived in 1865, which we call MWE. Now write the quantum Schrödinger wave equation again:

$$\left(\frac{\hbar^2}{2m}\nabla^2 + j\hbar\frac{\partial}{\partial t} - U\right)\Psi(\mathbf{r}, t) = 0$$

where Ψ is the wave function of probability wave, m is the particle mass, U is the potential energy function. This is the wave equation derived in 1926, which we call SE. Finally, write the Dirac quantum wave equation:

$$\left(j\hbar\frac{\partial}{\partial t} + j\hbar c\alpha \cdot \nabla - \beta m_0 c^2\right)\Psi(\mathbf{r}, t) = 0 \tag{12}$$

This is the wave equation derived in 1928, which we call DE.

Quantum scientist, Prof. Yongde Zhang believes that DE, as a single electron wave function equation, only takes into account the external electromagnetic field action, does not consider the electromagnetic action of the electron itself, so it is still an approximate equation. Secondly, it cannot be generalized to photons, because Dirac assumes that the particle position is an observable quantities. For photons, this assumption is not valid, and the photon problem cannot be described. Photons have no real representation of coordinates; Although a quasi-wave function description containing mechanical variables is sometimes imposed on the photons, it does not have a normal physical interpretation of the wave function (the modulus square is the probability density). A photon can, however, have a momentum representation, which is sufficient for practical purposes. ... It is not virtually possible to define a self-consistent wave function for photons, and there is no probability wave equation for photons, which there is for electrons.

The corresponding wave of the electron is the probability wave, so the wave equation of the electron is SE; The wave function of the electron is the $\Psi(\mathbf{r}, t)$ in SE, which is a probability wave function. Thus, SE fully represents and explains the properties of the electrons. But for photons things are not so simple; One idea is that the Maxwell wave equation is the wave equation of a free-state photon, because the corresponding wave of a photon can only be an electromagnetic wave. However, it is wrong to say that the two are "completely the same", because it is precisely because Maxwell electromagnetic wave theory can not explain the photoelectric effect, Einstein proposed the light quantum theory and explained the photoelectric effect perfectly. Although the photon's scalar nature and the photon flow's statistical essence both make it a probabilistic wave. However, there is no special probability wave equation for photons, so it is difficult to define the wave function for photons. So M.Lanzagorta says: "the photon cannot be localized also implies that we cannot define a consistent wave function for the photon."

Lanzagorta also states: "The photon is a non-localizable particle"; Also, "the photon cannot be localized"; The meaning of these words is consistent with that stated by Prof. Yongde Zhang. Further, Lanzagorta pointed out the difference between photons and ordinary microscopic particles, such as electrons. "It is a mathematically impossible to build a continuity equation using localization probability distributions that satisfy Einstein's special relativity "; it is impossible to write down a wave equation for the photon. It is impossible to write down a wave equation for the photon. It can be seen that studying photons is more difficult than studying electrons.

PHOTON NUMBER EQUATION AND PHOTON NUMBER MEASUREMENT

Classical optics also recognizes photons and mentions such particles in the discourse, but never considers the number of photons. This is because the number of photons contained in a beam of light (natural light or laserlight) is so large that talking about 10,000 photons is like talking about a drop of water in the ocean. Quantum optics, on the other hand, think down to the behavior of one or two photons. Recent developments in quantum informatics (QIT) have made it even more important to consider the one-photon, two-photons problem. For example, the concepts of "one-photon Quantum Radar" and "two-photon Quantum Radar" using entangled states appear in QR theory. Although they have not yet been realized in practice, the novelty and boldness of their imagination are quite surprising. For example, quantum communication technology uses a single photon string (single photon series), and also uses entangled states, which is said to have been successful;

This also gives people a sense of disbelief.

In QM there is so-called many-body theory, which involves, for example, writing rigorous quantum field equations when you consider many electrons, and then trying to solve them. The multi-photon problem is also a multi-body theory problem. In short, the multi-particle quantum theory is more complex and rigorous than the single-particle quantum theory. But that doesn't mean it's easy to master when it comes to single particles. In fact, functional devices with large numbers of particles (such as classical radar, which emits electromagnetic waves) are easier to implement; And devices that work with only one (or two) particles, such as single-photon quantum radars, are extremely difficult or impossible to succeed with.

So how many photons are there in a beam (microwave beam or laser beam)? So far, no one has measured them. This may be calculated or estimated; For example, there is a claim of 10^{18} photons, but we don't know on what. Now this paper deduces the equation related to the number of photons (n); The momentum of a single photon is given above

$$p = \frac{hf}{c}$$

The electromagnetic wave quanta can write

$$p = hk$$

The vector from is written as

$$\boldsymbol{p} = h\boldsymbol{k} \tag{13}$$

where $\boldsymbol{p}$ is the photon momentum vector; $\boldsymbol{k}$ is the wave vector.

If the number n of photons of a beam is, then the photon momentum equation should be written

$$\boldsymbol{p} = nh\boldsymbol{k} \tag{14}$$

This gives the photon number equation:

$$n = \frac{p}{hk} \tag{14a}$$

When the frequency is known, the magnitude of k is determined; However, it is no longer the momentum p of a single photon, but the momentum of the beam. Recently, there are reports abroad that Canadian scientists have proposed a new technique for measuring the momentum of light, referred to as the macroscopic momentum of light beams, which is very helpful to our topic.

Now, says scientists at the University of British Columbia in Canada: "We have not been able to determine how this momentum is transformed or moved. Since light carries so little momentum, the sensitivity of the equipment we have has not been sufficient to solve this problem." However, a new technique could finally help solve this 150-year-old conundrum, they say: "We cannot directly measure the momentum of a photon, so our approach is to detect its effect on the mirror by 'listening' to the elastic wave passing through it. We were able to trace the momentum of the light pulse itself from the signature of these waves, which opens the door to finally defining and simulating how light momentum exists inside a material."

I believe that the new result is actually the design of a new device designed to measure the weak interaction between photons. This provides a new approach to way to measure the number of photons in a beam of light, which we are considering. And it doesn't make sense to say that each photon is massless but that a large number of photons in a beam have momentum. Although the macroscopic momentum of a beam should be caused by the photon's dynamic mass, this scientific result should urge us to consider whether the photon's static mass is a problem.

PHOTON STATIC MASS AND PROCA WAVE EQUATION

Photons and neutrinos are two types of particles that still excite mystery. Whether they have a small, non-zero resting mass has been the subject of much debate. Traditional physical theories such as Maxwell's electromagnetic Theory and Special relativity (SR) suggest that photons have no resting mass, i.e., $m_0 = 0$; For this reason, photons are called "massless particles" to distinguish them from "massless particles" like electrons; The latter are also called matter particles.

Although efforts to measure the rest mass of photons have never stopped, and theories like quantum electrodynamics (OED) assume that the rest mass of photons is not zero; People still assume that photons are massless particles. Now it seems that this may not only be a barrier to understanding photons, but also one of the reasons for the lack of self-consistency in basic physical theories.

Particle physics usually assumes that the Lorentz-Einstein formula for mass velocity is true:

$$m = \frac{m_0}{\sqrt{1-v^2}} \quad (15)$$

where v is the particle velocity, c is the speed of light, m_0 is the rest mass at $v=0$. Physics textbooks never say that the above formula doesn't apply to photons, so let's give it a try; If you take $m_0 = 0$, $v = c$, you get $m = 0/0$; Becoming any size is not acceptable. The problem can only be caused by the following three aspects: (1) The formula of mass-velocity is wrong; (2) The photon's static mass is not zero; (3) The speed of the photon is not c. Obviously, any of these three things are inconsistent with special relativity; In fact, Einstein's theory could not explain his discovery of photons.

Another embarrassment in fundamental physics is that Schrödinger's quantum wave equation (SE) and Dirac's quantum wave equation (DE), both of which are important components of quantum mechanics, have particle mass m, whereas Maxwell's wave equation has no mass. Of course, in Maxwell's day (1865) there was no concept of wave-particle duality; But we do now, so what? SE and DE are very successful wave equations that describe the motion of electrons; Contrast this with the following: electrons are microscopic particles with rest mass; SE and DE have mass parameters m in the equation; These two points have internal logical consistency and are also excellent interpretations of wave-particle duality. ... But there is no mass parameter in ME. Does this mean that Maxwell equations are not accurate enough and need to be corrected? ... In addition, since the introduction of optical fiber in 1970, SE has been successfully illustrated; This suggests that SE can be used to analyze the participation of photons in physical processes, but there is a question: what m is in SE? ... Finally, the volatility of a microscopic particle depends on its statistics, and what the wave function represents is only a wave of probability. This principle holds true for electrons; It's a paradox for photons—although photons are the result of quantization of electromagnetic fields and waves, electromagnetic waves are not probability waves. Thus, the phrase "wave function modulus squared is probability density" does not apply to photons; Are photons still microscopic particles?

It must be noted that if a photon is a particle with a resting mass, all of these paradoxes are countable and the system is far more consistent than it otherwise would be. There are many physicists who do not believe that photon $m_0=0$, such as R.Lakes, a professor at the University of Wisconsin who has been working on measuring the rest mass of photons, once said firmly: "The photon is massive!"

In 1936, A.Proca proposed a new set of electromagnetic field equations, Proca assumed that the rest mass of photons $m_0 \neq 0$. This section deals with this problem. And we will derive the wave equation specific to photons.

The process of deriving the classical Maxwell electromagnetic wave equation (ME) is as follows: starting from the curl equation and incorporating the action of other equations into the derivation, the classical electromagnetic wave equation expressed by the intensity of the electric field can be derived from $\nabla \times \boldsymbol{E} = -\dfrac{\partial \boldsymbol{B}}{\partial t}$, we obtain:

$$\nabla^2 \boldsymbol{E} - \varepsilon\mu \frac{\partial^2 \boldsymbol{E}}{\partial t^2} = \nabla\left(\frac{\rho}{\varepsilon}\right) + \mu\frac{\partial}{\partial t}(\rho\boldsymbol{v}) \tag{16}$$

On the other hand, from the beginning, the classical electromagnetic wave equation expressed by the strength of the electromagnetic field can be derived from: $\nabla \times \boldsymbol{H} = \boldsymbol{J} + \dfrac{\partial \boldsymbol{D}}{\partial t}$, we obtain:

$$\nabla^2 \boldsymbol{H} - \varepsilon\mu \frac{\partial^2 \boldsymbol{H}}{\partial t^2} = -\nabla \times (\rho\boldsymbol{v}) \tag{17}$$

It can be seen that these two equations do not have a symmetric form. Only when the space is passive ($\rho=0$), the two equations have formal symmetry, so that the wave function $\Psi(\mathbf{r},t)$ can be used to express the sum uniformly, and a simplified formula $\left(\nabla^2 - \varepsilon\mu\dfrac{\partial^2}{\partial t^2}\right)$ $\Psi(\mathbf{r},t)=0$ can be obtained.

In a similar way, it can be derived from two curl equations of Proca equations. Now we write the Proca field equations:

$$\nabla \cdot \mathbf{D} = \rho - \kappa^2 \varepsilon \Phi \tag{18}$$

$$\nabla \cdot \mathbf{B} = 0 \tag{19}$$

$$\nabla \times \mathbf{H} = \mathbf{J} + \frac{\partial \vec{D}}{\partial t} - \frac{\kappa^2}{\mu}\mathbf{A} \tag{20}$$

$$\nabla \times \mathbf{E} = -\frac{\partial \mathbf{B}}{\partial t} \tag{21}$$

where $\mathbf{A}$ is the magnetic vector potential, Φ is the electrical marker potential, and the coefficient is K:

$$K = \frac{c}{\hbar}m_0 \tag{22}$$

So, relative to ME, two of the 4 equations have changed, and are related to m_0. Now, ME becomes a special case of the Proca system of equations (PE) at $m_0=0$. Since mass is introduced into the basic field equation, the derived wave equation will also be m relevant, so that some irrational phenomena in the physical theory can be solved. In a similar way, Zhi-Xun Huang derived the Proca wave equation expressed by the electric field intensity $\mathbf{E}$:

$$\nabla^2 \mathbf{E} - \varepsilon\mu\frac{\partial^2 \mathbf{E}}{\partial t^2} = \nabla\left(\frac{\rho}{\varepsilon}\right) + \mu\frac{\partial \rho}{\partial t} - \kappa^2 \frac{\partial \mathbf{A}}{\partial t} \tag{23}$$

where $\mathbf{A}$ is the magnetic vector potential, satisfied; $\mathbf{B} = \nabla \times \mathbf{A}$. For free space ($\rho=0$), it can be proved that:

$$\nabla^2 \mathbf{E} - \varepsilon\mu\frac{\partial^2 \mathbf{E}}{\partial t^2} - \kappa^2 \mathbf{E} = 0 \tag{24}$$

Similarly, we can prove

$$\nabla^2 \mathbf{H} - \varepsilon\mu\frac{\partial^2 \mathbf{H}}{\partial t^2} - \kappa^2 \mathbf{H} = 0 \tag{25}$$

The above two expressions are symmetric, they are Proca wave equation (PWE); If $K=0$, it is converted to the electromagnetic wave equation for photons without rest mass.

In short, Proca theory is useful for the study of photons and expands the idea. Although photons and electrons are very different; But if photons also have particles of rest mass (even very small), they have probability probabilistic properties like electrons. We can see this in the words of the Nobel Prize Laureate in Physics in 1933, who said, "With the introduction of light quantums, quantum mechanics must abandon the requirement of causality." ... The laws of physics represent the odds of an event occurring—our senses and instruments are imperfect, we can only sense the average, so our laws of physics deal with the odds. ... Even so, it is not wrong to find the "photon probability wave equation", but such an equation did not exist in the past. If we accept that photon wave is statistical, then they are indeed different from classical waves (such as mechanical waves and sound waves), and they do not seem to be the same as electromagnetic waves.

Since Proca himself did not deduce the wave equation, I make up for this gap and give PWE; But how to apply PWE in practice remains to be studied. As for other effects after adopting the Proca theory, it is only emphasized here that the phase velocity (v_p) and group velocity (v_g) of Proca wave are related to the angular frequency, showing the dispersion effect of electromagnetic waves in vacuum:

$$v_p = \frac{c}{\sqrt{1-\left(\frac{\omega_c}{\omega}\right)^2}} \tag{26}$$

$$v_g = c\sqrt{1-\left(\frac{\omega_c}{\omega}\right)^2} \tag{27}$$

where c is the speed of light in vacuum, and is the characteristic angular frequency:

$$\omega_c = \kappa c = \frac{m_0 c^2}{\hbar} \tag{28}$$

Obviously, when $\omega = \omega_c$, $v_p = \infty$, $v_g = 0$; When $\omega > \omega_c$, for a finite value greater than the speed of light, and for a finite value less than the speed of light. So, in the Proca context, the wave speed is not the speed of light even in free space conditions, and the size of the difference depends on the frequency. Of course, the difference is small, because the actual value is large. When $\frac{\omega}{\omega_c}=10$, $v_p=1.005\,c$, $v_g=0.995\,c$, the difference between phase velocity and group velocity is only 0.5%. In fact $\frac{\omega}{\omega_c} \gg 10$ (e.g. $10^6 \sim 10^{10}$), so the difference between the phase velocity, group velocity, and the speed of light is very small. Nevertheless, this is fundamentally different from the conventional theory ($v_p = v_g = c$, in free space).

To establish a quantitative concept of rest mass for microscopic particles, here are some data-electron rest mass $m_e = 9.10938188 \times 10^{-28}$g, which is the international recommended value in 1998; However, one value of the upper limit of the photon rest mass is 2×10^{-50}g, and the other is 1.2×10^{-51}g. It can be seen that the photon's rest mass, if not zero, is much smaller than that of the electron.

THE CONCEPT OF WAVE-PARTICLE DUALITY OF LIGHT MAY BE PROBLEMATIC

The basic concept in physics is that matter is made up of particles. In this way, separateness seems to "override" the continuity embodied by fields and waves. But you can't get rid of waves; It is true, for example, that light is made up of clusters of photons, But light is also a wave, and the properties of light waves are characterized by continuity parameters such as frequency, wavelength, phase and amplitude. Lose the "light wave" and there is no light. For example, "quantum" (energy quantum, light quantum) is a discrete concept, but the energy hf of each quantum, and the frequency f is a continuous concept, can be measured very accurately... It is clear, therefore, that since all particles are quanta of corresponding fields, the most fundamental forms of matter can be considered to be fields, with their related and corresponding waves. In this way, continuity again seems to "overwhelm" the separateness. Therefore, any theory based on the "one overriding the other" is not a good theory.

The theory of photons proposed by Einstein in 1905 not only successfully explained the photoelectric effect, but also made people realize that radiation of any frequency (wavelength) is particular, and the energy and momentum of particles are

$$E = hf \tag{29}$$

$$p = h/\lambda \tag{30}$$

where h is Planck constant ($=6.62606876 \times 10^{-34}$J·S). In 1924, Louis de Broglie proposed that wave-particle duality is not limited to light radiation, and is equally unavoidable when describing the behavior of matter particles such as electrons. He pointed out that if a particle (energy, momentum) is incident, it must carry a wave (first called phase wave, later called matter wave), its frequency and wavelength are respectively E p

$$f = E/h \tag{29a}$$

$$\lambda = h/p \tag{30a}$$

On the surface, formula (29a) and formula (30a) are the same as (29) and (30), but in fact they are different concepts. Einstein is based on the theory that waves have particle-like properties, which has been supported by experimental phenomena. De Broglie's theory that particles have waviness was based on electrons took several years to prove experimentally (the electron diffraction experiment of C. Davisson and I. H. Germer in 1928).

After the birth of quantum mechanics, it was suggested that its two systems (Heisenberg's matrix mechanics and Schrödinger's wave mechanics) represented exactly different directions for particles and waves, exacerbating the contradiction between the two images. Schrödinger later argued that the fundamental thing is a wave, and that the wave packet formed by superposition of the eigensolutions of the wave equation is a particle. N.Bohr pointed out that the particles were stable, but the wave packet gradually spread out (fat) as it propagated, so Schrödinger's statement was incorrect.

In May 1927, W.Heisenberg said that Bohr had made him realize that "uncertainty in experimental observations is related to wave-particle duality" and does not arise only from discontinuous particles or continuous waves. In September, N.Bohr gave a lecture on his idea of complementarity—that waves and particles have contradictory images while complementing each other. Bohr argued that what appears as a particle in one context appears as a wave in another. However, some physicists (such as de Broglie) believe in "wave-particle association", that is, being both a wave and a particle at the same time.

The fifth Solvy Conference was held in October 1927, with the central topic being "electrons and photons" and the development of quantum mechanics. At the meeting, de Broglie proposed the "waveguide theory" (also known as the double solution theory), which believes that the wave equation of quantum mechanics has two solutions, one is the continuous wave function ψ, representing the monochromatic plane wave; The other is a singular solution, where the singularity represents the particle. The idea that the wave guides the particles, as if the particles were straddling the wave, was not supported. In 1955, de Broglie published his final formula for the theory of double solutions:

$$u = Fe^{js/\hbar}$$

And think that the above formula represents the particle structure, and the exponential term represents the wave. During this period, de Broglie was concerned with nonlinear equations, with the study of solitons and solitons at the time, and argued that "particles are localized peaks in waves".

N.Bohr had proposed a principle 90 years earlier. It has to do with subatoms such as electrons and photons, which seem to behave like particles in some experiments and waves in others. According to Bohr, these particle and wave properties are complementary, and no experiment can show both properties at the same time. According to Bohr, the photon concept implies an unexpected dilemma: the particle picture cannot be reconciled with the effect of interference, which can only be explained by fluctuations. Moreover, the dilemma is exacerbated by the fact that interference is the only means of providing the idea of frequency and wavelength. The situation was not yet clarified, and Broglie pointed out that wave-particle duality was also unavoidable in explaining the behavior of matter particles. Bohr points out that there are opposites in the dilemma, each of which relates to an essential aspect of experience. So what is to be done? Bohr argues that what is encountered is the opposition revealed by observations of microscopic objects made by different experimental devices, a situation that has no precedent in classical physics and can be called "complementarity" or "complementarity". In theory, quantum mechanics makes it possible to describe complementarity by writing the commutation relation of a pair of conjugate variable operators (p, q) after replacing the physical variables with operators:

$$qp - pq = j\frac{h}{2\pi} \qquad (31)$$

Here $j = \sqrt{-1}$. At the same time, specify the limits of causal analysis with the uncertainty relation:

$$\nabla q \cdot \nabla p = \frac{h}{4\pi} \qquad (32)$$

Bohr argues that the complementarity view is appropriate both to encompass the individuality of quantum phenomena and to illuminate other unique aspects of experience. Whereas classical physics does not consider interference with object by means of observation in scientific experiments, this kind of neglect is not allowed in the study of quantum physics—for example, when electrons are observed under a gamma microscope, gamma photons interfere with the electrons, meaning that the object is associated with the instrument and loses its independent reality. N.Bohr had long argued that the two major requirements of classical physics (natural phenomena subject to strict causality and the description of object phenomena according to the laws of space-time) could not be met at the same time. Since no clear distinction can be drawn between phenomena and instruments, it is not surprising that the methods of quantum mechanics are limited to statements of statistical regularities; Bohr thinks that complementary descriptions can be used to cope with this situation.

In Bohr's view, since the simultaneous measurement of position and momentum for a microscopic particle is mutually exclusive, one can only gain knowledge of the object in a complementary sense. That is, position and momentum are a complementary pair of observable quantities in quantum mechanics. In fact "complementarity" can be generally expressed as follows: for every dynamical degree of freedom there is a pair of complementary observables.

All of the above theories seem profound on the surface, but they are in fact problematic, as I pointed out earlier. In 2013, Huang Zhixun proposed that for the wave and radiation with very low frequency (wavelength is very long), the concept of particle property is difficult to establish. For example, if you take $f=100Hz$, then $\lambda = c/f = 3 \times 10^6 m$; And suppose that in microwave $f=10GHz=10^{10}Hz$, then $\lambda = 3 \times 10^{-2} m = 3cm$. In both cases, the wavelength of the former is extremely long, and it is impossible to imagine that such waves can correspond to "particles." The wavelength of the latter is in the order of centimeters, and it is still difficult to imagine the corresponding "particles", but perhaps it is possible to start thinking about particles. Obviously, the wave-particle duality requires certain conditions, and this condition is that the wavelength is small enough; But the current theory does not give a boundary for wavelength (or frequencies). In other words, complete self-consistency in theory is not achieved. This alone shows that there is nothing wrong with the fundamental concepts of physics.

My criticism, though simple, points directly to the fundamental flaw in the concept of wave-particle duality. Unfortunately, my ideas has so far escaped the attention of the mainstream physics community.

PROFESSOR ZHIYUAN LI'S OUTSTANDING CONTRIBUTION TO THE STUDY OF WAVE-PARTICLE DUALITY

In recent years, Chinese physicist Zhiyuan Li has persisted in studying the problem of wave-particle duality. His theoretical analysis is both serious and profound, and his experimental design is unique. Now, it is Professor Li who takes a clear stand against the wave-particle duality theory. He has been studying the problem of wave-particle duality for 20 years. He points out that in the past it was widely believed that wave-particle duality was governed by the Heisenberg uncertainty relation. In the 1980s and 1990s, a number of scholars proposed atomic interferometer designs based on quantum optics and atomic physics methods, in which the detection of particle paths and interference fringes is not bound by the uncertainty relationship, but the quantum entanglement between the two still ensures that the microscopic particles obey the wave-particle duality principle. The application and extension of the concept of quantum entanglement in atomic interferometers has also led to the emergence of physical concepts such as "delayed selection" and "quantum erase" that go against people's intuitions.

This paper briefly introduces Li's research methods and uses two figures (Figure 1 and Figure 2) as an auxiliary illustration.

Li's research method is to admit that the time-tested and unambiguous system of quantum mechanics operating norms is also applicable to photon and atom interferometers, and then rigorously solve the Schrödinger equation to obtain the exact solution of the wave function evolution, and then further analyze the particle information and wave information of microscopic particles. Quantitatively calculate the values of path discernability D and interference fringe visibility V, and finally see whether the interferometer's operating results follow or violate the standard wave-particle duality principle.

Professor Li realized that the evolution of the wave function was the only thing that could be predicted based on SE, according to the standard operating rules of quantum mechanics. Photons and other microscopic particles follow SE as they move through the Mach-Zehnder interferometer shown in Figure 1:

$$j\hbar\frac{\partial}{\partial t}\Psi(\mathbf{r}, t)=H(\mathbf{r}, t)\Psi(\mathbf{r}, t) \qquad (33)$$

The mainstream view today is that when BS2 in an interferometer system is a quantum beam splitter, the Hamiltonian of the system can be simply written as $H(\mathbf{r})=a_w H_{in}(\mathbf{r})+a_p H_{out}(\mathbf{r})$, then the wave function of the system is $\psi=a_w \psi_w+a_p \psi_p$. The measurement results of the detector X and Y are show as the linear superposition of two abstract quantum states: particle property and wave property. At this time, the microscopic particles are in the quantum superposition state of partial particle state and a partial wave dynamics. This is exactly the conclusion obtained by several international teams carrying out "quantum delayed selection experiment". In contrast, the early view of quantum physics was that microscopic particles were either in full particle states ψ_p or in full wave dynamics ψ_w, one or the other.

However, this analysis in the face of everyday experience. It is hard to imagine in actual experiments splitter BS2 the macroscopic objects in "in the mood" $[H(\mathbf{r}, t)=H_{in}(\mathbf{r})]$ and the "state" $[H(\mathbf{r}, t)=H_{out}(\mathbf{r})]$ of quantum superposition. For this reason, Li proposed a more natural physical explanation in mid-2016: the time-modulated Hamiltonian model. In this model, BS2 is in a time-modulated state of "entering state" and "coming out state", meaning that one moment is in "entering state" and the other is in "coming out state", and the two cannot co-exist in time coordinates. Math in such a state can be expressed as $H(\mathbf{r}, t)=H_{in}(\mathbf{r})$ and $H(\mathbf{r}, t)=H_{out}(\mathbf{r})$ $H(\mathbf{r}, t)=H_{mod}(\mathbf{r}, t)=a(t)H_{in}(\mathbf{r})+b(t)H_{out}(\mathbf{r})$, and the corresponding wave function for the two kinds of particles and volatility of abstract quantum state time modulation state. $\psi=\psi_m(t)=a(t)\psi_w+b(t)\psi_p$. The physical starting point of the above two models is not the same, and the physical image and physical meaning are also very different.

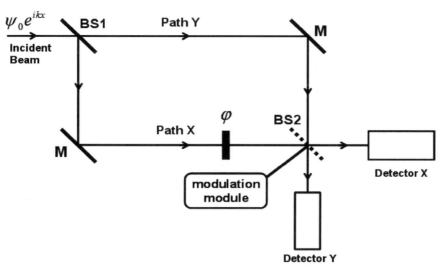

Fig. 1 Delay selection experiment based on Mach-Zehnder dual-arm interferometer

Importantly, the theoretical results clearly show that it is impossible to observe both the wave and particle properties of microscopic particles based on such an interferometer delay selection experiment scheme, because the path discernability and the visibility of interference fringes meet the condition $V^2+D^2 \leq 1$. Is it possible, then, to break through the limitations of the wave-particle duality principle so that $1 \ll V^2+D^2 \leq 2$?

Professor Li further analyzed the propagation and evolution of the wave function in the interferometer, and found that at the spatial intersection of the two particle beams, the wave functions will overlap, which should in principle produce interference fringes. But this simple physical picture was completely ignored in the past few years. Herefore, he proposed a new type of weak measurement interferometer, replacing BS2 with another interference fringe weak measurement device (such as using weak scattering and weak absorption techniques), while obtaining high-contrast interference fringes, the microscopic particles can continue to propagate in the original direction (99% efficiency), which is detected and recorded by the particle detectors X and Y. This new interferometer uses two sets of measuring instruments acting simultaneously, without interfering with each other, and in principle can observe the wave (V=1) and particle properties (D=0.99) of the microscopic particles at the same time, $V^2+D^2 \gg 1 \to 2$. This simple device can be applied to photons, electrons, atoms, molecules, and quasticles in condensed matter, etc. The key is to design and manufacture a weak measurement detector that can record interference fringes without changing the information of particle motion path. This analysis also shows that the delay-selection experiment belongs to the strong measurement interferometer in principle, and the functional device BS2 to detect particle properties has a huge destructive interference to the wave function. The two sets of instruments repel each other, so it is impossible to detect particle properties and wave properties at the same time.

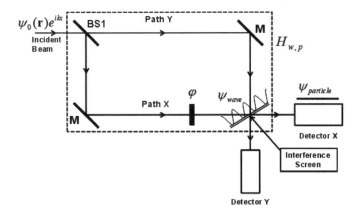

Fig. 2 Scheme of Mach-Zehnder photon interferometer for weak measurement.

Now a complete quantum mechanical modeling, calculation and analysis is needed.

In 2019, Professor Li started from SE, wave function and other basic elements to establish a reasonable physical model and conduct quantum mechanical solutions. First of all, it is determined that the atom interferometer is a quantum scattering or propagation problem, rather than a quantum eigenstate problem. Therefore, the whole process of space-time evolution of the atomic wave function must be calculated and solved, starting from the incident wave function, tracking its changes through all atomic optical devices such as beam splitter, phase shifter, path detector, and finally arriving at the interference fringe detector. Secondly, the atom is a complex system, containing the atomic centroid and the outer electrons, so the evolution of the atomic centroid wave function and the internal electronic state must be considered at the same time. Third, to evaluate the wave-particle duality of an atom, it is necessary to aim at the atom's centroid rather than its inner electrons. Therefore, it is necessary to analyze, calculate and measure the path information and interference fringes of the atomic center of mass, but not the path information and interference fringes of the internal electrons.

In his 2019 paper, he distinguished between strong measurement path detectors, which worked for early atomic interferometers, and weak measurement path detectors, which act directly on an atom's centroid to obtain precise information about its position. Introducing the interaction the total Hamiltonian is $H(R) = H_{free}(R) + H_{int}(R)$ Hamiltonian, it is simply. To solve the propagation and evolution of the atomic wave $\delta\varphi$ function, it is found that the influence of the path detector on the centroid wave function of the incident atomic beam can be described simply and reasonably by introducing random phase offset. The wave functions of the two propagation paths are finally overlapped on the interference fringe observation screen and the total wave function is written as

$$\psi(r) = a(r)\psi_{1,\text{tran}}(r) + b(r)\psi_{2,\text{tran}}(r) = \psi_0 e^{jkl}[a_1 a(r)e^{j\varphi_1} e^{j\delta\varphi_1} + [a_2 b(r)e^{j\varphi_2} e^{j\delta\varphi_2}] \quad (34)$$

$$I(r) = \overline{A|\psi(r)|^2} = A[a^2 + b^2 + 2ab\ \overline{a_1 a_2 \cos(\varphi_1 - \varphi_2 + \delta\varphi_1 - \delta\varphi_2)}] = A[a^2 + b^2] \quad (35)$$

This shows that when the path detector acts directly on the atom's center of mass, the overstrong interaction leads to a large random phase shift, which destroys the quantum coherent superposition of the two path propagation wave functions, making the interference fringe completely disappear.

Another class of path detectors are weak measurement detectors, including microwave microcavity path detectors and Bragg grating path detectors, which obtain deterministic path information through direct strong measurement interactions with the internal electronic states of atoms (indirect interactions for atomic centroids, H_{int}^{iqs} (R, r) described by the Hamiltonian). In order to obtain the effect of H_{int}^{iqs} (R,r), it is necessary to conduct fully quantum-mechanical modeling, analysis and calculation of the atom-electron-detector interaction system. For the microwave microcavity weak measurement path detector, the Hamiltonian of the system is $H_S = H_{acm} + H_{photon} + H_{electron} + H_{int}^{p.e.} + H_{int}^{a.e.} + H_{int}^{a.p.}$. It is generally believed in quantum physics circles that the interference fringes disappear due to the quantum entanglement between the path detector and the atomic path information, thus maintaining the wave-particle duality principle. However, in practice, the information of the interference fringe detector and the path detector are artificially associated by coincidence counting operation. Then the obtained measurement results of interference fringes already contain the path information of atoms, that is, the interference fringes are modulated by the electronic state transition information or the evolution information of the particle path detector. Therefore, non-physical quantum entanglement and quantum erasure phenomena are artificially generated.

Zhiyuan Li believes that the quantum entanglement effect claimed by some people in the past is to observe the wave-particle duality of the internal electronic states of atoms, and the test of the truly concerned atomic wave-particle duality does not work. In other words, the electrons inside the atom satisfy the wave-particle duality principle, while the atomic center of mass violates the wave-particle duality principle. The theoretical analysis results show that the wave and particle properties of atoms can be observed at the same time on this weakly measured atomic interferometer device, as long as the correct physical model and analysis method are adopted.

In 2020, Professor Xuewen Chen and his doctoral team designed and built the weak measurement photon interferometer on the basis of Professor Li's theoretical research, and tested the wave-particle duality of single photons. They demonstrated the simultaneous observation of the wave and particle properties of single photons with conclusive evidence, and its comprehensive index broke through the limitations of the wave-particle duality principle of classical quantum mechanics. The experimentally observed interference fringes show not only the fluctuations of those photons that are scattered, but also the fluctuations of the photons that enter the interferometer from the single photon source (the object of wave-particle duality detection). Therefore, the Chinese scientists' experimental results show that the wave and particle properties of photons can be observed simultaneously.

DISCUSSION

Professor Li's rigorous thinking and profound theoretical analysis are excellent works. We noticed that he started from the original thinking methods and operating tools of quantum mechanics, that is to say, he did these studies not to oppose QM, but to find a new quantum mechanics.

Now we can make a comparison. QM holds that all microscopic particles (whether they have mass or not) have wave-particle duality, sometimes appearing as particles (which have definite orbits) and sometimes as waves (which can produce interference fringes); This depends on the experimental method of the observer. But it is not possible to observe both at the same time, in fact, the root point is a quantum relationship that is both mutually exclusive and complementary, and any experiment will lead to uncertainty about their conjugate variables; Therefore, the complementarity principle is consistent with the uncertainty relation. ... But Einstein, the inventor of the theory of photons, has long recognized the paradox that light is both a wave and a particle. But by rejecting the uncertainty principle, he could not accept Bohr's complementarity theory, which saw uncertainty relations as an example and consequence of the complementarity principle.

In the author's opinion, Professor Li's theory and experiment only prove that it is possible to observe the particle properties and the volatility of photons at the same time, but not that these two properties must exist in the object at the same time in the microscopic world. To prove that the latter is a universal law of the objective world seems to be far away. Therefore, his work does not represent a repudiation of the principle of complementarity, quantum non-locality, quantum entanglement, the uncertainty principle, quantum statistical interpretation, etc. That is to say, a scientist who admits in advance the basic principles and modes of operation of QM will not conclude at the end of his research work that "quantum mechanics is wrong".

For the sake of caution, I put a few questions to Prof. Li: "(1) It is said that in 2007 and 2011, A. Stiinberg directed the publication of two papers opposing the 'principle of complementarity'. If this is true, then you are not the first to oppose the complementarity principle? (2) If the complementarity principle is dead, what next? What is your opinion? The complementarity principle is linked to the Heisenberg uncertainty principle, so is it a threat to quantum mechanics? (3) de Broglie argues that "particles ride on waves". what do you think? (4) Your thoughts and activities have always been within the framework of the QM, not bent on destroying it, as Einstein did. This is evident from your clever use of quantum weak measurements. (5) I have a personal suggestion for using nonlinear SE to help understand this problem."

Zhiyuan Li gave the following reply: "There are many people in history who have opposed the Copenhagen orthodox interpretation of quantum mechanics. I am just only a Chinese scholar who has thought about this. However, all of them have failed to shake the foundation of quantum mechanics... I'm not a big fan of the principle of complementarity, especially when combined with two-particle quantum physics, which naturally gives rise to bizarre, counterintuitive physical explanations of two-electron, two-photon quantum entanglement and so on. I am a committed materialist and firmly opposed to any physical theory that leads to idealism."

Before that, however, Professor Li had spoken of his thoughts, which I recorded: "How do microscopic particles move? Newton mechanics and Schrödinger equation (SE). Is quantum mechanics a complete statistical description of microscopic particles (massless photons and massless electrons, protons, atoms, molecules)? Is there a deeper, unknown law behind it that describes the behavior of individual particles? ... In today's highly developed technology of quantum physics, quantum optics, atomic physics and quantum regulation, these are not only philosophical questions, but also scientific questions that can be tested."

On the surface, Professor Li seems inclined to revert to the EPR paper and to J.Bell's earlier ideas. But I don't think so. His point is to re-examine many of the fundamental questions with modern technology. Moreover, he could not have been unaware of two things, one is that Schrödinger's equation (SE) was derived from Newton mechanics, not from relativistic mechanics. Second, J.Bell proposed the Bell inequality from the original intention of the EPR paper, but after Aspect's precise experiments in supporting of QM were published, Bell completely switched sides and sided with QM.

Now I suggest that further analysis should begin with the introduction of nonlinearities to the Schrödinger equation (SE), which we earlier studied in 2016. LSE can be written as:

$$j\hbar \frac{\partial \Psi}{\partial t} = \hat{H}\Psi$$

where $\hat{H}$ is the linear operator, the Hamiltonian of the system. If a nonlinear term is added to it, the nonlinear Schrödinger equation (NLSE) can be obtained; take

$$\hat{H} = \frac{\hbar^2}{2m}\nabla^2 + U \qquad \text{(LSE)} \qquad (36)$$

$$\hat{H} = \frac{\hbar^2}{2m}\nabla^2 + U - \beta|\Psi|^2 \qquad \text{(NLSE)} \qquad (37)$$

where β is the nonlinear coefficient; Take $U=0$, the NLSE is:

$$j\hbar \frac{\partial \Psi}{\partial t} + \alpha \nabla^2 \Psi + \beta |\Psi|^2 \Psi = 0 \qquad (38)$$

In formula $\alpha = -\hbar^2/2m$; It can be proved that NLSE can have solitary wave solutions, indicating that SE can strengthen its particle image by introducing nonlinearity.

We know that in SE, the kinetic energy operator $\left[-\frac{\hbar^2}{2m}\nabla^2\right]$ is already the embodiment of the particle property, in which m represents the particle mass. Therefore, SE is not only the basic equation of quantum wave mechanics, but also the equation that has reflected the duality of wave and particle. Now, the nonlinear term is introduced, and the isolated wave solution is obtained under the action of the dispersion effect. This not only overcomes the problem of SE's wave packet divergence, but also better presents particle properties in the wave equation ... Perhaps what we have said above can explain the results of recent experiments?

CONCLUSION

A physicist once asked, while giving an academic talk, "What is a photon? What does it look like? What is the size? How does it move? How to interact with matter? What are the space-time details?"

Here, we try to answer. It is well known that an electron has a rest mass and that it has a size (radius $r \leq 10^{-17}$cm), not a point particle. Mainstream physicists — to the photon has no rest mass, then it will not inch, become a point particle. However, any theory in physics that is based on point particles is not a good theory.

This paper argues that the hypothesis of "photon has no rest mass" causes the lack of self-consistency of the theory. For mass photons, the Proca equations proposed in 1936 can be used to replace the Maxwell equations. We derive a new wave equation for electromagnetic waves and photons, called Proca wave equation (PWE). In PWE there are terms containing particle mass parameter, which is consistent with Schrödinger wave equation and Dirac wave equation. This improved the theoretical relationship and draws a clear line between massive photons and point particles.

In short, some strange phenomena of photons (such as photon self-interference, homomorphic photon interference, single photon passing through double slit at the same time, quantum post-selection, quantum entanglement, etc.) can not be explained by traditional classical and deterministic methods. As R.Feynman said, we do not know why nature is the way it is, but we must accept it as it is; The important thing is that theory and experiment must agree.

The author believes that when strange phenomena appear in the microcosmic world, if they have been proved to exist by experiments, or even have been applied in practice, then we should acknowledge and accept them. Feynman is absolutely right that the only thing scientists can do is to be loyal to nature.

It must be admitted that there are still many contradictions and confusion regarding the photon and the wave-particle duality of light, and further study is undoubtedly right. But for things in the microcosmic world, perhaps some ambiguity that needs to be dealt with without undue attention to this. Photons belong to the quantum world, where the distinction between a particle and wave becomes blurred, and the notion of "size" becomes vague and meaningless. We can describe the behavior of photons mathematically, but we can't visualize them as regular images ... Now the particles of the wave are both present in separate states and may be observed simultaneously. This is the wonder of nature, the wonder that fascinates us!

REFERENCES

[1] Suplee C. Physics in 20th Century[M]. New York: Abrans, 1999.
[2] Speziali P. Albert Einstein–Michel Besso Correspondance, 1903–1955[M]. Paris: Hernmann, 1942.

[3] Stacy R. Essentials of biological and medical physics[M]. New York: Mc Graw Hill, 1955.

[4] Dirac P. Directions in physics[M]. Hoboken: John Wiley, 1964.

[5] Lanzagorta M. Quantun Radar[M]. San Rafael: Morgan and Claypool Pub., 2012.

[6] 黄志洵. 关于电磁波特性的一组新方程[J]. 中国传媒大学学报(自然科学版), 2019, 26(05): 1-6+18.

[7] 黄志洵. 影响物理学发展的8个问题[J]. 前沿科学, 2013, 7(03): 59-85.

[8] 黄志洵. 非线性薛定谔方程及量子非局域性[J]. 前沿科学, 2016, 10(02): 50-62.

Circular Section Metal Wall Cut-Off Waveguide Accurate Calculation when Used as a Standard Attenuator[*]

INTRODUCTION

In microwave measurement technology, the parameters to be measured are frequency, power, impedance, attenuation, phase shift, etc., and attenuation measurement occupies a prominent position. Microwave attenuation measurement methods, an important method is the replacement method, including high frequency replacement method (microwave replacement method), medium frequency replacement method, low frequency replacement method (audio replacement method), the essence is to compare the unknown attenuation with the known precise value, so as to measure the value of the unknown attenuation. Therefore, a high precision standard attenuator is required. The attenuator of WBCO (cut-off waveguide attenuator) is commonly used for IF substitution. Because the receiving electrode is usually a mobile structure (mounted on a piston), it is also called piston attenuator. In addition, because of its highest accuracy, it can be used as a national measurement standard, also known as the primary standard of attenuation.

It must be pointed out that the accuracy of the microwave attenuation measurement system is lower than that of the cutoff attenuator used as the standard; For example, if the accuracy of the latter is 5×10^{-5}, the former is $10^{-3}\sim10^{-4}$. In China, the microwave attenuation measurement system produced by microwave instrument factories in the 1980s had an error of no more than one thousandth. Of course, the accuracy of 5×10^{-5} standard attenuator is difficult to do, but in 1980 by the Chinese Academy of Metrology (NIM) research group completed the project reached this level, known as the first attenuation standard or National attenuation basic standard. —Below this is the Secondary attenuation standard, its standard cut-off attenuator accuracy is 5×10^{-4}, then the accuracy of the entire microwave attenuation measurement system is 10^{-3}, that is, the error is not more than one thousandth.

From 1966 to 1980, the American Bureau of Standards (NBS), the National Institute of Physics Labaratory (NPL), and NIM all built a first-stage attenuator working in the medium frequency (30MHz), all using a cut-off waveguide attenuator. Professor Mengzong Song of NIM is the chief designer, and engineers Yongjian Xia and Ronggen Mo are responsible for the structural design. The machining problem of precision circular waveguide is solved. Zhixun Huang is responsible for theoretical calculation and error analysis. In addition, there

---

[*] The paper was originally published in *Global Journal of Science Frontier Research: A Physics and Space Science*, 2024, 24 (1), and has since been revised with new information.

are other researchers involved in this project, after several years of efforts finally successful, technical indicators reached the international advanced level. This paper is a comprehensive theoretical summary by the author, including my new equations, which are contributions to the basic waveguide theory.

STANDARD ATTENUATORS IN MICROWAVE ATTENUATION MEASUREMENT

The working principle of the cutoff waveguide attenuator is the exponential decline characteristic of the evanescent field in the waveguide when the waveguide cutoff point is below ($f<f_c$, $\lambda>\lambda_c$), and a circular waveguide is placed along the direction z:

$$E = E_0 \, e^{-\alpha z}$$

The field strength at $z=0$ (starting point) is E_0. If $z=l$, the field strength is E_l, there is:

$$E_l = E_0 \, e^{-\alpha l} \tag{1}$$

If the attenuation generated by a waveguide of length l is defined as A, then there is

$$A = \ln \frac{E_0}{E_l} = \alpha l \quad (\text{Np}) \tag{2}$$

For example, when $A=20\text{dB}=2.3\text{Np}$, $e^{-A}=0.1$. This means that when the design is $\alpha=1\text{dB/mm}$, the field strength decreases to 1/10 of the starting value after only 20mm along the direction z. This is a fast rate of descent, indicating that large attenuation, such as more than 100dB, can be obtained only with a shorter cut-off waveguide.

When the electromagnetic wave propagates in the metal wall waveguide, there will be a cutoff phenomenon, that is, only the frequency $f>f_c$ wave, above the cutoff frequency, can propagate. At the frequency of $f<f_c$, the evanescent field is characteristic, which is a reactive energy storage field. Such devices, known as cut-off waveguides (WBCO), have prominent uses in electronic instrument design. In 1942, E.Linder gave a formula for calculating the attenuation constant of the cutoff waveguide:

$$\alpha = \frac{2\pi}{\lambda_c} \sqrt{1-\left(\frac{\lambda_c}{\lambda}\right)^2} \tag{3}$$

The cut-off wavelength λ_c is determined by the size of the waveguide cross section. We know that length and time are the two basic units of metrology; It is now associated with two basic units, which determines that it can obtain α high accuracy (if the waveguide machining size is very precise). ... Of course, Linder's formula is not very accurate, so improving the formula is a work for future generations. Both R. Grranthan in 1948 and

Circular Section Metal Wall Cut-Off Waveguide Accurate Calculation when Used as a Standard Attenuator

R.Rauskolb in 1962 made efforts, but the accuracy of their formulas was not high enough to meet the requirements for the design and construction of first-order standard attenuators. The problem was solved by me.

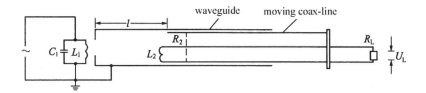

Fig. 1 Example of Output Attenuator for a Meter Wave Standard Signal Generator

To visualize the actual operating state of the cut-off waveguide, Figure 1 shows the output circuit of a standard signal generator capable of obtaining a voltage output in the microvolt (μV) range. When the receiving electrode (L) mounted on the coaxial line is moved, the output voltage U is uniformly and continuously changed for use by the experimenter. Here, l represents the length of the actual operating cut-off waveguide.

The main mode of the circular waveguide, strictly written as HE_{11} mode, indicates that the effect of the finite conductivity of the wall cannot be ignored, and the H_{11} mode (TE_{11} mode) under ideal conditions will be slightly affected by the mode coupling. But generally speaking, the main mode in the circular waveguide is H_{11} mode (TE_{11} mode); If its inner diameter is $2a$, the cutoff wavelength of the mode is

$$\lambda_{c.11} = \frac{2\pi}{k_{11}} a = 3.412 a \tag{4}$$

Set λ to the operating wavelength (depending on the signal source), so in order to obtain the H_{11}-mode evanescent field must be satisfied:

$$a < \frac{\lambda}{3.412} \tag{5}$$

Here are the transverse dimensions of the circular waveguide used in the primary attenuation standards built by the highest metrological institutions in several countries: $2a$ =3.2cm(NIM); $2a \approx 5.06$cm(NPL); $2a \approx 8.12$cm(NBS).

The relation A-l equation of the cutoff attenuator is not linear at all locations and has A nonlinear segment at smaller l (and therefore smaller A). In order to maintain high accuracy, the application can avoid this section, that is, avoid the initial attenuation section of about 20dB, that is, maintain the linear relationship $A = \alpha l$. The laser length measurement technology can be accurately measured, so the focus of the study turns to the attenuation constant α. It must be pointed out that the outstanding contribution of the author is the monograph *Introduction to the Theory of Waveguide Cutoff Below*, published in 1991, which makes the cutoff waveguide theory into a complete system, and derives some new equations and formulas, and gives the solution methods. In the book, it is pointed out that for microwave attenuation measurement, a branch of metrology, in the 30 years from 1950 to 1980, the accuracy of the cutoff attenuator standard was increased from 5×10^{-3} to close to 5×10^{-5}, because it is basically a calculation standard, and the attenuation constant is mainly determined by the basic unit (length and frequency). In the case of the increasing level of machining and surface treatment technology, it is required to theoretically derive the attenuation constant formula with higher accuracy (higher than 5×10^{-5}). For the HE_{11} mode in the metal-walled circular waveguide, the formula is derived as:

$$\alpha_{11} = \frac{2\pi}{\lambda_c}\sqrt{\frac{1}{2}\left[1-\left(\frac{\lambda_c}{\lambda}\right)^2\varepsilon_r-J_{11}\right]+\frac{1}{2}\sqrt{\left[1-\left(\frac{\lambda_c}{\lambda}\right)^2\varepsilon_r-J_{11}\right]^2+J_{11}^2}} \quad (Np/m) \qquad (6)$$

among

$$J_{11} = \frac{g\tau}{a} = \mu_{rc}\left[1+\frac{1}{k_{11}^2-1}\left(\frac{\lambda_c}{\lambda}\right)^2\right]\frac{\tau}{a} \qquad (7)$$

where μ_{rc} is the relative magnetic permeability of the wall metal; ε_r is the relative dielectric constant of the filling medium; τ is the skin depth of the inner wall, $\tau=(\pi\mu\sigma f)^{-1/2}$, where σ is the RF conductivity of the wall metal, $\mu=\mu_{rc}\mu_0$ is the magnetic permeability of the wall metal. The calculation shows that the formula derived by the author is the most accurate one of the same kind (the accuracy is 1×10^{-6}), which can meet the calculation needs of establishing high precision national attenuation standard.

It can be obtained from formula (2):

$$dA = \frac{d\alpha}{\alpha}A+\alpha\, dl \qquad (8)$$

The above formula indicates the method of giving electrical performance indicators for standard cutoff attenuators, which is very important. The first term on the right side of the equation is caused by the circular cut-off waveguide, and the second term is caused by the length measuring (displacement measuring) mechanism. Since the 1970s, due to the development of laser length measurement technology, the second item has been greatly reduced (such as displacement resolution up to 0.0001dB), so the main task is to reduce $d\alpha/\alpha$. The basic parameters of the first order cutoff attenuation standard developed successfully by various countries are derived from two main factors: the uncertainty of the radius of the circular waveguide (da/a) and the uncertainty of the RF conductivity of the waveguide wall material ($d\sigma/\sigma$). The operating mode is $H_{11}(TE_{11})$, and the operating frequency is 30MHz.

Now look at where the US and China have reached. According to literature [11], the

NBS of the United States in 1960 was as follows: the average inner diameter of the cut-off waveguide was 81.21015mm (at 20°C), the diameter machining uniformity was ±0.762μm, the diameter measurement uncertainty was ±1.27mm, the radio frequency conductivity of the tube wall was $1.2825\times10^{-7}(\Omega\,\text{m})^{-1}$, and the uncertainty was $d\sigma/\sigma=\pm5\%$. For such a waveguide, the attenuation constant $\alpha=0.393701$ dB/mm (at 20°C). The partial error and the uncertainty of pipe diameter caused by $\pm3\times10^{-5}$, the small change of temperature in the constant temperature room caused by $\pm1\times10^{-5}$, and the uncertainty of pipe wall conductivity caused by $\pm2.6\times10^{-5}$. NBS technical report gives $\alpha=0.393701$ dB/mm. $d\alpha/\alpha=\pm1\times10^{-4}$, that is, one part in 10,000.

In addition, according to literature [4], the situation of NIM in China in 1980 was as follows: average inner diameter of cutoff waveguide 31.96254mm (at 20°C), non-uniformity of pipe diameter processing ±0.5μm, uncertainty of pipe diameter measurement ±0.6μm, radio frequency conductivity of pipe wall $=1.6034\times10^{-7}(\Omega\,\text{m})^{-1}$, uncertainty ±4%. In this case, the attenuation constant $\alpha=0.99995775$ dB/mm (20°C); After synthesizing the partial errors, the NIM technical report gives $d\alpha/\alpha=5\times10^{-5}$, that is, 5 parts per 100,000.

It can be seen that the attenuation standard established by China's NIM is better than that established by the NBS of US. As for a lot of theoretical analysis work is not NBS.

HISTORICAL BACKGROUND: FROM THOMSON EQUATION TO CMS EQUATION

British scientist Joseph Thomson (1856-1940) won the Nobel Prize in Physics in 1906 for his discovery of the electron. In 1893 J. Thomson published a book: *Notes on New Research in Electromagnetics*—a continuation of Professor Clerk Maxwell's *Treatises on Electricity and Magnetism*. It fully affirms the realizability of transmitting electromagnetic waves in a circular metal-walled tube (that is, a circular waveguide). Why is Thomson so early can come to the right conclusion? This is mainly because his mathematical starting point is correct: first analyze the electrical vibration of the cylindrical cavity inside the conductor, that is, solve the two-dimensional wave equation. This made him the first scientist in history to predict waveguides.

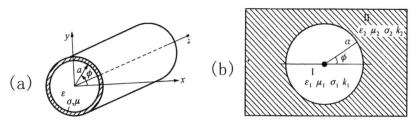

Fig. 2 Circular waveguide

For the reader's ease of understanding, Thomson's analysis is presented in today's notation, and its system of units is changed to SI. A medium cylinder with radius of (a) is buried in a uniform infinite conductor space (ε_c, μ_c), and the cylindrical coordinate system (r, ϕ, z) is taken for analysis (FIG. 2). Thomson showed that the following partial differential equations held in the medium:

$$\frac{\partial^2 H_z}{\partial x^2} + \frac{\partial^2 H_z}{\partial y^2} = \frac{1}{V^2} \frac{\partial^2 H_z}{\partial t^2}$$

He called V "the speed of electrodynamic action" propagating through a medium, which is actually the speed of electromagnetic waves. The following partial differential equation holds in the conductor:

$$\frac{\partial^2 H_z}{\partial x^2} + \frac{\partial^2 H_z}{\partial y^2} = \sigma \mu_c \frac{\partial^2 H_z}{\partial t}$$

where σ is electrical conductivity. The above two formulas can be written as

$$\nabla_t^2 H_z = \frac{1}{v^2} \frac{\partial^2 H_z}{\partial t^2} \quad \text{(in medium)}$$

$$\nabla_t^2 H_z = \sigma \mu_c \frac{\partial^2 H_z}{\partial t} \quad \text{(in conductor)}$$

This was Thomson's starting point. When you take a cylindrical coordinate system, then

$$\frac{\partial^2 H_z}{\partial r^2} + \frac{1}{r} \frac{\partial H_z}{\partial r} + \frac{1}{r^2} \frac{\partial^2 H_z}{\partial \phi^2} = \frac{1}{V^2} \frac{\partial^2 H_z}{\partial t^2} \quad \text{(in medium)}$$

$$\frac{\partial^2 H_z}{\partial r^2} + \frac{1}{r} \frac{\partial H_z}{\partial r} + \frac{1}{r^2} \frac{\partial^2 H_z}{\partial \phi^2} = \sigma \mu_c \frac{\partial^2 H_z}{\partial t} \quad \text{(inside the conductor)}$$

Thomson assumed H_z changes according to the law of $\cos m\phi \cdot e^{j\omega t}$, which in fact introduced the steady state, monochromatic simple harmonic concept proposed by Helmholtz (1821-1894) in 1860. In this way, two ordinary differential equations are obtained:

$$\frac{d^2 H_z}{dr^2} + \frac{1}{r} \frac{dH_z}{dr} + \left(\frac{\omega^2}{V^2} - \frac{m^2}{r^2} \right) H_z = 0 \quad \text{(in medium)} \tag{9}$$

$$\frac{d^2 H_z}{dr^2} + \frac{1}{r} \frac{dH_z}{dr} - \left(j\omega \mu_c + \frac{m^2}{r^2} \right) H_z = 0 \quad \text{(in conductor)}$$

where m is a positive integer. Let $\gamma_c = \sqrt{j\omega \mu_c \sigma}$, then the equation in the conductor is

$$\frac{d^2 H_z}{d(\gamma_c r)^2} + \frac{1}{\gamma_c r} \frac{dH_z}{d(\gamma_c r)} - \left[1 + \frac{m^2}{(\gamma_c r)^2} \right] H_z = 0 \tag{10}$$

The equation in the medium is Bessel equation, while in the conductor it is Bessel equation with virtual variable.

Obviously, the equation in the medium has only the following special solutions:

$$H_z^I = A \cos m\phi \cdot J_m\left(\frac{\omega}{V}r\right) \cdot e^{j\omega t} \tag{11}$$

where A is the constant. The other particular solution of the general solution (Neumann function) does not exist, because it is necessary to ensure that the field at $r=0$ is finite. For the equation in the conductor, there may be two special solutions are $I_m(\gamma_c r)$ and $\mathbf{K}_m(\gamma_c r)$, but the bigger, r I_m bigger, this does not meet the physical reality, so there is only one special solution:

$$H_z^{II} = B \cos m\phi \cdot \mathbf{K}_m(j\gamma_c r) \cdot e^{j\omega t} \tag{12}$$

where B is the constant. Therefore, the change in the medium field is according to the Bessel function; Within the conductor, the field is modified by the second class Bessel function. The latter description is what makes him unique.

Thomson derived the characteristic equations of circular waveguide H_{mn} modes, and it is surprising that this work succeeded at the end of the 19th century. His method was to obtain two different equations with two boundary conditions, combine them to eliminate the two arbitrary constants, and finally obtain a transcendental equation, called the characteristic equation (a term later used in the 20th century, but not by Thomson). The two boundary conditions are as follows:

① At the interface between the medium and the conductor, the "magnetic field line" parallel to the surface is continuous, which can be written as

$$H_z^I \big|_{r=a} = H_z^{II} \big|_{r=a}$$

② The interface between the medium and the conductor is continuous with r the vertical "electric strength". In the present expression, it can be written as

$$E_\phi^I \big|_{r=a} = E_\phi^{II} \big|_{r=a}$$

The first boundary condition causes

$$A\, J_m\left(\frac{\omega}{V}a\right) = B\, \mathbf{K}_m(j\gamma_c a)$$

However, in order to apply the second boundary condition, we must first write the expression, i.e

$$E_\phi^I \approx -\frac{1}{j\omega\varepsilon}\frac{dH_z^I}{dr} \quad \text{(in medium)}$$

$$E_\phi^{II} \approx -\frac{1}{\sigma}\frac{dH_z^{II}}{dr} \quad \text{(in conductor)}$$

This way, we can write

$$\frac{dH_z^I}{dr} = A \cos m\phi\, J_m'\left(\frac{\omega}{V}r\right) e^{j\omega t}\frac{\omega}{V}$$

$$\frac{dH_z^{II}}{dr} = B \cos m\phi\, K_m'(j\gamma_c r) e^{j\omega t} \cdot j\gamma_c$$

By the second boundary condition, we get

$$A\frac{1}{j\omega\varepsilon}\frac{\omega}{V}J_m'\left(\frac{\omega}{V}a\right) = B\frac{1}{\sigma}\cdot j\gamma_c\, K_m'(j\gamma_c a)$$

Simultaneous solution, and set $\mu = \mu_0$, $\mu_c = \mu_{rc}\mu_0$, then

$$\frac{\omega}{Va}\frac{J_m'\left(\frac{\omega}{V}a\right)}{J_m\left(\frac{\omega}{V}a\right)} = \frac{\mu_{rc}}{j\gamma_c a}\frac{K_m'(j\gamma_c a)}{K_m(j\gamma_c a)} \tag{13}$$

This is the characteristic equation of the circular waveguide when the wall conductivity is finite, that is, the equation (86) in Thomson's book.

Thomson pointed out that when "the wavelength of the electric vibration can be compared with the diameter of the cylinder ($2a$)", that is, when the wavelength is very short and the frequency is very high, it is $\gamma_c a$ very large, then

$$K_m(j\gamma_c a) = (-1)^m e^{-\gamma_c a}\sqrt{\frac{\pi}{2\gamma_c a}}$$

Thus there is

$$K_m'(j\gamma_c a) = jK_m(j\gamma_c a)$$

At this time, the right end of equation (13) becomes $\mu_{rc}/\gamma_c a$, which is a small quantity, so when the frequency is very high (i.e. microwave), the following equation is obtained as an approximate solution of the equation:

$$J_m'\left(\frac{\omega}{V}a\right) \tag{14}$$

This means that the tangential component of the electric field strength at the interface is zero. At this time, take $m = 1$ to obtain the equation of "maximum period of electric vibration":

$$\frac{\omega}{V}a = 1.841$$

From this, the cut-off wavelength of the H_{11} mode is

Circular Section Metal Wall Cut-Off Waveguide Accurate Calculation when Used as a Standard Attenuator

$$\lambda_{c.11} = 0.543 \times 2\pi a = 3.142a \tag{15}$$

This is slightly larger than half a circumference (πa). The above concept is completely correct, but Thomson did not have a "mode" at the time.

Thomson pointed out that if $\lambda > \lambda_c$ (the operating wavelength is greater than the cut-off wavelength), "the electrical vibration is not attenuated", which is the electromagnetic wave transmission. However, if we carefully examine the right side of the characteristic equation, we find that there is a small imaginary number term, "which marks the gradual reduction of vibration". That is to say, when the finite conductivity of the wall is taken into account, there is a small attenuation in the propagation of the electromagnetic wave.

The "wave whose wavelength is comparable to the diameter of a cylinder" predicted by Thomson turned out to be the microwave. The circular waveguide he predicted was realized 43 years later (i.e. 1936) by scientists at Bell Laboratories (BTL) in the United States. This shows that the predictions of science can be made much earlier than the development of technology, and demonstrates the power of applied mathematics. It is commendable that he deals with the finite conductive wall at the beginning, thus making the discussion close to reality and giving people a deeper impression. Under microwave conditions, he pointed out that his characteristic equation can be solved according to the ideal conductive wall, so as to find the cutoff frequency and cutoff wavelength, and the error is not large. In this way, Thomson correctly solved the problem of calculating the sum of the principal modules. In addition, he pointed out that the finite conductive wall must cause a small attenuation of the guided wave. His shortcomings lie in: (1) In the derivation process, it is assumed that the conductor σ is very large and the medium $\sigma=0$; (2) From the beginning, it is stipulated that it is a type of magnetic wave mode. Two points (1) and (2) determine that his analysis is not very strict, especially the second point is the most important.

Thomson started with the H_z scalar wave equation. However, when the conductivity is finite, the electric and magnetic fields must be superimposed to satisfy the boundary conditions at $r=a$. Therefore, the resultant field is neither a transverse magnetic field nor a transverse electric field, but a hybrid mode. A TM or TE mode can exist alone only if the wall conductivity $\sigma = \infty$.

From 1893 to 1936, a long gap was formed in the field of waveguide theory. It was not until 1936 that J. Arson and three others published a characteristic equation that solved the problem of both assuming finite wall conductivity and mixed mode analysis. This is known as the Carson-Mead-Schelkunoff equation, and since both are finite wall analyses, it should be possible to derive J. Thomas's equation from the CMS equation. On account of

$$j\gamma_c a = j\sqrt{j\omega\mu_c\sigma} a = \sqrt{-j\omega\mu_c\sigma} a = a\omega\sqrt{\frac{\mu_c\sigma}{j\omega}}$$

So when $\sigma \gg \omega\varepsilon$, there is

$$j\gamma_c a = a\omega\sqrt{\mu_c \varepsilon_c} = ak$$

where ream $k = \omega\sqrt{\mu_c \varepsilon_c}$, let

$$v = a\sqrt{k^2 + \gamma^2}$$

So when $k^2 \gg \gamma^2$, there is

$$v \approx j\gamma_c a$$

So the right end of the characteristic equation is

$$\frac{\mu_{rc}}{ak}\frac{K'_m(ak)}{K_m(ak)} = \frac{\mu_{rc}}{v}\frac{K'_m(v)}{K_m(v)}$$

On the other hand, if the medium is a vacuum (ε_0, μ_0), then

$$\frac{\omega}{V}a = \omega\sqrt{\mu_0 \varepsilon_0}\, a = k_0 a$$

where $k_0 = \omega\sqrt{\mu_0 \varepsilon_0}$, let

$$u = a\sqrt{k_0^2 + \gamma^2}$$

When $k_0^2 \gg \gamma^2$, we have

$$u = \frac{\omega}{v}a$$

So Thomson's characteristic equation can also be written as

$$\frac{1}{u}\frac{J'_m(u)}{J_m(u)} = \frac{\mu_{rc}}{v}\frac{K'_m(v)}{K_m(v)} \tag{16}$$

But mathematics knows

$$K_m(jx) = (-1)m\left(\frac{\pi}{2}j^{m+1}\right)H_m^{(1)}(x)$$

where $H_m^{(1)}$ is the first kind of order Hankel function, so there is m

$$\frac{K'_m(jx)}{K_m(jx)} = \frac{H_m^{(1)'}(x)}{H_m^{(1)}(x)}$$

The characteristic equation can be obtained as

$$\frac{1}{u}\frac{J'_m(u)}{J_m(u)} = \frac{\mu_{rc}}{v}\frac{H_m^{(1)'}(v)}{H_m^{(1)}(v)} \tag{17}$$

Obviously, this is part of the CMS equation, or it can be immediately derived from the CMS equation. It can also be seen that Thomson's equation does not contain waveguide propagation constant γ, which will undoubtedly greatly reduce the application value of the equation.

A NEW METHOD FOR ACCURATE SOLUTION OF CMS EQUATION

In the development of waveguide theory, a prominent research direction is the characteristic equation method, which is remarkably effective for cylindrical guided wave systems and can be applied in other cases. The essence of the method is to derive the eigenequation containing the longitudinal propagation constant ($\gamma = \alpha + j\beta$), which is actually the eigen value equation. For cylindrical guided wave structures, the initial derivation was for metal-walled circular waveguides, but it was soon found that this method could be generalized to other guided wave systems.

Figure 2(b) shows a vacuum cylinder (ε_0, μ_0) embedded in an infinitely large conducting medium (ε_c, μ_c). For the sake of universality, the macro parameters of region I are $\varepsilon_1, \mu_1, \sigma_1$ and the macro parameters of region II are $\varepsilon_2, \mu_2, \sigma_2$. In this way, a generalized mathematical physical model can be constructed, which is applicable not only to metal wall circular waveguides, but also to a series of other guided waveguides, such as single-line surface waveguides, dielectric tube circular waveguides, dielectric rod circular waveguides, and optical fibers. The generalized characteristic equation is derived by J.Carson et al., in cylindrical coordinate system, the field vector is written as

$$\boldsymbol{E} = E_r \boldsymbol{i}_r + E_\phi \boldsymbol{i}_\phi + E_z \boldsymbol{i}_z$$

$$\boldsymbol{H} = H_r \boldsymbol{i}_r + H_\phi \boldsymbol{i}_\phi + H_z \boldsymbol{i}_z$$

where i is the unit vector. Starting from the two curl equations ($\nabla \times \boldsymbol{H} = j\omega\varepsilon\,\boldsymbol{E}$, $\nabla \times \boldsymbol{E} = -j\omega\mu\,\boldsymbol{H}$) in Maxwell equations, the relationship between the transverse field component and the longitudinal field component can be found out when electromagnetic wave propagates along the cylinder. First specify the following symbols:

$$h_1 r = r \sqrt{\omega^2 \varepsilon_1 \mu_1 + \gamma^2}$$

$$h_2 r = r \sqrt{\omega^2 \varepsilon_2 \mu_2 + \gamma^2}$$

Ignoring the mathematical derivation, the CMS equation can be obtained:

$$-\left[\frac{\mu_1 J'_m(u)}{u\, J_m(u)} - \frac{\mu_2 H'_m(v)}{v\, H_m(v)}\right]\left[\frac{k_1^2 J'_m(u)}{\mu_1 u\, J_m(u)} - \frac{k_2^2 H'_m(v)}{\mu_2 v\, H_m(v)}\right] = m^2 \gamma^2 \left(\frac{1}{u^2} - \frac{1}{u^2}\right)^2 \quad (18)$$

where $u = h_1 a$, $v = h_2 a$. The above equation is the generalized characteristic equation of cylindrical wave. The cylindrical function in the formula is a transcendental function with infinite roots for each value, so the solution of the above equation is a discrete spectrum, and a value can be obtained for each normal wave type, so it is an eigen value equation.

Carson et al.'s paper used 8th-order determinants to find non-zero solutions under Bezout condition when 8 constants were treated as variables, while we use 4th-order determinants. This is because we only take cosm ϕ or sinm ϕ for analysis; This can be done because these two types of modes are polarially degenerate. However, the derivation is an eigen-valued equation, and the degeneracy model of the eigen-valued problem can be ignored, so the derivation method can be greatly simplified.

For the most common non-ideal conductive wall circular waveguide, region 1 is air and region 2 is conductive medium, so the characteristic equation becomes

$$-k_0^2 \left[\frac{1}{u} \frac{J_m'(u)}{J_m(u)} - \frac{\mu_{rc}}{v} \frac{H_m'(v)}{H_m(v)} \right] \left[\varepsilon_{r1} \frac{J_m'(u)}{u J_m(u)} - \left(\varepsilon_{r2} + \frac{\sigma}{j\omega\varepsilon_0} \right) \frac{1}{v} \frac{H_m'(v)}{H_m(v)} \right] = m^2 \gamma^2 \left(\frac{1}{v^2} - \frac{1}{u^2} \right)^2 \quad (19)$$

it is

$$u = a\sqrt{\mu_{r1}k_0^2 + \gamma^2}$$

$$v = a\sqrt{k_2^2 + \gamma^2}$$

Zhixun Huang and Jin Pan propose a theoretical analysis and a new numerical solution, see below.

Suppose the region I is the non-dissipative medium and the region II is conductor, then

$$\varepsilon_1 = \varepsilon_{r1}\varepsilon_0; \mu_1 = \mu_0; \mu_2 = \mu_c = \mu_{rc}\mu_0$$

$$\varepsilon_2 = \varepsilon_c = \varepsilon_{r2}\varepsilon_0 + \frac{\sigma}{j\omega} = \varepsilon_0 \left(\varepsilon_{r2} + \frac{\sigma}{j\omega\varepsilon_0} \right)$$

where $\varepsilon_2(\varepsilon_c)$ is the complex permittivity. Then the wave numbers of two regions become:

$$k_1^2 = \omega^2 \varepsilon_{r1}\varepsilon_0\mu_0 = \varepsilon_{r1} k_0^2; k_2^2 = \omega^2 \varepsilon_c \mu_c = \mu_{rc} k_0^2 \left(\varepsilon_{r2} + \frac{\sigma}{j\omega\varepsilon_0} \right)$$

so we have:

$$-k_0^2 \left[\frac{1}{u} \frac{J_m'(u)}{J_m(u)} - \frac{\mu_{rc}}{v} \frac{H_m'(v)}{H_m(v)} \right] \left[\varepsilon_{r1} \frac{J_m'(u)}{u J_m(u)} - \left(\varepsilon_{r2} + \frac{\sigma}{j\omega\varepsilon_0} \right) \frac{1}{v} \frac{H_m'(v)}{H_m(v)} \right] - m^2\gamma^2 \left(\frac{1}{v^2} - \frac{1}{u^2} \right)^2 = 0 \quad (20)$$

Equation(20) is the basic propagation equation for circular waveguides, the field components in the guide for the (m, n) mode are proportional to $\exp[j(\omega\varepsilon - \gamma z - m\phi)]$. But

$$v^2 = a^2 k_2^2 + a^2\gamma^2 = a^2 k_1^2 \left[\left(\frac{k_2}{k_1} \right)^2 - 1 \right] + u^2$$

Equation (20) can be written in the form:

$$F(u)=0 \quad (21)$$

The algorithm for solving equation (21) is based upon the Newton-Raphson method, which is based upon the preliminary determination by trial of an approximate root u_0:

—First guess u_0,

—Next guess u_1, $u_1 = u_0 - [F(u_0)/F'(u_0)]$,

—Third guess u_2, $u_2 = u_1 - [F(u_1)/F'(u_1)]$.

This process may be repeated until the desired degree of approximation is attained:

$$u_n = u_{n-1} - \frac{F(u_{n-1})}{F'(u_{n-1})}$$

where n is the calculation number, not mode index. And now we may have:

$$\frac{u_n - u_{n-1}}{u_n} < 1 \times 10^{-7}$$

In common case, it has been assumed that the bounding conductor is lossfree, but in practice, this assumption will not be true. Because of the finite conductivity of the surrounding conductor, the electric field will penetrate into the metal, and the resistance losses thereby incurred will cause the coupling of wave modes, except the circular symmetrical modes in the circular waveguide. When $m \neq 0$, this case yields waves of the type HE mn ($E_z \neq 0$) and EH mn ($H_z \neq 0$), named with the first derivation given by Carson-Mead-Schelkunoff. We must bear firmly in mind, that the hybrid wave has six field components. So we can strictly distinguish between the hybrid modes HE^{mn} and the transverse electric modes H_{mn}.

Putting $m = 1$, from the CMS equation on HE mode, we have:

$$F(u) = -k_0^2 \left[\frac{1}{u} \frac{J_1'(u)}{J_1(u)} - \frac{\mu_{rc}}{v} \frac{H_1'(v)}{H_1(v)} \right] \left[\varepsilon_{r1} \frac{J_1'(u)}{u J_1(u)} - \left(\varepsilon_{r2} + \frac{\sigma}{j\omega\varepsilon_0} \right) \frac{1}{v} \frac{H_1'(v)}{H_1(v)} \right] - m^2 \gamma^2 \left(\frac{1}{v^2} - \frac{1}{u^2} \right)^2 \quad (22)$$

But

$$J_1'(u) = J_0(u) - \frac{1}{u} J_1(u); \quad H_1'(v) = H_0'(v) - \frac{1}{v} H_1(v)$$

Then

$$F(u) = -k_0^2 \left[\varepsilon_{r1} \left(\frac{1}{u} \frac{J_0(u)}{J_1(u)} - \frac{1}{u^2} \right)^2 - \left(\frac{1}{u} \frac{J_0(u)}{J_1(u)} - \frac{1}{u^2} \right) \left(\varepsilon_{r1} \mu_{rc} + \varepsilon_{r2} + \frac{\sigma}{j\omega\varepsilon_0} \right) \times \right.$$

$$\frac{1}{v} \left(\frac{H_0(v)}{H_1(v)} - \frac{1}{v} \right) + \mu_{rc} \left(\varepsilon_{r2} + \frac{\sigma}{j\omega\varepsilon_0} \right) \frac{1}{v^2} \left(\frac{H_0(v)}{H_1(v)} - \frac{1}{v} \right)^2 \Bigg] -$$

$$\left(\frac{u^2}{a^2} - \varepsilon_{r1} k_0^2 \right)^2 \left(\frac{1}{v^2} - \frac{1}{u^2} \right)^2 \quad (23)$$

$$F'(u) = -k_0^2 \frac{2\varepsilon_{r1}}{u^2} \left(\frac{J_0(u)}{J_1(u)} - \frac{1}{u} \right) \left[\frac{2}{u^2} - 1 - \left(\frac{J_0(u)}{J_1(u)} \right)^2 \right] + k_0^2 \left(\varepsilon_{r1} \mu_{rc} + \varepsilon_{r2} + \frac{\sigma}{j\omega\varepsilon_0} \right) \times$$

$$\left\{\frac{1}{uv}\left[\frac{2}{u^2}-1-\left(\frac{J_0(u)}{J_1(u)}\right)^2\right]\left(\frac{H_0(v)}{H_1(v)}-\frac{1}{v}\right)+\left(\frac{J_0(u)}{J_1(u)}-\frac{1}{u}\right)\frac{1}{v^2}\times\right.$$

$$\left[\frac{1}{v}\frac{H_0(v)}{H_1(v)}-\left(\frac{H_0(v)}{H_1(v)}\right)^2-1+\frac{1}{v^2}\right]-\left(\frac{J_0(u)}{J_1(u)}-\frac{1}{u}\right)\frac{1}{v^3}\left(\frac{H_0(v)}{H_1(v)}-\frac{1}{v}\right)\right\}-$$

$$k_0^2\mu_{rc}\left(\varepsilon_{r2}+\frac{\sigma}{j\omega\varepsilon_0}\right)\frac{2u}{v^3}\left(\frac{H_0(v)}{H_1(v)}-\frac{1}{v}\right)\left[\frac{2}{v^2}-1-\left(\frac{H_0(v)}{H_1(v)}\right)^2\right]-$$

$$2\left(\frac{1}{v^2}-\frac{1}{u^2}\right)\frac{u}{a^2}\left(\frac{1}{v^2}-\frac{1}{u^2}\right)+2\left(\frac{u^2}{a^2}-\varepsilon_{r1}k_0^2\right)\left(\frac{1}{u^3}-\frac{u}{v^4}\right)]\quad(24)$$

Now, in our practical condition $f \ll f_c$, the attenuation constant α is very large, so we have

$$u_0=a\sqrt{k_0^2+\gamma_0^2}\approx a\gamma_0$$

Putting

$$\gamma_0=\alpha_0\left(1+j\frac{\tau}{2a}\right)$$

Using Linder's formula:

$$\alpha_0=\frac{2\pi}{\lambda_c}\sqrt{1-(\lambda_c/\lambda)^2}$$

Then

$$u_0=\frac{2\pi}{\lambda_c}\sqrt{1-(\lambda_c/\lambda)^2}\left(a+j\frac{\tau}{2}\right)\quad(25)$$

So that the $F(u_0)$ and $F'(u_0)$ can be obtained by computer analysis. The skin depth of the walls of the guide is given by

$$\tau=\frac{1}{\text{Re}(j\omega^2\mu_c\varepsilon_c)}$$

For the most important mode HE_{11}, the cutoff wavelength may be written in the form

$$f_c=\frac{2\pi a}{1.84118378}=3.4125791\,a$$

The calculations will be made for the brass guide. It will be assumed that the original data are given by:

$u_0=4\times10^{-7}$H/m (permeability of free space),

$\varepsilon_0=8.854187818\times10^{-12}$F/m (permittivity of free space),

$\varepsilon_{r1}=1.000537$ (relative permittivity of air),

$\mu_{rc} = 0.99998$ (relative permeability of brass),

$\sigma = 1.6034 \times 10^7$ $1/\Omega \text{m}$ (RF conductivity of guide wall),

$a = 1.598125 \times 10^{-2} \text{m}$ (inner radius of circular guide),

$f = 30 \text{MHz}$ (signal frequency).

Now, we calculate the α and β:

$\varepsilon_{r1} = 1$, $\alpha = 0.9999593$ dB/mm, $\beta = 0.0827073$ rad/m

$\varepsilon_{r1} = 1.000537$, $\alpha = 0.9999593$ dB/mm, $\beta = 0.0827073$ rad/m

$\varepsilon_{r1} = 2$, $\alpha = 9999440$ dB/mm, $\beta = 0.0827096$ rad/m

It shows how the attenuation constant α and phase constant β depend on the relative permittivity of medium in the guide. It is seen that the difference between vacuum and air should not be sufficient large to be observed.

The attenuation constant of 30MHz WBCO attenuator standard built in NIM is 0.99995775 dB/mm. As a numerical example, we calculated the attenuation constant of same attenuator, the final result is 0.99995930 dB/mm. This value is larger than that mentioned above, the correction will be 1.55 part in 10^6.

USING THE SURFACE IMPEDANCE PERTURBATION METHOD TO DERIVE AN ACCURATE ATTENUATION CONSTANT FORMULA: THE HUANG EQUATION

Zhixun Huang has tried to directly derive the calculation formula of the attenuation constant of the precise cut-off waveguide by using other analytical methods instead of starting from the characteristic equation of the cylindrical structure, and has succeeded. We use the surface impedance theory of the waveguide. Let's start with the Maxwell curl equation, let

$$\varepsilon_c = \varepsilon + \frac{\sigma}{j\omega} = \varepsilon\left(1 + \frac{\sigma}{j\omega\varepsilon}\right) \tag{26}$$

we have

$$\nabla \times \boldsymbol{H}_c = j\omega\varepsilon_c \boldsymbol{E}_c$$

The subscript c stands for conductor. The planar metal surface impedance is defined as

$$z_s = \sqrt{\frac{\mu_c}{\varepsilon_c}} \tag{27}$$

Plug in the formula (26) and get

$$z_s = \sqrt{\frac{j\omega\mu_c}{\sigma + j\omega\varepsilon}} \tag{28}$$

There are good conductors

$$Z_s \approx \sqrt{\frac{j\omega\mu_c}{\sigma}} = \sqrt{\frac{\omega\mu_c}{\sigma}} e^{j\pi/4} \quad (29)$$

In 1948, Leontovich deduced the relationship between tangential electric field and tangential magnetic field in metal as follows.

$$E_c \approx \sqrt{\frac{\omega\mu_c}{\sigma}} e^{j\pi/4} (i_n \ H_c) \quad (30)$$

where i_n is the unit vector perpendicular to the metal surface; Take the axis to the metal surface as the origin point to the coordinates inside the metal, then it can be written

$$E_c = E_s \ e^{-\gamma_c r}$$

where subscript s represents surface. So we can derive

$$E_s \approx \sqrt{\frac{\omega\mu_c}{\sigma}} e^{j\pi/4} (i_n \ H_s) \quad (31)$$

Because σ is very large and E_s is small, the magnetic field is mainly in the inner surface of the conductor. It can now be concluded that

$$E_s \approx z_s (i_n \ H_s) \quad (32)$$

The approximate sign means that the displacement current in the conductor is ignored. The above equation is a concise form of the surface boundary condition of a nonideal conductor, where the surface impedance is the scale coefficient in this vector equation and is scalar. A good conductor is small, and therefore E_s is small.

Leontovich condition describes the relationship between the surface electromagnetic field components of a non-ideal conductor with good conductivity. Despite the small E_s size of a good conductor, the calculations considered are obviously much better than those considered when the conductivity is infinite. It can be seen that Leontovich's condition already implies the concept of surface impedance. The application of Leontovich condition is as follows: ① When electromagnetic wave passes through an object, the complex refractive index is $n = n' + jn''$, requiring $n'' \gg 1$; ② Let the skin layer be τ, $\tau \ll \lambda/2\pi$; ③ Let it be the radius of curvature of the surface of the object, $\tau \ll r$ (a surface with little curvature can treat the wave of the adjacent surface as a plane wave).

Metal-walled circular waveguides are not flat metals and are more complicated to deal with. On the inner surface of such a waveguide, since the inner radius is a, the coordinates of any point on the cylinder are (a, ϕ, z), and the tangential vector field of the point is

$$E_s = E_\phi i_\phi + E_z i_z$$
$$H_s = H_\phi i_\phi + H_z i_z$$

Define the impedance dyadic of the circular waveguide as

$$\ddot{Z} = Z_\varphi i_\varphi i_\varphi + Z_z i_z i_z \tag{33}$$

where z_φ is the circumferential surface impedance; z_z is the axial surface impedance; $i_\varphi i_\varphi$ is a dyadic circumferential unit vector; $i_z i_z$ is the dyadic axial unit vector. The electric field vector and the magnetic field vector of the inner surface of the circular waveguide are related by the following formula:

$$\ddot{Z}_s = Z(H_s \ i_r) \tag{34}$$

This is also a way of writing surface boundary conditions, which can be shortened to Leontovich conditions under certain conditions. On account of

$$H_s \times i_r = (H_\varphi i_\varphi \ H_z i_z) \times i_r = H_z i_\varphi - H_\varphi i_z$$

Therefore,

$$\ddot{Z}(H_s \ i_r) = \begin{bmatrix} 0 & 0 & 0 \\ 0 & z_\varphi & 0 \\ 0 & 0 & z_z \end{bmatrix} \begin{bmatrix} H_z \\ -H_\varphi \end{bmatrix}$$

The first row and column of the first matrix to the right of the equation are zero, so the remaining factors can be used:

$$\ddot{Z} \cdot (H_s \ i_r) = \begin{bmatrix} z_\varphi & 0 \\ 0 & z_z \end{bmatrix} \begin{bmatrix} H_z \\ -H_\varphi \end{bmatrix} = z_\varphi H_z i_\varphi - z_z H_\varphi i_z$$

hence

$$E_s = Z_\varphi H_z i_\varphi - Z_z H_\varphi i_z \tag{35}$$

Thus writable

$$E_\varphi = Z_\varphi H_z$$

$$E_z = Z_z H_\varphi$$

hence

$$Z_\varphi = \frac{E_\varphi}{H_z}\bigg|_{r=a} \tag{36}$$

$$Z_z = -\frac{E_z}{H_\varphi}\bigg|_{r=a} \tag{37}$$

where Z_φ is the circumferential surface impedance of the inner wall of the circular waveguide; Z_z is the axial surface impedance of the inner wall of the circular waveguide, which is generally mode dependent. For certain patterns, they may not differ much. In short, for circular waveguide and cylindrical resonator, the definition of surface impedance must be expanded to two. Since both are complex:

$$Z_\varphi = R_\phi + jX_\phi \tag{38}$$

$$Z_z = R_z + jX_z \tag{39}$$

So there are four real quantities that need to be determined. If the two impedances are normalized to the vacuum wave impedance (=376.62), the symbol is z_{00} Ω

$$\tilde{Z}_\varphi = \frac{Z_\varphi}{Z_{00}} \tag{40}$$

$$\tilde{Z}_z = \frac{Z_z}{Z_{00}} \tag{41}$$

The above statement shows that it is necessary to distinguish between the two surface impedances for non-planar metals. The ambiguity of the surface impedance reflects the non-unity of the field relation of the surface. In the simple theory, the difference between the two kinds of surface impedance in the inner wall of the circular waveguide is ignored

$$\tilde{Z}_\varphi \approx \tilde{Z}_z \tag{42}$$

In fact, the normalized surface impedance of the flat metal Z_f (i.e. Z_s) is taken instead of the two. This approach loses its rigor, but simplifies the derivation considerably.

The surface impedance analysis of metal-walled waveguides must lead to the theory of coupled modes. Given the surface impedance is equivalent to given the tangential field. Strictly speaking, as long as the loss of the waveguide wall is considered, there is no separate wave or magnetic wave in the circular waveguide except H_{0_n} wave and E_{0_n} wave, because the surface impedance makes the wave and magnetic wave coupling. The strict theory when the coupling is not ignored is the coupled wave theory. The solution to the problem of wave propagation in an ideal waveguide with a given shape is known, and the mode is assumed M_0. When the wall of the waveguide is a non-ideal conductor, the field in the waveguide is given by:

$$\psi = M_0 + \sum_{i=1}^{\infty} C_r M_i \tag{43}$$

In the formula M_i is the orthogonal module. For non-ideal waveguides, there is no longer a pure, single normal mode. As indicated in the above equation, a new factor is introduced into the orthogonal normalizing function set—the self-coupling coefficient C of the mixed mode. It is no longer possible to classify TM and TE modes in the waveguide. The cause of C can be described from different angles, one of which is the surface impedance. To get a better sense of this, consider the E_{mn} (EH *mn*, to be precise) mode groups in circular waveguides. Let the longitudinal electric field component in the ideal waveguide be

$$E_z = J_m(hr)\cos m\phi$$

Let's just write down the time phase factor of the field. For a non-ideal conductive wall waveguide, the surface impedance will couple a certain amount of H-wave energy, resulting in

$$H_z = c_i\, J_m(hr)\sin m\phi$$

In the case of ideal waveguide. H_ϕ is

$$H_\phi^E = -j\frac{k_0}{h_0} J'_m(hr)\cos m\phi$$

The coupling is caused by H waves

$$H_\phi^H = C_i[-j\frac{\beta_0 m}{h_0^2 r} J_m(hr)\cos m\phi]$$

hence

$$H_\phi\big|_{r=a} = \left[-j\frac{k_0}{h_0} J'_m(ha) - jC_i\frac{\beta_0 m}{h_0^2 a} J_m(ha)\right]\cos m\phi$$

On the other hand

$$E_z\big|_{r=a} = J_m(ha)\cos m\phi$$

Combine the above types can be obtained

$$\tilde{Z}_z = \frac{J_m(ha)}{-j\frac{k_0}{h_0} J'_m(ha) + jC_i\frac{\beta_0 m}{h_0^2 a} J_m(ha)} \quad (44)$$

Expand the pair by Taylor series, and take the first term, and get $J_m(ha)\, h_0 a$

$$J_m(ha) \approx a(\delta h) J'_m(h_0 a)$$

Plug in the above formula and get

$$\tilde{Z}_z \approx \frac{a\delta h}{j\frac{k_0}{h_0} + jC_i\frac{\beta_0 m}{h_0^2 a}\delta h} \quad (45)$$

If we ignore the first term in the right denominator, we get

$$\tilde{Z}_z \approx \frac{h_0^2 a}{jC_i\beta_0 m} \quad (46)$$

That is

$$C_i \approx \frac{h_0^2 a}{\beta_0 m}\frac{1}{j\tilde{Z}_z} \quad (47)$$

make

$$C_\tau = \frac{1}{C_i} \tag{48}$$

we have

$$C_\tau \approx \frac{\beta_0 m}{h_0^2 a} j\tilde{Z}_z \tag{49}$$

Because Z_z is small, C_τ is also a small amount; And when $Z_z \to 0$, then $C_\tau \to 0$. This means that all E waves in the circular waveguide are stable. In addition, when $m=0$, $C_\tau=0$. Therefore, the purity of E_{0_n} mode is not impaired by Z_z existence.

In a similar way, we can lead the circumferential surface impedance of the exit mode E_{mn}

$$\tilde{Z}_\varphi = \frac{j\dfrac{\beta_0 m}{h_0^2 a} J_m(ha) + jC_i \dfrac{k_0}{h_0} J'_m(ha)}{C_i J_m(ha)} \tag{50}$$

In addition to the approximation relation of Bessel function, $J'_m(ha) \approx J'_m(h_0 a)$ is taken, so it is obtained

$$\tilde{Z}_\varphi = \frac{\beta_0 m}{h_0^2 a} \frac{1}{C_i} + j \frac{k_0}{h_0} \frac{1}{a \cdot \delta h} \tag{51}$$

If we ignore the second term on the right, we get

$$C_i \approx \frac{\beta_0 m}{h_0^2 a} \frac{1}{\tilde{Z}_\varphi} \tag{52}$$

Thus obtained

$$C_\tau = \frac{h_0^2 a}{\beta_0 m} (j\tilde{Z}_\varphi) \tag{53}$$

So z_ϕ to 0, $C_\tau \to 0$.

The case of H_{mn} (strictly HE mn) pattern groups is discussed below. Let the longitudinal magnetic field component in the ideal waveguide be

$$H_z = J_m(hr) \cos m\phi$$

For non-ideal waveguides, the inner wall surface impedance will be coupled to a certain wave energy, resulting in E

$$E_z = C_i J_m(hr) \sin m\phi$$

Follow the previous way to extrapolate

$$\tilde{Z}_\varphi = j\frac{k_0}{h_0}\frac{\mathbf{J}'_m(ha)}{\mathbf{J}_m(ha)} - jC_i\frac{\beta_0 m}{h_0^2 a} \tag{54}$$

$$\tilde{Z}_z = \frac{C_i \mathbf{J}_m(ha)}{-j\frac{\beta_0 m}{h_0^2 a}\mathbf{J}_m(ha) + jC_i\frac{k_0}{h_0}\mathbf{J}'_m(ha)} \tag{55}$$

The characteristic equation can also be derived from the surface impedance analysis. For example, for the H_{mn} (strictly HE mn) pattern group, from the formula (54) and the formula (55), make them equal, and specify ($\gamma_0 = j\beta_0$), thus it can be deduced:

$$\frac{k_0^2}{h_0^2}\left[\frac{\mathbf{J}'_m(ha)}{\mathbf{J}_m(ha)}\right]^2 + j\frac{k_0}{h_0}\left(\tilde{Z}_\varphi + \frac{1}{\tilde{Z}_z}\right)\left[\frac{\mathbf{J}'_m(ha)}{\mathbf{J}_m(ha)}\right] + \left[\left(\frac{m\gamma_0}{h_0^2 a}\right)^2 - \frac{\tilde{Z}_\varphi}{\tilde{Z}_z}\right] = 0 \tag{56}$$

The above equation is called the surface impedance characteristic equation in order to distinguish it from the characteristic equation established by the field matching method. Make

$$\frac{k_0}{h_0}\frac{\mathbf{J}'_m(ha)}{\mathbf{J}_m(ha)} = y \tag{57}$$

we have

$$y^2 + j\left(\tilde{Z}_\varphi + \frac{1}{\tilde{Z}_z}\right)y + \left[\left(\frac{m\gamma_0}{h_0^2 a}\right)^2 - \frac{\tilde{Z}_\varphi}{\tilde{Z}_z}\right] = 0 \tag{58}$$

So it can be solved

$$y = -\frac{j}{2}\left(\tilde{Z}_\varphi + \frac{1}{\tilde{Z}_z}\right) \pm \frac{1}{2}\sqrt{-\left(\tilde{Z}_\varphi + \frac{1}{\tilde{Z}_z}\right)^2 - 4\left[\left(\frac{m\gamma_0}{h_0^2 a}\right)^2 - \frac{\tilde{Z}_\varphi}{\tilde{Z}_z}\right]} = 0 \tag{59}$$

Take a plus sign (HE mn pattern group) or a minus sign (EH mn pattern group). Let $z_\varphi \ll 1$, approximate equation for HE mn can be obtained:

$$\frac{k_0}{h_0}\frac{\mathbf{J}'_m(ha)}{\mathbf{J}_m(ha)} \approx -j\tilde{Z}_\varphi + j\frac{m\gamma_0}{h_0^2 a}\tilde{Z}_z \tag{60}$$

The above formula can be solved by perturbation method.

Now let's see how we derive the attenuation constant. When propagating with an ideal conductive wall, the field component is proportional to $e^{j\omega t - \gamma_0 z}$, $\gamma_0 = \sqrt{h_0^2 - k_0^2}$. When the waveguide wall is not ideally conductive, the field component is proportional to $e^{j\omega t - \gamma z}$, $\gamma = \sqrt{h^2 - k_0^2}$. If the non-ideal waveguide is regarded as a perturbation of the ideal waveguide, it can be proved

$$\gamma=\sqrt{\gamma_0^2+2h_0\delta h+(\delta h)^2}$$

When $\lambda>\lambda_c$, $\gamma_0=\alpha_0$, therefore

$$\gamma=\sqrt{\alpha_0^2+2h_0\delta h+(\delta h)^2}$$

So the problem is finding δh, it is a complex quantity, $\delta h=h_1+jh_2$. Thus it can be proved that

$$\alpha=\text{Re}=\gamma\frac{2\pi}{\lambda_c}\sqrt{\frac{1}{2}\left(\frac{\lambda_c}{2\pi}\right)^2 A+\frac{1}{2}\sqrt{\left(\frac{\lambda_c}{2\pi}\right)^4(A^2+B^2)}} \tag{61}$$

$$A=\alpha_0^2+2h_0h_1+h_1^2-h_2^2$$

$$B=2h_2(h_0+h_1)$$

In the perturbation method, δh is a function of the surface impedance and the waveguide geometry. In order to derive the H_{11} wave, the derivation is a little more complicated. The normalized value of the wave surface impedance is given. The above formula can be expressed as the binary simultaneous equations of, and a new equation is obtained after elimination τ_s:

$$\left[\frac{k_0}{h_0}\frac{J'_m(ha)}{J_m(ha)}\right]^2+j\left[\frac{k_0}{h_0}\frac{J'_m(ha)}{J_m(ha)}\right]\left(z_\phi+\frac{1}{z_z}\right)+\left[\left(\frac{\beta_0 m}{h_0^2 a}\right)^2-\frac{z_\phi}{z_z}\right]=0 \tag{62}$$

we have

$$\frac{k_0}{h_0}\frac{J'_m(ha)}{J_m(ha)}=-\frac{j}{2}\left(z_\phi+\frac{1}{z_z}\right)\left[1-\sqrt{1+\frac{2}{z_\phi+\frac{1}{z_z}}\left(\frac{\beta_0 m}{h_0^2 a}-\frac{z_\phi}{z_z}\right)}\right] \tag{63}$$

By using function expansion, it can be obtained approximately

$$\frac{J'_m(ha)}{J_m(ha)}=\frac{h_0}{k_0}\left\{\frac{-jz_\phi-\left(\frac{\beta_0 m}{h_0^2 a}\right)^2 jz_z}{1+z_\phi z_z}\right\}\left\{1+z_z\frac{z_\phi-\left(\frac{\beta_0 m}{h_0^2 a}\right)^2 z_z}{(1+z_\phi z_z)^2}\right\}$$

Finally get

$$\alpha=\frac{2\pi}{\lambda_c}\sqrt{\frac{1}{2}(M-J)+\frac{1}{2}\sqrt{(M-J)^2+J^2\left(1-J+\frac{J^2}{4}\right)}} \tag{64}$$

where J represents the influence of the finite conductivity (or skin depth) of the waveguide material. If the term J^3, J^4, is ignored, there is

$$\alpha = \frac{2\pi}{\lambda_c}\sqrt{\frac{1}{2}(M-J)+\frac{1}{2}\sqrt{(M-J)^2+J^2}} \qquad (65)$$

it is

$$\alpha = \frac{2\pi}{\lambda_c}\sqrt{\frac{1}{2}\left[1-\left(\frac{\lambda_c}{\lambda}\right)^2-J\right]+\frac{1}{2}\sqrt{\left[1-\left(\frac{\lambda_c}{\lambda}\right)^2-J\right]^2+J^2}} \qquad (66)$$

where J is related to the wave pattern:

$$J_{01} = \frac{\mu_r \tau}{a}\left(\frac{\lambda_c}{\lambda}\right)^2$$

$$J_{11} = \frac{g\tau}{a}$$

$$g = (1+0.4186]\mu_r\left(\frac{\lambda_c}{\lambda}\right)^2$$

Formula (66) is the main result of this section. It's called Huang's equation. In the derivation process, Leontovich condition is assumed to be true, which naturally has errors; Bessel function operation and final formula shape sorting, there are also errors. However, the formula (66) is more accurate than the formulas listed in foreign literature, and its accuracy can reach 10^{-7}.

In formula (66), if the term J^2 is zero, then

$$\alpha = \frac{2\pi}{\lambda_c}\sqrt{1-\left(\frac{\lambda_c}{\lambda}\right)^2-J} \qquad (67)$$

This is Rauskolb's formula, where $J = g\frac{\tau}{a}$. If we take $g = 1$ and $J = g/a$, we get Grantham and Freeman's formula

$$\alpha_{11} = \frac{2\pi}{\lambda_c}\sqrt{1-\left(\frac{\lambda_c}{\lambda}\right)^2-\frac{\tau}{a}} \qquad (68)$$

Therefore, we can see that the formulas in this paper can summarize the previous formulas.

In summary, for HE_{11} modes in a metal-walled circular waveguide, Huang's equation is:

$$\alpha_{11} = \frac{2\pi}{\lambda_0}\sqrt{\frac{1}{2}\left[1-\left(\frac{\lambda_c}{\lambda}\right)^2-J_{11}\right]+\frac{1}{2}\sqrt{\left[1-\left(\frac{\lambda_c}{\lambda}\right)^2\varepsilon_r-J_{11}\right]^2+J_{11}^2}} \qquad (69)$$

$$J_{11} = \frac{g\tau}{a} = \mu_{rc}\left[1+\frac{1}{k_{11}^2-1}\left(\frac{\lambda_{c\varepsilon}}{\lambda}\right)^2\right]\frac{\tau}{a} \qquad (70)$$

And $\lambda_c = \dfrac{2\pi a}{k_{11}}$, where $k_{11} = 1.84118378$; The formula α_{11} derived by me is the most accurate of its kind, and can fully meet the needs of establishing high-precision national attenuation standards.

As mentioned earlier, for microwave attenuation measurement, between 1950 and 1980, the accuracy of the cutoff attenuation standard was increased from 5×10^{-3} to close to 5×10^{-5}, because it is basically a calculation standard, and the attenuation constant is mainly determined by the basic unit (length and time). In the case of the increasing level of machining and surface treatment technology, it is required to theoretically derive the attenuation constant formula with higher precision. This gives rise to the so-called Huang equation.

GENERAL THEORY OF METAL-WALLED CIRCULAR WAVEGUIDES LINED WITH DIELECTRIC LAYERS: HUANG ZENG EQUATION

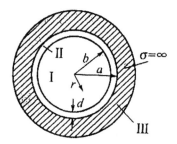

Fig. 3 Metallic wall circular waveguide lined with dielectric layer

In 1943, H.Buchholz published the paper "Circular Waveguides Filled with Dielectric Matter", which first put forward the problem of the influence of dielectric filled waveguides. In 1950, H.Wachowski and R. Beam published an article entitled "Dielectric Rod Waveguide with Shielding". In the above paper, the characteristic equation (assuming the conductivity of the waveguide material $\sigma = \infty$) when the inner wall of the circular waveguide is artificially added with a uniform dielectric layer (FIG. 3) is derived, which is called the BWB equation. In 1957, Unger proposed a perturbed approximate solution to the BWB equation. These works are called "artificial dielectric layer circular waveguide theory". In 1949, J.Brown published a paper entitled "Correction of Attenuation Constant of Piston Type Attenuator", which analyzed the normal mode of circular waveguides with uniform oxide layer on the inner wall, and also assumed that the conductivity of the waveguide material $\sigma = \infty$; The eigen equation derived under this assumption is called Brown equation, and its shape is different from that of BWB equation. In 1993, Zhixun

Huang and Cheng Zeng published the most general formula, called the Huang Zeng equation.

In early applications, the lined waveguide is mainly used as a long-distance transmission tool in the micro wave and infrared bands, that is, the application of the lined uniform dielectric layer to obtain a small attenuation of the main low-order mode. Later, in the study of mode suppression and how to reduce the radar reflection cross section (RCS), the opposite application of low-loss long-distance transmission has appeared, that is, the use of lining the medium layer to obtain a large attenuation of the main low-order mode; The theoretical basis is that since the internal radiation of the low-order mode in the terminal short-circuit waveguide is the main contribution to RCS, the RCS can be greatly reduced as long as the lined dielectric layer can effectively inhibit the main low-order mode.

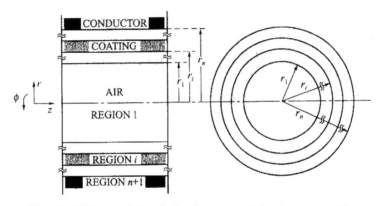

Fig. 4 Multi-layer dielectric lined metal wall circular waveguide

When assuming the physical model, Zhixun Huang and Cheng Zeng stipulate that the dielectric lining of the metal-walled circular waveguide can be multi-layered, that is, multilayered coated circular waveguide, as shown in FIG. 4. A difficult mathematical analysis results in the following characteristic equation:

$$k_0^2 \left(\frac{\varepsilon_{r1}}{h_1} \frac{J_m'(u)}{J_m(u)} - \frac{\varepsilon_{r2}}{h_2} \frac{F_e'(q)}{F_e(q)} \right) \left(\frac{\mu_{r1}}{h_1} \frac{J_m'(u)}{J_m(u)} - \frac{\mu_{r2}}{h_2} \frac{F_h'(q)}{F_h(q)} \right)$$

$$= \eta_b^2 + \left(R_e(q) R_h(q) \eta_a^2 - \frac{1}{k_0^2} P_2^2(q) \eta_a^2 \eta_b^2 + \frac{8\mu_{r2}\varepsilon_{r2}\eta_a\eta_b}{\pi^2 abh_2^4} \right) \frac{1}{R_e(q)R_h(q)} \quad (71)$$

$$\eta_b = \eta_1, \quad \eta_a = \eta_2$$
$$u = h_1 b, \quad q = h_2 b, \quad v = h_3 a$$

$$F_e(q) = \frac{\varepsilon_{r2}}{h_2} Q_2(q) - \frac{\varepsilon_{r3}}{h_3} \frac{H_m^{(2)'}(v)}{H_m^{(2)}(v)} P_2(q) \quad (72)$$

$$R_e(q) = \frac{\varepsilon_{r2}}{h_2} P_2'(q) - \frac{\varepsilon_{r3}}{h_3} \frac{H_m^{'(2)}(v)}{H_m^{(2)}(v)} P_2(q) \qquad (73)$$

among

$$P_i(h_i r_{i-1}) = J_m(h_i r_{i-1}) \, N_m(h_i r_i) - J_m(h_i r_i) \, N_m(h_i r_{i-1}) \qquad (74)$$

$$Q_i(h_i r_{i-1}) = J_m(h_i r_{i-1}) \, N_m'(h_i r_i) - J_m'(h_i r_i) \, N_m(h_i r_{i-1}) \qquad (75)$$

$$P_i'(h_i r_{i-1}) = J_m'(h_i r_{i-1}) \, N_m(h_i r_i) - J_m(h_i r_i) \, N_m'(h_i r_{i-1}) \qquad (76)$$

$$Q_i'(h_i r_{i-1}) = J_m'(h_i r_{i-1}) \, N_m'(h_i r_i) - J_m'(h_i r_i) \, N_m'(h_i r_{i-1}) \qquad (77)$$

Our new equation can be used to deal with the situation of artificially adding dielectric layer to the inner wall of the circular waveguide, and to analyze the effect of forming oxide layer in the precision cutoff attenuator.

CALCULATION OF THE INFLUENCE OF OXIDE LAYER ON THE INNER SURFACE OF CIRCULAR WAVEGUIDE

Microwave attenuation measurement technology mainly uses IF instead of the method, that is, the standard cut-off waveguide attenuator as the core, to form a comparable standard at IF (30MHz). If its accuracy reaches 5×10^{-5}, the overall microwave attenuation measurement accuracy can reach $10^{-3} \sim 10^{-4}$.

In short, the circular cut-off waveguide is the heart of the cut-off attenuation standard, and its processing conditions meet the requirements of high precision and low roughness to ensure that the electrical index is qualified. The entire attenuation standard should be placed in a laboratory with constant temperature and humidity. Taking the standard of the National Institute of Metrology (NIM) as an example, the accuracy is $dA = \pm(5 \times 10^{-5} + 0.0002)$dB; This means that i.e. $d\alpha/\alpha = \pm 5 \times 10^{-5}$, $\alpha \cdot dl = 0.0002$dB. Other indicators are: measuring range 0~120dB, resolution ±0.0001dB(0.1mdB). Accordingly, two examples of accuracy data can be calculated: ±0.0007dB/10dB (i.e., $\pm 7 \times 10^{-5}$); ±0.0052dB/100dB (i.e. $\pm 5.2 \times 10^{-5}$). The high resolution length measurement guarantees a reduction of dl, which is due to the use of laser length measurement technology.

Below the national standards, there are so-called secondary standards, which are distributed in various industrial sectors and assume the task of regional standards. The fixed-point same frequency comparison method (based on the medium frequency substitution method) is adopted, which is based on the cut-off attenuator as the core standard component. For example, the domestic T0-7 attenuation standard measuring instrument, its total index is: frequency 1GHz~12.5GHz (fixed intermediate frequency 30MHz after conversion), attenuation range 100dB, attenuation measurement accuracy

Circular Section Metal Wall Cut-Off Waveguide Accurate Calculation when Used as a Standard Attenuator

5×10^{-4}. Specifically for the standard cutoff attenuator, the circular cutoff waveguide is made of H59 brass, the diameter is $2a = 31.962$mm, the diameter machining accuracy is $\pm 5\mu m$, and the internal surface roughness Ra is $0.2\mu m$. Corresponding attenuation constant $\alpha = 1$dB/mm; The length measurement technique used is raster, not laser, but also achieves high resolution—reading resolution of 0.002dB(2mdB).

For the cut-off waveguide attenuators with accuracy up to 5×10^{-5}, the oxidation layer on the inner wall of the guided wall is a problem that needs to be considered. The thickness of this oxide layer is small, but it is a factor in error. As mentioned earlier, the relevant equations are the starting point for our analysis. For example, Brown's equation assumes that the oxide film is a pure medium and its dielectric constant is $\varepsilon_0 \sim \infty$. Brown's approximate analysis shows that the error of the propagation constant for H_{11} mode in the cutoff region is

$$\frac{d\gamma_{11}}{\gamma_{11}} = (1 \sim 1.8)\frac{d}{a} \tag{78}$$

where a is the inner diameter of circular waveguide; d is the oxide film thickness. Therefore, Brown believes that if the attenuation constant error is required to be $<1\times10^{-5}$, the oxide film thickness is required to be $d<10^{-4}$mm$(0.1\mu m)$. These can be called "circular cut-off waveguide theories for naturally occurring thick layers of oxide film".

The characteristics of this section are as follows: ① The characteristic equation of the metal-walled circular waveguide with dielectric layer is discussed; ② Brown equation was derived from Buchholz-Wachowski-Beam equation (BWB equation); ③ The perturbation solution (H_{mn} mode) of Brown equation is discussed. ④ The case of H-mode below the truncated frequency is further obtained; ⑤ The reasons for the inconsistency between the two sets of theoretical results are analyzed. ⑥ The H_{11}-mode cut-off waveguide is numerically calculated, the influence of oxide layer is numerically analyzed, and the limit of tolerable oxide layer thickness is obtained.

If we follow Figure 3 for discussion, we have $a = b+d$, $\rho = b/a < 1$, $x_1 = h_1 a$, $x_2 =, h_2 a$ $\rho x_1 = h_1 b$, $\rho x_2 = h_2 b$. Here is the radial propagation constant:

$$h_1^2 = +k_0^2\gamma^2$$
$$h_2^2 = k_2^2 + \gamma^2 \approx \varepsilon_r k_0^2 + \gamma^2 \tag{79}$$

where γ is the axial propagation constant: k_2 is a parameter that characterizes the performance of the medium is:

$$k_2 = \sqrt{\omega^2 \varepsilon_2 \mu_2 - j\omega\mu_2\sigma_2}$$

where ε_2 is the dielectric constant of the dielectric layer μ_2 is the permeability of the dielectric layer, $\mu_2 = \mu_r\mu_0$ σ_2 is the conductivity of the dielectric layer. For the actual medium, $\sigma_2 \approx 0$, so there are

$$k_2 = \omega\sqrt{\varepsilon_2\mu_2} = k_0\sqrt{\varepsilon_r\mu_r}$$

take the medium $\mu_r = 1$, so get

$$k_2 \approx \sqrt{\varepsilon_r}k_0 \tag{80}$$

Unger gives the form of the BWB equation as:

$$m^2\left[\frac{1}{x_1^2}-\frac{1}{x_2^2}\right]^2 - \rho^2\frac{x_2^2-x_1^2}{x_2^2-\varepsilon_r x_1^2}\left[\frac{1}{x_1}\frac{J'_m(\rho x_1)}{J_m(\rho x_1)} + \frac{\varepsilon_r W_m(\rho x_2)}{\rho x_2^2 U_m(\rho x_2)}\right]$$

$$\cdot\left[\frac{1}{x_1}\frac{J'_m(\rho x_1)}{J_m(\rho x_1)} + \frac{1}{\rho x_2^2}\frac{V_m(\rho x_2)}{Z_m(\rho x_2)}\right] = 0 \tag{81}$$

$$U_m(\rho x_2) = J_m(\rho x_2)N_m(x_2) - N_m(\rho x_2)J_m(x_2)$$

$$V_m(\rho x_2) = \rho x_2^2[J'_m(\rho x_2)N_m(x_2) - N'_m(\rho x_2)J'_m(x_2)]$$

$$W_m(\rho x_2) = \rho x_2^2[J_m(x_2)N'_m(\rho x_2) - J'_m(\rho x_2)N_m(x_2)]$$

$$Z_m(\rho x_2) = x_2[J'_m(x_2)N_m(\rho x_2) - J_m(\rho x_2)N'_m(x_2)]$$

Now we want to transform the formula, for example

$$\rho^2\frac{x_2^2-x_1^2}{x_2^2-\varepsilon_r x_1^2} = \rho^2\frac{h_2^2-h_1^2}{h_2^2-\varepsilon_r h_1^2} = \rho^2\frac{(\varepsilon_r-1)k_0^2}{(1-\varepsilon_r)\gamma^2} = \rho^2\frac{k_0^2}{-\gamma^2} = \frac{k_0^2 b^2}{k_z^2 a^2}$$

again, let's deal with the following items:

$$\frac{\varepsilon_r}{\rho x_2^2}\frac{W_m(\rho x_2)}{U_m(\rho x_2)} = \frac{\varepsilon_r}{x_2}\frac{J_m(x_2)N'_m(\rho x_2) - J'_m(\rho x_2)N_m(x_2)}{J_m(\rho x_2)N_m(x_2) - N_m(\rho x_2)J_m(x_2)}$$

$$-\frac{\varepsilon_r}{h_2 a}\frac{J'_m(h_2 b)N_m(h_2 a) - J_m(h_2 a)N'_m(h_2 b)}{J_m(h_2 b)N_m(h_2 a) - J_m(h_2 a)N_m(h_2 b)}$$

it is

$$P_m(h_2 r) = J_m(h_2 r)N_m(h_2 a) - J_m(h_2 a)N_m(h_2 r)$$

where r represents radial coordinates. Then there is

$$P_m(h_2 b) = J_m(h_2 b)N_m(h_2 a) - J_m(h_2 a)N_m(h_2 b)$$

$$P'_m(h_2 b) = J'_m(h_2 b)N_m(h_2 a) - J_m(h_2 a)N'_m(h_2 b)$$

Therefore available

$$\frac{\varepsilon_r}{\rho x_2^2}\frac{W_m(\rho x_2)}{U_m(\rho x_2)} = -\frac{\varepsilon_r}{h_2 a}\frac{P'_m(h_2 b)}{P_m(h_2 b)}$$

it is

$$Q_m(h_2 r) = J_m(h_2 r)N'_m(h_2 a) - J'_m(h_2 a)N_m(h_2 r)$$

the same can be proved:

$$\frac{1}{\rho x_2^2}\frac{V_m(\rho x_2)}{Z_m(\rho x_2)} = -\frac{1}{h_2 a}\frac{Q'_m(h_2 b)}{Q_m(h_2 b)}$$

After this kind of sorting, the BWB equation becomes

$$\left[\frac{1}{h_1}\frac{J'_m(h_1 b)}{J_m(h_1 b)} - \frac{\varepsilon_r}{h_2}\frac{P'_m(h_2 b)}{P_m(h_2 b)}\right] \cdot \left[\frac{1}{h_1}\frac{J'_m(h_1 b)}{J_m(h_1 b)} - \frac{\mu_r}{h_2}\frac{Q'_m(h_2 b)}{Q_m(h_2 b)}\right]$$

$$= \frac{m^2}{b^2}\frac{k_z^2}{k_0^2}\left[\frac{1}{h_1^2} - \frac{1}{h_2^2}\right]^2 \quad (82)$$

the above formula is strict when the waveguide material is ideally conductive ($\sigma = \infty$). We tried to derive the Brown equation from the BWB equation, and now we have

$$h_1 a = h_1 b + h_1 d$$

$$h_2 a = h_2 b + h_2 d$$

Suppose $d \ll b$, cylindrical functions can be expanded by variables. $h_1 b$ and $h_2 b$. The Taylor series shows:

$$f(x_0 + \Delta) = f(x_0) + f'(x_0)\Delta + \frac{1}{2!}f''(x_0)\Delta^2 + \ldots$$

linear expansion is preferable if Δ small:

$$f(x_0 + \Delta) = f(x_0) + f'(x_0)\Delta$$

Therefore there may be

$$J_m(h_2 a) = J_m(h_2 b) + h_2 d \, J'_m(h_2 b)$$

$$J'_m(h_2 a) = J'_m(h_2 b) + h_2 d \, J''_m(h_2 b)$$

The second type of Bessel function also has a similar relationship, so it can be proved that:

$$\frac{P'_m(h_2 b)}{P_m(h_2 b)} \approx -\frac{1}{h_2 d} \quad (83)$$

Now we need to utilize the Wronsky relation of the integer order Bessel function:

$$J_m(x)N'_m(x) - J'_m(x)N_m(x) = J_{m+1}(x)N_m(x) - J_m(x)N_{m+1}(x)$$

$$= J_m(x)N_{m-1}(x) - J_{m-1}(x)N_m(x) = \frac{2}{\pi x}$$

and

$$J_m(x)N''_m(x) - J''_m(x)N_m(x) = -\frac{2}{\pi x^2}$$

$$J_m'(x)N_m''(x) - J_m''(x)N_m'(x) = \frac{2}{\pi x}\left(1 - \frac{m^2}{x^2}\right)$$

It can therefore be proved that:

$$\frac{Q_m'(h_2 b)}{Q_m(h_2 b)} \approx h_2 d\left(1 - \frac{m^2}{h_2^2 b^2}\right) \tag{84}$$

The following symbols are also introduced:

$$u = h_1 b = b\sqrt{k_0^2 + \gamma^2} \tag{85}$$

$$q = h_2 b = b\sqrt{k_2^2 + \gamma^2} \tag{86}$$

and consider the following relationship:

$$k_z^2 = -\gamma^2$$

This style (82) can be arranged as

$$-\left[k_0^2 \frac{\mathbf{J}_m'(u)}{\mathbf{J}_m(u)} + \frac{buk_2^2}{\mu_1 dq^2}\right] \cdot \left[\frac{\mathbf{J}_m'(u)}{\mathbf{J}_m(u)} - \frac{\mu_r u d}{b}\left(1 - \frac{m^2}{q^2}\right)\right] = m^2 \gamma^2 u^2 \left(\frac{1}{u^2} - \frac{1}{q^2}\right)^2$$

can also be written as

$$-\left[\frac{k_0^2}{\mu_0}\frac{\mathbf{J}_m'(u)}{\mathbf{J}_m(u)} + \frac{buk_2^2}{dq^2 \mu_2}\right] \cdot \left[\mu_0 \frac{\mathbf{J}_m'(u)}{\mathbf{J}_m(u)} - \frac{\mu_r u d}{b}\left(1 - \frac{m^2}{q^2}\right)\right] = \frac{m^2 \gamma^2}{u^2}\left[1 - \frac{u^2}{q^2}\right]^2 \tag{87}$$

this is Brown's characteristic equation. The error only comes from ignoring small quantities of higher order during linear expansion, which requires very little (the oxide film is very thin). Brown's equation is correct, but it cannot be used for film thickness.

Equation (87) can also be written as

$$\left[\frac{k_0^2}{\mu_0}\frac{\mathbf{J}_m'(u)}{\mathbf{J}_m(u)} + \frac{buk_2^2}{dq^2 \mu_2}\right] \cdot \left[\mu_0 \frac{\mathbf{J}_m'(u)}{\mathbf{J}_m(u)} - \frac{\mu_r u d}{b}\left(1 - \frac{m^2}{q^2}\right)\right] = \frac{m^2 k_z^2}{u^2}\left[1 - \frac{u^2}{q^2}\right]^2 \tag{88}$$

Now the perturbation solution of Brown's equation is required; Multiply the left side of the formula and expand it, multiply the left and right side by d/b, omit the $(d/b)^2$ term, and set $\delta = d/b$ to obtain the approximate equation:

$$k_2^2 \frac{\mu_0}{\mu_2}\left[\frac{\mathbf{J}_m'(u)}{\mathbf{J}_m(u)}\right] + k_0^2 \delta \left[\frac{\mathbf{J}_m'(u)}{\mathbf{J}_m(u)}\right]^2 - k_2^2 \frac{u^2}{q^2}\delta\left(1 - \frac{m^2}{q^2}\right)$$

$$-m^2 k_z^2 u^2 \delta \left(\frac{1}{u^2} - \frac{1}{q^2}\right)^2 = 0 \tag{89}$$

It is actually

$$f(u) = 0 \tag{90}$$

now we're going to approximate it. For the sake of comparison with Brown's original

Circular Section Metal Wall Cut-Off Waveguide Accurate Calculation when Used as a Standard Attenuator

text, using one of his symbols, it is actually:

$$h_{mn} = \dot{}j\gamma_{mn} \tag{91}$$

hence

$$h_{mn}^2 = k_0^2 - \frac{u_{mn}^2}{b^2} \tag{92}$$

perturbation is for $\delta=0$ (no oxide film present). Case of $\delta=0$:

$$J'_m(u) = 0$$

let the root be

$$u_{mn} = k_{mn}$$

when a thin oxide film is present, perturbations are generated:

$$u_{mn} = +k_{mn} \, du_{mn} \tag{93}$$

now we're going to do the Newton-Raphson process, which is

$$du_{mn} \approx -\frac{f(k_{mn})}{f'(k_{mn})} \tag{94}$$

However, it can be proved that:

$$\frac{dh_{mn}}{h_{mn}} = -\frac{u_{mn}}{b^2 h_{mn}^2} \, du_{mn} \tag{95}$$

and

$$\frac{dh_{mn}}{h_{mn}} = \frac{d\gamma_{mn}}{\gamma_{mn}} \tag{96}$$

So when $u_{mn} = k_{mn}$,

$$\frac{d\gamma_{mn}}{\gamma_{mn}} = -\frac{k_{mn}^2}{b^2 h_{mn}^2} \, du_{mn} = \frac{k_{mn}}{b^2 h_{mn}^2} \frac{f(k_{mn})}{f'(k_{mn})} \tag{97}$$

So let's figure du_{mn} out first. From formula (89), when $J'_m(u) = 0$,

$$f(u) = -k_2^2 \frac{u^2}{q^2}\left(1 - \frac{m^2}{q^2}\right)\delta - m^2 k_z^2 u^2 \left[\frac{1}{u^2} - \frac{1}{q^2}\right]^2 \delta$$

Thus obtained

$$f(k_{mn}) = -k_2^2 \frac{k_{mn}^2}{q_{mn}^2}\left(1 - \frac{m^2}{q_{mn}^2}\right)\delta - \left(k_2^2 \frac{\delta}{q_{mn}^2}\right)\frac{m^2 k_z^2 k_{mn}^2 q_{mn}^2}{k_2^2}\left[\frac{1}{k_{mn}^2} - \frac{1}{q_{mn}^2}\right]^2$$

replace it with k_z^2, and arrange it h_{mn}^2

$$f(k_{mn}) = -k_2^2 \frac{\delta}{q_{mn}^2}\left\{k_{mn}^2\left(1 - \frac{m^2}{q_{mn}^2}\right) + \frac{m^2 h_{mn}^2 q_{mn}^2}{k_2^2 k_{mn}^2}\left(1 - \frac{k_{mn}^2}{q_{mn}^2}\right)^2\right\} \tag{98}$$

look below; $f'(k_{mn})$ Just take the first term of formula (89) (zero-order approximation) and set then there is $\mu_2 = \mu_0$,

$$f(u) = k_2^2 \frac{u}{q^2} \left[\frac{J_m'(u)}{J_m(u)} \right]$$

The expression is obtained by solving according to the differential rule obtain $f'(u)$, And then $J_m'(k_{mn})$ equal to 0, so that's it

$$f'(k_{mn}) = \frac{k_2^2}{q_{mn}^2} k_{mn} \frac{J_m''(k_{mn})}{J_m(k_{mn})}$$

Using the Bessel recurrence formula:

$$J_m''(k_{mn}) = -\left(1 - \frac{m^2}{k_{mn}^2}\right) J_m(k_{mn})$$

Finally get

$$f'(k_{mn}) = -\frac{k_2^2}{q_{mn}^2} k_{mn} \left(1 - \frac{m^2}{k_{mn}^2}\right) \tag{99}$$

thus obtained

$$\frac{d\gamma_{mn}^{(b)}}{\gamma_{mn}^{(b)}} \approx \frac{\delta}{h_{mn}^2 b^2} \left[k_{mn}^2 \left(1 - \frac{m^2}{q_{mn}^2}\right) + \frac{m^2 h_{mn}^2 q_{mn}^2}{k_2^2 k_{mn}^2} \left(1 - \frac{k_{mn}^2}{q_{mn}^2}\right)^2 \right] \frac{k_{mn}^2}{k_2^2 - m^2} \tag{100}$$

h_{mn}: obtained from formula (92)

$$h_{mn}^2 = k_0^2 - \frac{k_{mn}^2}{b^2} \tag{101}$$

It can be determined by the following formula: q_{mn}

$$q_{mn}^2 = k_2^2 b^2 + \gamma_{mn}^2 b^2 = k_2^2 b^2 - h_{mn}^2 b^2 = k_2^2 b^2 - k_0^2 b^2 + k_{mn}^2 \tag{102}$$

These results are consistent with Brown's.

The opposite formula (100) can be deformed. On account of

$$k_{mn}^2 - \frac{m^2 k_{mn}^2}{q_{mn}^2} = (k_{mn}^2 - m^2) + m^2 \left(1 - \frac{k_{mn}^2}{q_{mn}^2}\right) = (k_{mn}^2 - m^2) + m^2 k_{mn}^2 \left(\frac{1}{k_{mn}^2} - \frac{1}{q_{mn}^2}\right)$$

So the equation (100) can be written as

$$\frac{d\gamma_{mn}^{(b)}}{\gamma_{mn}^{(b)}} = \frac{\delta k_{mn}^2}{h_{mn}^2 b^2} + \frac{m^2 \delta}{h_{mn}^2 b^2} \left\{ k_{mn}^2 \left(\frac{1}{k_{mn}^2} - \frac{1}{q_{mn}^2}\right) + \frac{k_{mn}^2 q_{mn}^2}{k_2^2 k_{mn}^2} \left(1 - \frac{k_{mn}^2}{q_{mn}^2}\right)^2 \right\} \frac{k_{mn}^2}{k_{mn}^2 - m^2}$$

where the symbol (b) indicates that this result is obtained by expanding the reference radius (not). The second term on the right can be simplified to

$$\frac{m^2\delta}{h_{mn}^2 b^2}\left\{\frac{q_{mn}^2-k_{mn}^2}{q_{mn}^2}+\frac{h_{mn}^2(q_{mn}^2-k_{mn}^2)}{k_2^2 k_{mn}^2 q_{mn}^2}\right\}\frac{k_{mn}^2}{k_{mn}^2-m^2}=\frac{m^2\delta}{h_{mn}^2 b^2}\frac{q_{mn}^2-k_{mn}^2}{k_{mn}^2-m^2}\left\{\frac{k_2^2 k_{mn}^2+h_{mn}^2(q_{mn}^2-k_{mn}^2)}{k_2^2 q_{mn}^2}\right\}$$

however

$$q_{mn}^2 - k_{mn}^2 = k_0^2 b^2 (\varepsilon_r - 1) \tag{103}$$

It can therefore be proved that:

$$k_2^2 k_{mn}^2 + h_{mn}^2 (q_{mn}^2 - k_{mn}^2) = k_0^2 q_{mn}^2 \tag{104}$$

we plug in these results

$$\frac{d\gamma_{mn}^{(b)}}{\gamma_{mn}^{(b)}} = \frac{\delta k_{mn}^2}{h_{mn}^2 b^2} + \frac{m^2\delta}{h_{mn}^2 b^2} \cdot \frac{q_{mn}^2 - k_{mn}^2}{k_{mn}^2 - m^2} \cdot \frac{k_0^2}{k_2^2} \cdot \frac{\delta k_{mn}^2}{h_{mn}^2 b^2} + m^2\delta \frac{\varepsilon_r - 1}{k_{mn}^2 - m^2} \cdot \frac{k_0^2}{k_2^2} \cdot \frac{k_0^2}{h_{mn}^2}$$

However $k_0^2/k_2^2 = \varepsilon_r$, in addition, we defined:

$$G_{mn}^2 = \frac{h_{mn}^2}{k_0^2} = 1 - \frac{k_{mn}^2}{b^2 k_0^2} = 1 - \frac{\lambda^2}{\left(\frac{2\pi b}{k_{mn}}\right)^2} = 1 - \frac{\lambda^2}{\lambda_{c.mn}^2}$$

hence

$$\frac{d\gamma_{mn}^{(b)}}{\gamma_{mn}^{(b)}} = \frac{\delta k_{mn}^2}{h_{mn}^2 b^2} + m^2\delta \frac{\varepsilon_r - 1}{\varepsilon_r} \cdot \frac{1}{(k_{mn}^2 - m^2)} G_{mn}^2$$

however

$$F_{mn}^2 = \frac{h_{mn}^2 b^2}{k_{mn}^2} = \frac{b^2 k_0^2}{k_{mn}^2} - 1 = \left(\frac{2\pi b}{k_{mn}}\right)^2 \frac{1}{\lambda^2} - 1 = \frac{\lambda_{c.mn}^2}{\lambda^2} - 1 \tag{105}$$

So finally get

$$\frac{d\gamma_{mn}^{(b)}}{\gamma_{mn}^{(b)}} = \frac{\delta}{F_{mn}^2} + \frac{m^2\delta(\varepsilon_r - 1)}{\varepsilon_r(k_{mn}^2 - m^2)G_{mn}^2} \tag{106}$$

This expression is relatively comprehensive, Brown has not given such a clear expression.

ANALYSIS OF CIRCULAR CUT-OFF WAVEGUIDE AND H$_{11}$-MODE

Approximation theory still admits that positive scale can exist. Start with a clear formula, which is obtained from equation

$$\frac{d\gamma_{mn}^{(b)}}{\gamma_{mn}^{(b)}} = \frac{d/b}{\left(\frac{\lambda_{c.mn}}{\lambda}\right)^2 - 1} + \frac{d}{b} \cdot \frac{m^2}{k_{mn}^2 - m^2}\left(1 - \frac{1}{\varepsilon_r}\right) \cdot \frac{1}{1 - \left(\frac{\lambda}{\lambda_{c.mn}}\right)^2} \tag{107}$$

Let $m = n = 1$, get the formula applicable to H_{11} mode

$$\frac{d\gamma_{11}^{(b)}}{\gamma_{11}^{(b)}} = \frac{d/b}{\left(\frac{\lambda_{c.11}}{\lambda}\right)^2 - 1} + \frac{d}{b}\frac{1}{k_{11}^2 - 1}\left(1 - \frac{1}{\varepsilon_r}\right)\frac{1}{1 - \left(\frac{\lambda}{\lambda_{c.11}}\right)^2} \tag{108}$$

For the specific case of the cut-off waveguide attenuation standard, $\lambda \gg \lambda_c$, it is obtained

$$\frac{d\gamma_{11}^{(b)}}{\gamma_{11}^{(b)}} \approx -\frac{d}{b} - \frac{d}{b}\frac{1}{k_{11}^2 - 1}\left(1 - \frac{1}{\varepsilon_r}\right)\left(\frac{\lambda_{c.11}}{\lambda}\right)^2 \tag{109}$$

Obviously, the second term on the right side of the equation is small, so

$$\frac{d\gamma_{11}^{(b)}}{\gamma_{11}^{(b)}} \approx -\frac{d}{b} \tag{110}$$

The minus sign indicates that the oxidation film reduces the attenuation constant.

Now proceed from equation (24) of Brown's paper; Let $\mu_2 = \mu_0$, $m = n = 1$, then the formula becomes

$$\frac{d\gamma_{11}^{(b)}}{\gamma_{11}^{(b)}} = \frac{d}{b}\frac{1}{h_{11}^2 b^2}\frac{k_{11}^2}{k_{11}^2 - 1}\left\{k_{11}^2\left(1 - \frac{1}{q_{11}^2}\right) + \frac{h_{11}^2 q_{11}^2}{k_2^2 k_{11}^2}\left(1 - \frac{k_{11}^2}{q_{11}^2}\right)^2\right\}$$

Brown, however, "takes the unperturbed decay constant k_{11}/b", which has two caveats.

(1) This assumption is only true if the terms k_0 in the formula are ignored. By numerical verification, it is proved that it is allowed to work in the cut-off area.

(2) Brown lost a minus sign and should have taken it instead

$$h_{11}^2 b^2 = -k_{11}^2 \tag{111}$$

hence

$$\frac{d\gamma_{11}^{(b)}}{\gamma_{11}^{(b)}} = -\frac{d}{b}\frac{k_{11}^2}{k_{11}^2 - 1}\left\{1 - \frac{1}{q_{11}^2} + \frac{h_{11}^2 q_{11}^2}{k_2^2}\left(\frac{1}{k_{11}^2} - \frac{1}{q_{11}^2}\right)^2\right\}$$

$$= -\frac{d}{b}\frac{k_{11}^2}{k_{11}^2 - 1}\left\{1 - \frac{1}{q_{11}^2} - \frac{q_{11}^2}{b^2 k_2^2 k_{11}^2} + \frac{2}{b^2 k_2^2} - \frac{k_{11}^2}{b^2 k_2^2 q_{11}^2}\right\} \tag{112}$$

If $\varepsilon_r \gg 1$ is assumed, then $k_2^2 \gg k_0^2$, therefore desirable:

$$q_{11}^2 = \approx b^2 k_2^2 - b^2 k_0^2 + k_{11}^2 \; b^2 k_2^2 + k_{11}^2 \tag{113}$$

Hence there

$$\frac{q_{11}^2}{b^2k_2^2k_{11}^2}+\frac{k_{11}^2}{b^2k_2^2q_{11}^2}=\frac{b^2k_2^2+k_{11}^2}{b^2k_2^2k_{11}^2}+\frac{q_{11}^2-b^2k_2^2}{b^2k_2^2q_{11}^2}=\frac{1}{k_{11}^2}+\frac{2}{b^2k_2^2}-\frac{1}{q_{11}^2}$$

You get it when you plug it in

$$\frac{d\gamma_{11}^{(b)}}{\gamma_{11}^{(b)}} \approx -\frac{d}{b} \tag{114}$$

This result is correct, and Brown's original formula is:

$$\frac{dh_{11}}{h_{11}}=\frac{d}{b}\frac{1}{k_{11}^2-1}\left\{\frac{k_{11}^4-k_{11}^2+b^2k_2^2\left(k_{11}^2+1\right)}{k_{11}^2+b^2k_2^2}\right\}$$

this is wrong, because of the loss of a minus sign. Brown is wrong, but his conclusion is still correct on an order of magnitude scale.

We discuss the perturbation of BWB equation (H mn-mode) and H_{11}-mode cases. There is now

$$h_1b = h_1a - h_1d$$
$$h_2b = h_2a - h_2d$$

Suppose $d \ll a$, cylindrical functions can be expanded by variables h_1a, h_2a. Introduce the following symbols:

$$u = h_1a = a\sqrt{k_0^2+\gamma^2} \tag{115}$$

$$x_2 = h_2a = a\sqrt{k_2^2+\gamma^2} \tag{116}$$

And let $\delta = d/a$, similar analysis is obtained

$$k_2^2\frac{u}{x_2^2}\left[\frac{J_m'(u)}{J_m(u)}\right]+k_0^2\delta\left[\frac{J_m'(u)}{J_m(u)}\right]^2-k_2^2\frac{u^2}{x_2^2}\delta K-m^2h_{mn}^2u^2\delta\left[\frac{1}{u^2}-\frac{1}{x_2^2}\right]^2=0 \tag{117}$$

The term δ^2 was omitted from the derivation, where K is

$$K=\frac{J_m'(u)}{J_m(u)}+\left(1-\frac{m^2}{x_2^2}\right) \tag{118}$$

Take formula (117) as

$$f(u)=0 \tag{119}$$

It can be proved that:

$$\frac{d\gamma_{mn}^{(a)}}{\gamma_{mn}^{(a)}}=-\frac{k_{mn}}{a^2h_{mn}^2}du_{mn} \tag{120}$$

It can also be proved that:

$$\frac{d\gamma_{mn}^{(a)}}{\gamma_{mn}^{(a)}}=\frac{m^2\delta}{h_{mn}^2a^2}\left\{1-\frac{k_{mn}^2}{x_{2.mn}^2}+\frac{h_{mn}^2x_{2.mn}^2}{k_2^2k_{mn}^2}\left[1-\frac{k_{mn}^2}{x_{2.mn}^2}\right]^2\right\}\frac{k_{mn}^2}{k_{mn}^2-m^2} \tag{121}$$

and

$$h_{mn}^2 = k_0^2 - \frac{k_{mn}^2}{a^2} \tag{122}$$

$$x_{2.mn}^2 = k_2^2 a^2 - k_0^2 a^2 + k_{mn}^2 \tag{123}$$

But the formula has been rearranged to become

$$\frac{d\gamma_{mn}^{(a)}}{\gamma_{mn}^{(a)}} = \frac{m^2\delta}{h_{mn}^2 a^2} \frac{x_{2.mn}^2 - k_{mn}^2}{k_{mn}^2 - m^2} \left\{ \frac{k_2^2 k_{mn}^2 + h_{mn}^2 \left(x_{2.mn}^2 - k_{mn}^2 \right)}{k_2^2 x_{2.mn}^2} \right\}$$

However ($x_{2.mn}^2 - k_{mn}^2$) can be written as:

$$x_{2.mn}^2 - k_{mn}^2 = k_0^2 a^2 \left(\varepsilon_r - 1 \right) \tag{124}$$

And know $k_0^2 / k_2^2 = 1/\varepsilon_r$, so the last can be obtained

$$\frac{d\gamma_{mn}^{(a)}}{\gamma_{mn}^{(a)}} = m^2 \delta \frac{(\varepsilon_r - 1)}{\varepsilon_r} \frac{1}{(k_{mn}^2 - m^2) G_{mn}^2} \tag{125}$$

and

$$G_{mn}^2 = \frac{h_{mn}^2}{k_0^2} = 1 - \left(\frac{\lambda}{\lambda_{c \cdot mn}} \right)^2 \tag{126}$$

$$\lambda_{c.mn} = \frac{2\pi a}{k_{mn}} \tag{127}$$

The formula can be written as

$$\frac{d\gamma_{mn}^{(a)}}{\gamma_{mn}^{(a)}} = \frac{d}{a} \frac{m^2}{k_{mn}^2 - m^2} \left(1 - \frac{1}{\varepsilon_r} \right) \frac{1}{1 - \left(\frac{\lambda}{\lambda_{c \cdot mn}} \right)^2} \tag{128}$$

Let $m = n = 1$, get the formula applicable to H_{11} modes:

$$\frac{d\gamma_{11}^{(a)}}{\gamma_{11}^{(a)}} = \frac{d}{a} \frac{1}{k_{11}^2 - 1} \left(1 - \frac{1}{\varepsilon_r} \right) \frac{1}{1 - \left(\frac{\lambda}{\lambda_{c.11}} \right)^2} \tag{129}$$

for the specific case of the cut-off attenuation standard, $\lambda \gg \lambda_c$, thus obtained

$$\frac{d\gamma_{11}^{(b)}}{\gamma_{11}^{(b)}} \approx -\frac{d}{a} \frac{1}{k_{11}^2 - 1} \left(1 - \frac{1}{\varepsilon_r} \right) \left(\frac{\lambda_{c.11}}{\lambda} \right)^2 \tag{130}$$

equation (130) shows that the artificial application of dielectric layer on the inner wall causes only a small effect and can be ignored.

So why are the results of the two theories inconsistent?

First, the formula should be written as

$$h_{mn}^{(b)} = \sqrt{k_0^2 - \frac{u_{mn}^2}{b^2}} = \sqrt{k_0^2 - \frac{k_{mn}^2}{b^2}} \tag{131}$$

the other formula should be written as

$$h_{mn}^{(a)} = \sqrt{k_0^2 - \frac{k_{mn}^2}{a^2}} \tag{132}$$

the difference between them can therefore be found:

$$h_{mn}^{(b)} - h_{mn}^{(a)} = \sqrt{k_0^2 - \frac{k_{mn}^2}{b^2}} - \sqrt{k_0^2 - \frac{k_{mn}^2}{a^2}}$$

however

$$\delta = \frac{d}{a} = \frac{a-b}{a} = 1 - \frac{b}{a}$$

hence

$$b^2 = a^2(1-\delta)^2 \tag{133}$$

Hence there

$$h_{mn}^{(b)} = \sqrt{k_0^2 - \frac{k_{mn}^2}{a^2(1-\delta)^2}} \approx \sqrt{k_0^2 - \frac{k_{mn}^2}{a^2}(1+2\delta)} - h_{mn}^{(a)} \sqrt{1 - \frac{2k_{mn}^2}{a^2 h_{mn}^{(a)^2}} \delta}$$

Again, we approximate, we get

$$h_{mn}^{(b)} \approx h_{mn}^{(a)} \left[1 - \frac{k_{mn}^2}{a^2 h_{mn}^{(a)^2}} \delta \right]$$

Thus obtained

$$h_{mn}^{(b)} - h_{mn}^{(a)} = -\frac{\delta k_{mn}^2}{a^2 h_{mn}^{(a)}} \tag{134}$$

On the other hand, formula (121) can be deformed and arranged into the following form:

$$\frac{dh_{mn}^{(a)}}{h_{mn}^{(a)}} = + - \frac{\delta k_{mn}^2}{a^2 h_{mn}^{(a)}} + \frac{\delta}{a^2 h_{mn}^{(a)^2}} \left\{ k_{mn}^2 \left(1 - \frac{m^2}{x_{2.mn}^2}\right) \frac{m^2 h_{mn}^2 x_{2.mn}^2}{k_2^2 k_{mn}^2} \left[1 - \frac{k_{mn}^2}{x_{2.mn}^2} \right]^2 \right\} \frac{k_{mn}^2}{k_{mn}^2 - m^2}$$

compared to the past formula, hence

$$dh_{mn}^{(a)} - dh_{mn}^{(b)} = -\frac{\delta k_{mn}^2}{a^2 h_{mn}^{(a)}} \tag{135}$$

Therefore, it is obtained from formula (134) and formula (135):

$$h_{mn}^{(a)} + dh_{mn}^{(a)} = h_{mn}^{(b)} + dh_{mn}^{(b)} \tag{136}$$

hence

$$\gamma_{mn}^{(a)} + d\gamma_{mn}^{(a)} = \gamma_{mn}^{(b)} + d\gamma_{mn}^{(b)} \tag{137}$$

That is to say, after the propagation constant is perturbed, the result is the same whether it is expanded a or expanded b. In other words, under first-order approximation, the results obtained by BWB equation are consistent with those obtained by Brown equation.

Consider the following formula:

$$\gamma_{11}^{(a)} + d\gamma_{11}^{(a)} = \gamma_{11}^{(b)} + d\gamma_{11}^{(b)} \tag{138}$$

the sum is the same, but the items are different. To give a number like this:

$$1.0001000 + 0.0000006 = 1.0000006 + 0.0001 \tag{139}$$

Even though the equation is true, the small quantity on the left is 6×10^{-7}, and the small quantity on the right is 1×10^{-4}. So the end result is different.

If the medium layer is not artificially laid, but caused by natural oxidation, the use of Brown analysis is obviously more reasonable, because just finished processing is the radius b, the propagation constant; $\gamma_{11}^{(b)}$. Oxidation expands the ideal conductive boundary to cause perturbation. From an experimental point of view, the propagation constant in the presence of an oxide layer can only be compared with that in the absence of oxidation $\gamma_{11}^{(b)}$, and cannot be compared with that in the absence of oxidation $\gamma_{11}^{(a)}$.

The conclusion is that the process of naturally forming an oxide layer on the inner surface of a copper or brass circular waveguide may cause an error of 10^{-5} orders of magnitude in the attenuation constant (depending on the thickness), which cannot be ignored for high accuracy standards! Brown's analysis had a few flaws at the end, but it was mostly correct. As for the sign of the error, all analyses indicate that the presence of an inner dielectric layer reduces the attenuation constant. There are no contradictions or differences to explain here.

We finally discuss the numerical calculation of the H_{11}-mode case where the operating frequency is much less than the cutoff frequency, or the operating wave is longer than the cutoff wavelength. Then, the H_{11}-mode can be numerically calculated by the formula (140) and written as

$$\frac{\Delta \alpha_{11}}{\alpha_{11}} \approx -\frac{d}{a} \tag{140}$$

hence

$$\left| \frac{\Delta \alpha_{11}}{\alpha_{11}} \right| \approx \frac{d}{b} \tag{141}$$

namely

$$d \approx b \left| \frac{\Delta \alpha_{11}}{\alpha_{11}} \right| \tag{142}$$

For example, if there is such a copper (or brass) waveguide with an internal radius b of 16000.00μm, there is

$$d \approx 1.6 \times 10^4 \left| \frac{\Delta \alpha_{11}}{\alpha_{11}} \right| \quad \mu m \tag{143}$$

It can be seen that when it is specified that $|\Delta \alpha_{11}|/\alpha_{11}$ is less than a certain value, it is also necessary to ensure that the thickness of the oxide film is less than a certain value. For example, $<5 \times 10^{-5}$, oxide film thickness $d \leq 0.8 \mu m$.

It can be seen that if we want to make a cutoff attenuation standard with an uncertainty of 5×10^{-5}, it is necessary to take appropriate anti-oxidation measures for copper (or brass) waveguides. Because even if the error assigned to this aspect is 1×10^{-5}, the oxide film thickness is still required to be $d < 0.16 \mu m$. In the scientific research of surface physics and surface chemistry, there are special instruments and methods for measuring the thickness of oxide film, such as the ellipsometer method (using the polarization of the metal and the oxide film to measure) and the film resistance method, which can be selected, and this paper will not be described.

CONCLUSION

Based on a research project I participated in at NIM in China in the past, based on the mathematical equations in the waveguide theory and a lot of mathematical analysis, this paper makes an original research on the cut-off waveguide theory used in first-order attenuation criteria, and gets a series of new achievements. They are unknown to the international scientific community, such as the later named equations (Huang equation, Huang Zeng equation), as well as some unique algorithms. This paper is a complete summary, a large number of mathematical analysis work has been elevated to the level of physical significance, so that this paper is not only a contribution to metrology, but also to the enrichment of basic scientific theories—providing new mathematical equations in teaching, and strengthening the waveguide theory in physics.

From the development and experimental results, NIM reached a high level of accuracy of 5×10^{-5} for the cut-off attenuation standard in 1980. This requires a theoretical calculation accuracy better than 1×10^{-6}. It is best to reach 1×10^{-7}. This paper shows how the author achieved this requirement through hard work, and vividly illustrates the twists and turns and hardships of scientific research work.

I would like to thank Professor Min Zhu for his help in writing this article.

REFERENCES

[1] Barlow H, Cullen A. Microwave measurements[M]. London: Constable, 1950.

[2] Huang Z X. An Introduction to the Theory of Waveguide Bolow Cutoff[M]. Beijing: Metrology Publishing House, 1981(1st pub.), 1991(2nd pub.).

[3] Thomson J. Notes on recant researches in electricity and magnetism, intended as a seguel to Prof. Maxwell's treatise on electricity and magnetism[M]. London: Dawsons of Pall Mall, 1893;reprinted 1968.

[4] 黄志洵. $H_{(11)}$ 模截止式衰减器的误差分析 [J]. 电子学报, 1963, (04): 128–141.

[5] 黄志洵. 一级衰减标准所用高精密圆截止波导金属壁内部氧化层的影响与分析 [J]. 计量学报, 1983, (04): 316–318+320.